T0325807

Building Support for Scholarly Practices in Mathematics Methods

A volume in
The Association of Mathematics Teacher Educators (AMTE) Professional Book Series
Christine Browning, *Series Editor*

Building Support for Scholarly Practices in Mathematics Methods

edited by

Signe E. Kastberg
Purdue University

Andrew M. Tyminski
Clemson University

Alyson E. Lischka
Middle Tennessee State University

Wendy B. Sanchez
Kennesaw State University

INFORMATION AGE PUBLISHING, INC.
Charlotte, NC • www.infoagepub.com

Library of Congress Cataloging-in-Publication Data

Names: Kastberg, Signe E., 1963- editor.
Title: Building support for scholarly practices in mathematics methods /
  edited by Signe E. Kastberg, Purdue University [and three others].
Description: Charlotte, NC : Information Age Publishing, Inc., [2017] |
  Series: Association of Mathematics Teacher Educators (AMTE) professional
  book series | Includes bibliographical references.
Identifiers: LCCN 2017035865 (print) | LCCN 2017043159 (ebook) | ISBN
  9781641130271 (E-book) | ISBN 9781641130257 (pbk.) | ISBN 9781641130264
  (hardcover)
Subjects: LCSH: Mathematics teachers–Training of. | Mathematics–Study and
  teaching.
Classification: LCC QA11.2 (ebook) | LCC QA11.2 .B8674 2017 (print) | DDC
  510.712–dc23
LC record available at https://lccn.loc.gov/2017035865

Cover photo provided by University of Washington College of Education

Printed in the United States of America

For mathematics teacher educators
whose creativity, commitment, and care
inspire scholarly inquiry and practice
in mathematics teacher education.

CONTENTS

SECTION I

PERSPECTIVES AND MATHEMATICS METHODS COURSES

SECTION II

USING PERSPECTIVES TO INFORM SCHOLARLY INQUIRY AND PRACTICE

SECTION III

LEARNING GOALS AND ACTIVITIES IN MATHEMATICS METHODS COURSES

SECTION IV
ACTIVITY DEVELOPMENT

SECTION V
ACTIVITIES AND IMPLEMENTATIONS

SECTION VI

LOOKING INWARD

SECTION VII

COMMENTARY

FOREWORD

Uniquely positioned as the lead organization and public voice for mathematics teacher education, the Association of Mathematics Teacher Educators (AMTE) established the production of a professional book series. This book, *Building Support for Scholarly Practices in Mathematics Methods*, is the third book of the series. Its focus on the practices of mathematics teacher educators (MTEs) furthers the mission of AMTE to improve mathematics teacher education as well as promoting AMTE's goals. Although all goals of AMTE's are advanced in this work, four goals are particularly prominent: research and scholarly endeavors in mathematics teacher education, equitable practices, effective mathematics teacher education programs and practices, and communication and collaboration among MTEs.

The work leading to the development of *Building Support for Scholarly Practices in Mathematics Methods* is described by the editors, Signe Kastberg, Andrew Tyminski, Alyson Lischka, and Wendy Sanchez, as an outgrowth of 5 years of ongoing scholarly inquiry centered on the investigation of MTEs' practices in mathematics methods courses. The development was launched in 2012 in a session presentation at the AMTE annual conference. This session focused on frameworks and activities used in methods courses. As a result of this session, the participants encouraged the editors to create a working group within the North American chapter of the Psychology of Mathematics Education (PME-NA) to continue collaboration of MTEs engaged in the exploration of mathematics methods. Thus, a PME-NA working group was established in 2012, and as a result, participating MTEs were now situated within a well-organized structure designed to sustain the ongoing collaboration. Further, they expanded the focus to include the study of

Building Support for Scholarly Practices in Mathematics Methods, pages xi–xii
Copyright © 2018 by Information Age Publishing
xi

residue—meaning that the focus of the scholarly inquiry now included the study of the "impact of an experience beyond methods courses."

The following two events are noteworthy given the influence of these events leading to conception of this book. In 2013, the editors presented a session at the AMTE annual conference entitled *Building a Theoretically Grounded Practice of Methods Instruction*. In 2015, the editors held a conference, Scholarly Inquiry and Practices Conference for Mathematics Education Methods (SIP; Sanchez, Kastberg, Tyminski, & Lischka, 2015), funded by the National Science Foundation. SIP was designed with an emphasis to support MTEs' engagement in conversations on theoretical perspectives. Further, the editors posit the significance of SIP leading to the organization and contents of this book. The editors employed the ongoing conversations on theoretical perspectives to set the stage of this book and noted that the conversations at SIP culminated in images of the variation in mathematics methods represented in the chapters in this book.

It is also pertinent to note that during the development of this book, AMTE was in the process of writing the newly released *Standards for Preparing Teachers of Mathematics* (SPTM). Although both SPTM and this book were in developmental stages at the same time, drafts of the AMTE Standards were available for review and influenced the elaboration of what "well-prepared beginning mathematics teachers" need to know and be able to do as presented within the pages of *Building Support for Scholarly Practices in Mathematics Methods*. The SPTM is a set of comprehensive standards describing a national vision for the initial preparation of all teachers, prekindergarten through grade 12, who teach mathematics. The standards advocate for practices that support candidates in becoming effective teachers of mathematics who guide student learning. In particular, one of the four standards, "Candidate Knowledge, Skills and Dispositions," focuses on the social contexts of mathematics teaching and learning. *Building Support for Scholarly Practices in Mathematics Methods* offers several examples of practices that promote equity and access in diverse classroom settings and help beginning teachers make connections with their students.

Collectively, the chapters in this book provide an initial work of the enactment of the SPTM as well as many other standards and prove to be an excellent resource, inspiring others to engage in examining their practices, share and collaborate with others, and continue to learn. This book is invaluable in highlighting the work of MTEs engaged in examining and researching their own practices as they focus on the development of beginning teachers of mathematics.

—**Christine D. Thomas**
Georgia State University
AMTE President 2015–2017

REFERENCE

Sanchez, W., Kastberg, S., Tyminski, A., & Lischka, A. (2015). *Scholarly inquiry and practices (SIP) conference for mathematics education methods*. Atlanta, GA: National Science Foundation.

PREFACE

This book is intended for mathematics teacher educators (MTEs) who teach prospective teachers (PTs) in mathematics methods courses. Through stories of practice and reports of research, it provides a focus on scholarly inquiry and practice (Lee & Mewborn, 2009) in mathematics methods courses for PTs. The chapters in this book arose from the work of the Scholarly Inquiry and Practices Conference on Mathematics Methods,[1] held in the fall of 2015 in Atlanta, Georgia. Over 50 MTEs were assembled to discuss ways in which theoretical perspectives influence teaching and research in mathematics methods courses.

Methods courses, in many ways, are the heart of teacher preparation. It is in these courses that PTs are asked to think about whom, how, and what they teach in the context of society. The content of mathematics methods courses has been shown to vary substantially across institutions (Taylor & Ronau, 2006). What is taught in mathematics methods courses is of interest to all stakeholders in mathematics education. Moreover, what is learned is even more important. What do PTs have the opportunity to learn through their mathematics methods courses, what do they learn, and what do they carry with them into their teaching practice? Further, how can MTEs build scholarly inquiry and practice (Lee & Mewborn, 2009) that explores this variation across mathematics methods courses in order to learn from each other? This book explores these questions by unpacking the ways in which MTEs use theoretical perspectives to inform their construction of goals, activities designed to address those goals, facilitation of activities, and ways in which MTEs make sense of experiences PTs have as a result.

The 22 chapters in the book are organized in seven sections that highlight how MTEs' theoretical perspectives inform their scholarly inquiry and practice (Lee & Mewborn, 2009). The final section provides insight as we look backward to reflect, and forward with excitement, moving with the strength of the variation we found in our stories and the feeling of solidarity that results in our understandings of purposes for and insight into teaching mathematics methods. This work reflects the efforts of the Scholarly Inquiry and Practices Conference participants. We appreciate their willingness to share stories of practice and embark upon research inquiry that extended the conversations from the conference. In particular, we thank Rochelle Gutiérrez, Elham Kazemi, and Martin Simon for anchoring the discussions about perspectives that launched the conference. Thank you to the University of Washington College of Education for the wonderful book cover photo. Christine Browning, the series editor, has been instrumental in guiding this book to publication. We are most grateful to Fran Arbaugh who saw the power in the work of MTEs at the Scholarly Inquiry and Practices Conference and encouraged us to continue our efforts to create this book.

—**Signe E. Kastberg**
Andrew M. Tyminski
Alyson E. Lischka
Wendy B. Sanchez

NOTE

1. This material is based upon work supported by the National Science Foundation under Grant No. 1503358. Any opinions, findings, and conclusions or recommendations expressed in this material are those of the author(s) and do not necessarily reflect the views of the National Science Foundation.

REFERENCES

Lee, H., & Mewborn, D. (2009). Mathematics teacher educators engaging in scholarly practices and inquiry. In D. Mewborn & H. Lee (Eds.), *Scholarly practices and inquiry in the preparation of mathematics teachers*. In M. Strutchens (Series Ed.) (pp. 1–6). San Diego, CA: Association of Mathematics Teacher Educators.

Taylor, M., & Ronau, R. (2006). Syllabus study: A structured look at mathematics methods courses. *AMTE Connections, 16*(1), 12–15.

SECTION I

PERSPECTIVES AND MATHEMATICS METHODS COURSES

CHAPTER 1

SETTING THE STAGE

Explorations of Mathematics Teacher Educator Practices

Signe E. Kastberg
Purdue University

Andrew M. Tyminski
Clemson University

Alyson E. Lischka
Middle Tennessee State University

Wendy B. Sanchez
Kennesaw State University

When I (Signe) was assigned to teach a mathematics methods course for the first time, I reached out to Wendy Sanchez, whom I knew to be deeply committed to work with prospective mathematics teachers (PTs). As we talked, I wondered about the content of methods. My experiences teaching mathematics learners involved supporting their constructions of mathematics

Building Support for Scholarly Practices in Mathematics Methods, pages 3–10
Copyright © 2018 by Information Age Publishing

content, notions of self as a mathematics learner, and views of mathematics communities. Yet as I faced the prospect of working with PTs to support their constructions of mathematics teaching and learning, I wondered: What do I teach, if I am not teaching mathematics? *This probably gives away too much and puts me at risk for judgments about what I should have known or done, yet I am willing to take the risk to situate the work of authors in this book.*

Wendy shared ideas and resources, notably AMTE monographs (for example, Lee & Mewborn, 2009). Yet my efforts to create activities and ask questions seemed like a patchwork. I wanted to be intentional in developing curriculum and practices and then to reconstruct the curriculum and practices in engagement with PTs. Coherence among activities and emergence of productive pedagogies were elusive. These challenges were coupled with limited access to "exemplars of practice" (LaBoskey, 2007) containing images of ways of mathematics teacher educator (MTE) knowing in research literature (with notable exceptions such as Jaworksi, 2008).

Andrew Tyminski, Wendy Sanchez, Kelly Edenfield, and I began an enquiry (Schwab, Westbury, & Wilkof, 1978) of "the content of methods" through AMTE conference presentations and PME-NA working groups (Kastberg, Sanchez, Edenfield, Tyminski, & Stump, 2012; Kastberg, Sanchez, Tyminski, Lischka, & Lim, 2013). Discussions stemmed from AMTE monographs that challenged MTEs to engage in scholarly inquiry (Lee & Mewborn, 2009), explorations of "issues and practices through systematic data collection and analysis that yields theoretically-grounded and empirically-based findings" (p. 3), and the construction of scholarly practice and use of findings from "empirical studies of the teaching and learning of mathematics and the preparation of mathematics teachers" (p. 3).

The collaborations and research findings (Kastberg, Tyminski, & Sanchez, in press) motivated us to advocate for conscious use of theoretical perspectives in the development of curriculum and pedagogy for mathematics methods. We further sought explorations of MTEs' experiences and practices using methodologies such as autoethnography, self-study, and narrative inquiry in mathematics education research. To gain perspective, we searched for ways to enquire into and communicate about experiences of mathematics methods with MTE scholars. Scholarly Inquiry and Practices Conference for Mathematics Education Methods (SIP) (Sanchez, Kastberg, Tyminski, & Lischka, 2015), a conference dedicated to discussions of scholarly inquiry and practice (Lee & Mewborn, 2009), was the result of our efforts to create opportunities for enquiry and dialogue.

SIP was structured to support conversations of theoretical perspectives and associated *learning goals* for mathematics methods. These goals were then used to inform the construction of *activities* for mathematics methods, defined as situations MTEs provide for PTs to support development toward learning goals (Mewborn, 2000) through *experiences* or "ways in which the preservice teachers internalized those activities" (p. 31). Three broad and commonly used theoretical perspectives (sociopolitical, cognitive, and situative) were selected to frame conference discussions. Our meanings for these perspectives during and after SIP evolved through use of the perspectives as lenses to reinterpret constructing curriculum and pedagogy with PTs in mathematics methods courses. The conference participants taught us that these perspectives are not absolutes, but are situated, interpreted, and operationalized in different ways by MTEs. Use of theoretical perspectives alone or in concert served to enrich participants' discussions of curriculum and pedagogy of mathematics methods. Conversations at SIP culminated in images of the variation in mathematics methods represented in the chapters in this book.

In the two remaining sections, we first discuss variation in mathematics methods as a critical strength of MTEs' work. This discussion is undertaken in the time of accountability and standards that necessitates understanding that just as one size cannot fit all for mathematics learners, MTEs' practices and curriculum are derived with and for PTs and the contexts in which they will teach and live. We conclude with an overview of the book sections and brief introductions of the chapters.

VARIATION IN MATHEMATICS METHODS

Research has demonstrated mathematics methods courses vary in content, activities, and goals (Harder & Talbot, 1997; Kastberg et al., in press; Otten, Yee, & Taylor, 2015; Taylor & Ronau, 2006; Watanabe & Yarnevich, 1999). In response to such findings, Taylor and Ronau (2006) suggested "establishing a common framework offers the possibility of developing shared sets of lenses and a common language, allowing us to conduct a broad-based and open discussion about syllabi and about mathematics methods courses in general" (p. 15). The draft version of the AMTE Standards for Mathematics Teacher Preparation (Bezuk et al., 2016), too, have called for consistency in outcomes of mathematics teacher education programs.

> AMTE's goal is for the standards in this document to provide a clear, comprehensive vision for initial preparation of teachers of mathematics.... [W]e, in this document's standards, elaborate what beginning teachers of mathematics must know and be able to do as well as the dispositions they must have to increase equity, access, and opportunities for the mathematical success of each student. (p. 14)

AMTE's vision, however, does not require conformity and should not define how these outcomes are achieved within teacher preparation programs. It is important for our field to understand that outcome alignment should not come at the cost of variation and autonomy for MTEs. Variation exists because MTEs' work is psychological, social, temporal, and contextual and results in the evolution of curriculum and pedagogy of mathematics methods.

To understand why variation has been reported, take, for example, Marshall and Chao's chapter on mathematics autobiographies (Chapter 18). A quick review of Marshall and Chao's mathematics methods courses would reveal that both MTEs engage PTs in the construction of mathematics autobiographies and utilize Drake's (2006) autobiography story assignment. Yet a longer look would reveal Marshall's choice to invoke a journaling approach and Chao's use of photovoice (Wang & Burris, 1997). These choices help illustrate how their sociopolitical perspectives play out in relation to their personal histories and those of their students within the institutional and community contexts of their work. In short, the origin of variation in this case is human understanding of context and interpretation of needs, goals, and aims. When activities share the same name, such as rehearsal (see Arbaugh, Adams, Teuscher, Van Zoestin, & Wieman, Chapter 9) or clinical interviews (see Chao, Hale, & Behm Cross, Chapter 8), similarities are suggested and, looking broadly, common elements exist among them. Yet, as these authors' descriptions reveal, looking more closely, variation in focus, goals, and implementation also exists. Chapters in this book illustrate that reported variation in mathematics methods is the result of drawing from perspectives, conducting scholarly inquiry, and constructing scholarly practices identified by Lee and Mewborn (2009).

ORIENTATION AND OVERVIEW

The authors of chapters in this book have provided common elements to enable readers to build insight about MTEs' practices. Each chapter includes a description of theoretical perspectives used to inform authors' work. In addition, the terms *learning goal*, *activity*, and *experiences* described earlier in this chapter are used consistently throughout the chapters unless otherwise specified (for example, see Gutierrez, Gerardo, Vargas, & Irving, Chapter 10). The book is comprised of stories and reports of studies across institutional contexts that represent the teaching of mathematics methods "as a complex intellectual endeavor that unfolds in an equally complex sociocultural context" (Borko, Liston, & Whitcomb, 2007, p. 5).

Organization of the Book

The book is organized in seven sections. The first section sets the stage for the remaining sections by describing three theoretical perspectives (sociopolitical, cognitive, and situative) that structured discussions at the SIP conference. The three keynote speakers at the conference were selected based on their expertise and work within a particular theoretical perspective, and each contributed a chapter drawn from their presentation. Gutiérrez (Chapter 2) shares a view of the sociopolitical perspective; Simon (Chapter 3) describes challenges in mathematics teacher education from a cognitive perspective; and Kazemi (Chapter 4) describes and illustrates her view of the situative perspective. These chapters serve to orient readers to an interpretation of each perspective, rather than providing the only possible interpretation. These perspectives are representative, rather than exhaustive, of those that have been and can be used to frame the work of MTEs.

Chapters in Section II describe the affordances and constraints of utilizing perspectives in the design of scholarly inquiry and practice. Weston (Chapter 5) explores her use of the knowledge quartet (Rowland, 2008) while the work of Earnest and Amador (Chapter 6) and Harper, Herbel-Eisenmann, and McCloskey (Chapter 7) provide additional insights regarding the way in which perspectives inform scholarly inquiry and practice. Section III provides examples of learning goals and how MTEs' scholarly practices are developed in connection to these goals. Included in this section are contributions by Chao, Hale, and Cross (Chapter 8) exploring clinical interviews and Arbaugh et al. (Chapter 9) exploring rehearsals. In addition, Gutiérrez, Gerardo, Vargas, and Irving (Chapter 10) explore rehearsals for the development of political knowledge for teaching (Gutiérrez, 2013) and Kinach, Bismark, and Salem (Chapter 11) draw from a cognitive perspective to describe approaches to conceptual-development teaching. Chapters in Section IV provide descriptions of MTEs' work in the development, refinement, or adaptation of a mathematics methods course activity. Wessman-Enzinger and Salem (Chapter 12) illustrate one approach to designing activities for PTs using the context of integer operations. Lawler, LaRochelle, and Thompson (Chapter 13) describe and illustrate how activities used in methods courses can be revised to include learning goals from a sociopolitical perspective. The final chapter in this section, Ward (Chapter 14), illustrates how MTEs' perspectives coupled with PTs' experiences inform revisions of an activity. Chapters in Section V describe MTEs' enactments and course contexts and attend to how these factors influence PTs' learning outcomes. Virmani, Taylor, Rumsey, and colleagues (Chapter 15) examine a variety of ways to embed mathematics methods courses within the context of the K–8 classroom. Singletary, de Araujo, and Conner (Chapter 16), draw from the situative perspective in their discussion of PTs' reflections on transcripts of their teaching as opportunities for learning. The section also includes the aforementioned chapter on autobiography by Marshall and Chao (Chapter 18), along with an exploration by Harper, Sanchez, and Herbel-Eisenmann (Chapter 17) that examines the role of language across two university settings through a mathematical activity with sociopolitical underpinnings. Chapters in Section VI explore MTEs' use of perspectives in teaching and self-evaluation. Casey, Fox, and Lischka (Chapter 19) explore potential connections between theoretical perspectives and the ways MTEs evaluate video as a curricular resource. In Chapter 20, Smith, Taylor, and Shin explore MTEs' theoretical perspectives, learning goals, and activities, as well as alignment between perspectives, learning, and activities. The final entry in the section, by McCloskey, Lawler, and Chao (Chapter 21), describes how MTEs might explore progress toward their teaching goals using the metaphor of a mirror derived from the work of Guitérrez (2016). Section VII concludes with the closing chapter by Richard Kitchen, which provides insight as we look backward to reflect, and forward with excitement, moving with the strength of the variation we found in our stories and the feeling of solidarity that results in our understandings of purposes for and insight into teaching mathematics methods. Kitchen's discussion draws from his insights on the chapters to identify relationship building as a key structure critical to MTEs in mathematics methods.

The editors would like to thank all of the SIP conference participants and authors for their hard work and dedication in writing and refining their chapters through cycles of reviews and revisions. We invite readers to consider the ideas and issues raised by the authors and build upon them in their own work as we move forward in developing scholarly practices for teaching mathematics methods courses. We thank the National Science Foundation[1] for the funding the SIP conference that motivated the book. Finally, we thank Dr. Denise Spangler, who was the external evaluator on the grant but more importantly was a mentor to each of us at various points in our thinking before, during, and after the SIP conference.

NOTE

1. This material is based upon work supported by the National Science Foundation under Grant No. 1503358. Any opinions, findings, and conclusions or recommendations expressed in this material are those of the author(s) and do not necessarily reflect the views of the National Science Foundation.

REFERENCES

Bezuk, N., Bay-Williams, J., Clements, D., Martin, W. G., Aguirre, J., Boerst, T., . . . White, D. (2016). *AMTE Standards for Mathematics Teacher Preparation*. Retrieved from https://amte.net/sites/default/files/SPTM.pdf

Borko, H., Liston, D., & Whitcomb, J. (2007). Genres of empirical research in teacher education. *Journal of Teacher Education, 58*, 3–11.

Drake, C. (2006). Turning points: Using teachers' mathematics life stories to understand the implementation of mathematics education reform. *Journal of Mathematics Teacher Education, 9*(6), 579–608.

Gutiérrez, R. (2013). Why (urban) mathematics teachers need political knowledge. *Journal of Urban Mathematics Education, 6*(2), 7–19.

Gutiérrez, R. (2016). Nesting in nepantla: The importance of maintaining tensions in our work. In N. M. Joseph, C. Haynes & F. Cobb (Eds.), *Interrogating whiteness and relinquishing power* (pp. 253–282). New York, NY: Peter Lang.

Harder, V., & Talbot, L. (1997, February). *How are mathematics methods courses taught?* Paper presented at the Annual Meeting of Association of Mathematics Teacher Educators, Washington, DC. Retrieved from http://www.eric.ed.gov/PDFS/ED446936.pdf

Jaworski, B. (2008). Development of the mathematics teacher educator and its relation to teaching development. In B. Jaworski & T. Wood (Eds.), *The international handbook of mathematics teacher education: The mathematics educator as a developing professional* (Vol. 4, pp. 335–361). Rotterdam, The Neatherlands: Sense.

Kastberg, S., Sanchez, W., Edenfield, K., Tyminski, A., & Stump, S. (2012). What is the content of methods? Building an understanding of frameworks for mathematics methods courses. In L. R. Van Zoest, J.-J. Lo, & J. L. Kratky (Eds.), *Proceedings of the 34th annual meeting of the North American Chapter of the International Group for the Psychology of Mathematics Education* (pp. 1259–1267). Kalamazoo, MI: Western Michigan University.

Kastberg, S., Sanchez, W., Tyminski, A. M., Lischka, A., & Lim, W. (2013). Exploring mathematics methods courses and impacts for prospective teachers. In M. Martinez & A. C. Superfine (Eds.), *Proceedings of the 35th annual meeting of the North American Chapter of the International Group for the Psychology of Mathematics Education* (pp. 1349–1357). Chicago, IL: University of Illinois at Chicago.

Kastberg, S., Tyminski, A., & Sanchez, W. (in press). Reframing research on methods courses to inform mathematics teacher educators' practice. *The Mathematics Educator.*

LaBoskey, V. K. (2007). The methodology of self-study and its theoretical underpinnings. In J. Loughran, M. L. Hamilton, V. K. LaBoskey, & T. Russell (Eds.), *International handbook of self-study of teaching and teacher education practices* (pp. 817–869). Dordrecht, The Netherlands: Springer.

Lee, H., & Mewborn, D. (2009). Mathematics teacher educators engaging in scholarly practices and inquiry. In D. Mewborn & H. Lee (Eds.), *Scholarly practices and inquiry in the preparation of mathematics teachers* (pp. 1–6). San Diego, CA: Association of Mathematics Teacher Educators.

Mewborn, D. (2000). Learning to teach elementary mathematics: Ecological elements of a field experience. *Journal of Mathematics Teacher Education, 3*, 27–46.

Otten, S., Yee, S., & Taylor, M. (2015). Secondary mathematics methods courses: What do we value? In T. Bartell, K. Bieda, R. T. Putnam, K. Bradfield, & H. Dominguez (Eds.), *Proceedings of the 37th annual meeting of the North American*

Chapter of the International Group for the Psychology of Mathematics Education (pp. 772–779). East Lansing, MI: Michigan State University.

Rowland, T. (2008, July). *The knowledge quartet: A theory of mathematical knowledge in teaching.* Paper presented at the International Congress on Mathematics Education, Monteray, Mexico. Retrieved from http://tsg.icme11.org/document/get/405

Sanchez, W., Kastberg, S., Tyminski, A., & Lischka, A. (2015). *Scholarly inquiry and practices (SIP) conference for mathematics education methods.* Atlanta, GA: National Science Foundation.

Schwab, J. J., Westbury, I., & Wilkof, N. J. (1978). *Science, curriculum, and liberal education: Selected essays.* Chicago, IL: University of Chicago Press.

Taylor, M., & Ronau, R. (2006). Syllabus study: A structured look at mathematics methods courses. *AMTE Connections, 16*(1), 12–15.

Wang, C., & Burris, M., A. (1997). Photovoice: Concepts, methodology, and use for participatory needs assessment. *Health Education & Behavior, 24*(3), 369–387.

Watanabe, T., & Yarnevich, M. (1999, January). *What really should be taught in the elementary methods course?* Paper presented at the Annual meeting of the Association of Mathematics Teacher Educators, Chicago, IL. Retrieved from http://www.eric.ed.gov/PDFS/ED446931.pdf

CHAPTER 2

POLITICAL CONOCIMIENTO FOR TEACHING MATHEMATICS

Why Teachers Need It and How to Develop It[1]

Rochelle Gutiérrez
University of Illinois at Urbana-Champaign

Contrary to popular belief and research, addressing equity in mathematics education will not simply come once teachers understand the content they are to teach; when they find accessible, quality, or motivating activities and instructional strategies to use with students; or even when they develop meaningful relationships with students. Many teachers find their biggest struggle lies in understanding and negotiating the politics in their everyday practice. This is particularly true in mathematics, where teachers may expect their work to be straightforward—universal and culture free (Martin, 1997; Powell & Frankenstein, 1997). Teachers have not been trained to negotiate their local politics. Even teachers who have shown substantial

Building Support for Scholarly Practices in Mathematics Methods, pages 11–37
Copyright © 2018 by Information Age Publishing
11

success with students, especially ones who historically have been excluded from mathematics, suggest their knowledge of content, pedagogy, and students is not enough to maintain that success. Politics get in the way, their work is undermined, or they leave the profession.

Imagine if teachers were trained with as much skill and practice in dealing with the politics of teaching as they were with lesson planning, assessment, strategic instructional decisions, classroom management, connecting topics within mathematics, and relating to students. Instead of just carrying out local practices that are valued or have been in place for years, they might question whether those practices are in the best interest of students. They might be more inclined to engage in dialogue and influence others to consider new perspectives. Rather than stand by while new policies are being created that go against their sense of justice, they might advocate for their students or themselves, and perhaps more talented teachers might stay in the profession longer. In this chapter, I will argue (a) mathematics teaching is political, (b) mathematics teachers need political knowledge, (c) teacher education programs can develop political knowledge with teachers through particular activities, and (d) when mathematics teachers have opportunities to understand and deal with the politics of teaching, they are able to use that knowledge in their practice.

POLITICS OF TEACHING MATHEMATICS

All Teaching Is Political

Teaching has always been political, but we seem to be at an extreme point in history. We see talented and committed individuals reconsidering whether teaching will allow them to be the kinds of people they wanted to be when they entered this profession (Natale, 2014). As teachers are robbed of their ability to use professional judgment, even award-winning teachers are counseling the next generation of students to rethink teaching as a profession (Klein, 2014). Private and charter schools may be able to remain competitive because they can ask poor-performing students to leave or because they can simply close their doors if their school is no longer profitable (Seattle Education, 2015). Public school teachers know they must work with every student who walks through their doors. As such, part of teachers' work is creating a counternarrative to stories of students not having enough "grit" (Tough, 2016) or the view that teachers are slackers (Rosemond, 2004).

More and more, corporate America and billionaires with no expertise in education seek to control our schools. In 2015, Eli Broad and his foundation announced they are moving forward with a $490,000,000 plan to privatize the Los Angeles public schools (Blume, 2015). The goal is to create 260 new schools in the next 7 years and to launch a massive marketing campaign that will get families and the general public to embrace the idea that charter schools are the next great innovation for the nation. The Bill and Melinda Gates Foundation, as well as the Walton family, show similar interests in public education. The emphasis on charter schools is likely to intensify with the multimillionaire Betsy DeVos, a leader in the school choice movement, as the new Secretary of Education. Curriculum development corporations like Pearson have capitalized on the standards movement to expand to student assessments and all of the related products to support districts (Persson, 2015). With teachers' salaries and positions partly dependent on student test scores, Pearson is, in a very real sense, controlling who is allowed to stay in the profession.

Corporations are making huge profits by promoting new standards and ways of assessing them, yet the benefits to the public, and to students in particular, are not so clear. The Common Core State Standards are little more than the *Adding It Up* report (National Research Council, 2001) combined with the National Council for Teachers of Mathematics (NCTM) *Principles and Standards for School Mathematics* (NCTM, 2000), documents we already had in our professional community. In fact, the Common Core State Standards are a move away from the "equity principle," one of six key components of previous standards (NCTM, 2000) and a departure from the equity position statement (NCTM, 2008) that suggested teachers need to connect mathematics with students' cultural roots and history. Equity has been the focus of more NCTM presidents' messages than any other topic (Gojak, 2012), yet there is no mention of equity in the Common Core State Standards, and accommodations for "English/Language learners" are in an appendix, something only the tenacious teacher would find.

Content-specific education professors have always evaluated the work of prospective teachers (PTs) and helped decide who is qualified to become a teacher. Now, for-profit corporations control those decisions. Thirty-five states and the District of Columbia have adopted the edTPA, a teacher performance assessment managed by Pearson. Under this new paradigm, PTs pay $300 to upload evidence of planning, instruction, and assessment in hopes of being positively evaluated to become a teacher. As part of the process, they are required to document the kinds of textbooks used in the schools where they are student teaching, important information for a corporation that is seeking to market its products to those not already using them. So, in some ways, our PTs have become data collection agents for a for-profit corporation.

It is not always easy for PTs to understand both the upsides and downsides of new reforms. Take, for example, the Partnership for Assessment of Readiness for College and Careers (PARCC), one of two new national tests given to measure student learning and growth. The PARCC test seeks to better support students by offering a national standard and holding schools accountable for reaching it, thereby making it easier for parents anywhere in the country to judge the ability level of their students, regardless of the state or neighborhood in which they reside. There are many upsides to ensuring all students are held to high standards, as some fear our nation relies too heavily on social promotion (Balingit & St. George, 2016). However, most PTs do not realize that because the PARCC test was never normed on a national population before requiring states to use it, the test is not a valid measure of learning.[2] In fact, some educators have argued that schools are paying a corporation to norm the tests on the backs of their students (Gaines, 2015; Strauss, 2014) and are relinquishing upwards of 6 weeks of instruction to administer such tests. The first set of scores received by students was incredibly low, thereby justifying the need for states and districts to purchase additional materials from Pearson to raise those scores. The cycle often continues with more tests for students, little useful information for teachers about their students' learning, and more profit for corporations. I served on the PARCC item review committee at the high school level. When I raised the issue with Pearson officials in 2013 about consistently low student test scores across the nation and what this meant for students' futures, I was told that Pearson could not be held accountable for any decisions that school administrators made or what the public did with the test; Pearson was "just the people who make the tests." Their goal at that time was for the PARCC test to replace the ACT so that they would gain market share in testing for college. To some extent, their goal is already being realized, as colleges in Delaware, Kentucky, New Jersey, and Colorado are using PARCC scores in admission decisions and entry-level credit for courses. And, although the state of Illinois has recently stopped using the PARCC test (Rado, 2016), most states are still spending millions of dollars on Pearson-related products for PARCC testing. Where corporations might have had market share in textbook adoption, now they are poised to gain market share in college testing. Moreover, Pearson has recently expanded its markets to countries such as the Philippines with Affordable Private Education Center (APEC) secondary schools (Kamenetz, 2016) and intends to impact more than 200 million students worldwide by 2025 (Pearson.com). The increased influence of corporate America, high-stakes testing, and the deprofessionalization of teachers are all signs of an extreme point in the history of public education.

There is so much happening in the public sphere that it would be hard for a PT to keep track of it all or know how to make sense of it without guidance. Most teachers cannot understand how corporations or "philanthropists" could make money off of public schools. I list here just a few things that I have shared with my PTs. Pearson has a $32 million contract to administer tests with the state of New York and $500 million in Texas (Otterman, 2011; Phillips, 2014). California is spending $900 million on Common Core. PARCC and Smarter Balanced received $330 million from the U.S. Department of Education. There is a long history of errors in scoring or delays in reporting scores, design flaws, insufficient memory in systems for testing, and untimely reporting of scores. Students' test scores influence not only teacher's salaries but also students' chances of getting into the next level of schooling. For-profit corporations are in control of not only tests used to decide who stays in teaching but also who becomes a teacher in the first place (e.g., edTPA) and are collecting data about textbook use through this process. Pearson's EnVisionMATH has been found to exaggerate claims of impact and generalizability to students of all ability levels, while grossing a minimum of $320 million per year on this one product, with a potential revenue stream of $2 billion/year (Singer, 2014). Fueled by Race to the Top money, charter schools are popping up everywhere. (KIPP and other charter schools play by their own rules. Pearson owns Connections Academy, a group of virtual charter schools.) Corporations encourage new standards and new products for districts (yet little new content). Pearson places gag rules in test contracts to prevent teachers from raising questions about the tests. Pearson has been caught monitoring kids' social media to stop testing leaks (Strauss, 2015). Students who take PARCC mathematics and reading language arts tests will spend more time testing than aspiring lawyers who sit for the bar exam. And they will get nothing in return. Pearson was implicated in an FBI investigation for unfair bidding practices in a $1.3 billion deal to provide curriculum via iPads to students in the Los Angeles Unified School District (Singer, 2014).

Fortunately, there is a movement of growing resistance from parents, teachers, students, and journalists who are bringing together visions of education that move beyond testing and to highlight the lack of transparency and the attacks on public education. Researchers crunching large data are providing a picture of what is happening in public education, noting that, in the past decade, we have had nearly 2,500 charter schools that have received a total of $3.7 billion in federal funding but have closed or never opened their doors (Persson, 2015).

As the influence of corporate America intensifies, individuals are joining forces with others to reclaim this profession of ours. Their response is not that education should give up all testing. National tests have helped us understand which populations of students are being served well by the school system and which are not. Rather, individuals are finding resources on the internet such as Fair Test, Change the Stakes, New York Core, Saving Our Schools, Creating Balance in an Unjust World, Rethinking Schools, Teachers for Social Justice, TODOS Mathematics for All, and many local groups who are fighting for a definition of education that moves beyond standardized tests. Some teachers and principals are taking matters into their own hands by writing blogs to help distribute information to help families opt out of high-stakes tests (LaReviere, 2015). Others are writing letters to their students or to public officials that can create a wider public debate about not just testing but the nature of education and its place in our society (Goosetree, 2015; Lifshitz, 2015; Look-Ainsworth, 2015; Vilson, 2012).

Although these politics affect everybody, inner-city schools that lack the infrastructure or resources to carry out newer assessments or whose students need more support to reach learning goals based on new standards are more severely impacted. With edTPA and its associated text-heavy forms of evaluation, we may be discouraging or preventing individuals whose first language is not English from entering teaching. Given such politics, it is hard to imagine that we will be able to recruit and retain a large cadre of teachers of color into the profession. Regardless of where they work, PTs and mathematics teacher educators (MTEs) alike will need support to deepen their knowledge of the sociopolitical context of mathematics teaching and learning so that they can make informed decisions about their work (Association of Mathematics Teacher Educators [AMTE], 2017).

All Mathematics Teaching Is Political

How do the aforementioned politics relate to mathematics education in particular? I take as an example two schools—Railside and Union. Railside is a school in Northern California so noted for its success in mathematics that it has been studied by various researchers (Boaler, 2006; Boaler & Staples, 2008; Horn, 2004; Jilk, 2010; Nasir, Cabana, Shreve, Woodbury, & Louie, 2014); Union is a school in Chicago, also noted for its success (Gutiérrez, 1999, 2002a, 2014). Both schools serve low-income, largely Latin@/x[3] populations; both have had teachers who underwent extensive professional development for students to develop conceptual understanding over mere procedures; both have created a departmental community that held a common vision for advancement and a commitment to all students; both have used the Interactive Mathematics Program (Alper, Fendel, Fraser, & Resek, 1997) and showed clear signs of success. Their students have demonstrated the ability to make conjectures and defend their arguments publicly, attained higher test scores than peers in other schools, demonstrated higher classroom engagement overall, and produced a unimodal distribution of engagement from adolescents of different backgrounds. Students have worked in two languages, and a higher percentage of students took calculus (over 40% of the senior class at Union in the 1998–1999 school year).

Yet the efforts of both of these high school mathematics departments were derailed by district politics—a back-to-basic-skills movement in

Chicago and a teaching-to-the-test movement in Northern California. In both locations, highly successful teachers were demoralized and either succumbed to district mandates that went against their professional judgments or left their school or the profession altogether. These schools are not alone. We see pockets of success every day where teachers are working hard and are getting historically excluded students to see themselves as doers of mathematics and to perform well in coursework and on tests. So although the public and many mathematics education researchers seem to believe that the most difficult part about addressing issues of equity is how to get teachers to develop deep and flexible knowledge of mathematics or to adopt particular pedagogical practices, addressing equity is not a technical problem with a technical solution. Values, morals, and judgments all come into play, and these are the heart of politics.

Is it just mathematics teaching that is political, or is there actually something about mathematics as a discipline that is political? A number of researchers across the globe have begun to highlight the ways in which knowledge, power, and identity are interwoven with mathematics, something called the "sociopolitical turn" (Gutiérrez, 2010/2013;[4] Stinson & Bullock, 2015). Early examples that highlighted how power, identity, and knowledge relate to teaching, learning, and teacher education named these as "sociopolitical dimensions of mathematics education" (Valero & Zevenbergen, 2004); a "socio-political orientation" (Chronaki, 1999, p. 19); or simply "power" in mathematics education (Walkerdine, 1988; Walshaw, 2001). For example, Chronaki (1999) suggested that a "political view on mathematics education" should focus on "fostering of citizenship" (p. 19). In general, one distinction is that sociocultural dimensions tend to have enculturation as their goal, whereas sociopolitical dimensions concern themselves with emancipation. In writing about the sociopolitical turn, I chose not to hyphenate the word because I did not believe the social (issues of identity in particular) and the political (issues of power in particular) could be extracted from each other—there is no social without political and vice versa. In fact, sometime after 2010, when the sociopolitical turn was published, most researchers seem to have adopted the term sociopolitical instead of socio-political.

The way mathematics operates in our world and the politics that mathematics brings are important for MTEs to consider. On many levels, mathematics itself operates as Whiteness. Who gets credit for doing and developing mathematics, who is capable in mathematics, and who is seen as part of the mathematical community is generally viewed as White. School mathematics curricula emphasizing terms like Pythagorean theorem and pi perpetuate a perception that mathematics was largely developed by Greeks and other Europeans. Perhaps more importantly, mathematics operates with unearned privilege in society, just like Whiteness. Mathematics is viewed as so pure that it has become the discipline by which we measure other disciplines. See for example, the XKCD comic (n.d.) that depicts mathematicians so far removed from other disciplines that they hardly recognize other scientists.

We treat mathematics as if it is a natural reflection of the universe. When we identify mathematics in the world around us (e.g., Fibonacci sequences in pinecones, fractals in snowflakes), we convince ourselves that mathematics occurs outside of human influence. Rather than recognizing that we may see patterns we want to see (because we set the rules for finding them), we instead feel mathematics is a way of encoding the universe with eternal truths, a natural order of things that should not be questioned. And so mathematics is viewed as a version of the world that is proper, separate from humans, where no emotions or agendas take place.

Because of its perceived purity, we assume mathematics should be the basis for how we think about the world and what is important. Currently, mathematics operates as a proxy for intelligence. Society perpetuates the myth that there are some people who are good at mathematics and some who are not (Mighton, 2004). If you tell someone you are a mathematician or mathematics educator, often you are met with two reactions: confession (e.g., "I was never really good at mathematics") or adulation ("You must be really smart!"). As MTEs, we need to ask ourselves whether we are challenging that adulation or simply accepting it because we enjoy the benefits of increased status and economic gains. Are we really smart just because we do mathematics? As researchers, are we more deserving of large grants because we focus on mathematics education and not social studies or English? Is there something inherent in mathematics as a discipline and human activity that merits higher prestige and higher paychecks?

When we combine the belief that mathematics operates with no values, no judgments, no agenda, with the idea that it properly confers intelligence and importance in society, it can impact how one thinks of oneself. Beyond how well students do in mathematics courses or whether they can imagine themselves pursuing a STEM-based career, they are influenced by this notion of what counts as intelligent. If one is not viewed as mathematical, there will always be a sense of inferiority that can be summoned, especially because the average citizen will not necessarily question the role of mathematics in society. The effects are lasting. So many people are walking around in society who have experienced trauma, microaggressions from participating in math classrooms where the idea of being a successful person, being an intelligent person, is removing oneself from the context, not involving emotions, not involving the body, and being judged by whether one can reason abstractly. Those are all messages that we can unknowingly transmit. It is not just that teaching is political; mathematics is also political. Therefore, whether we recognize it or not, mathematics teaching is a highly political activity.

All Mathematics Teachers Need Political Knowledge to Be Successful

When we acknowledge a sociopolitical perspective on mathematics education, it raises questions about whether PTs are receiving the kinds of knowledge and skills they need. Many are being prepared as if once they develop "ambitious" teaching practices (Lampert et al., 2013) they will be rewarded for their efforts and their students will learn. As we saw in the cases of Union and Railside High, this reality does not exist. High-stakes education, Response to Intervention initiatives, Race to the Top campaigns, and the latest packaged reforms can keep us from acting on what is in the best interest of our students and their learning. In terms of preparing teachers to become professionals, there is nothing in edTPA that will assess whether PTs can successfully deconstruct the deficit messages about teachers, students, or public education in movies like *Waiting for Superman* (Guggenheim & Kimball, 2010) or *Won't Back Down* (Barnes & Hill, 2012). Nor can the edTPA identify teachers who can see limitations in the latest reform movements like "growth mindset" or "grit." On the surface, these movies and reforms address equity by helping students get a better education. However, the savvy educator understands that these movies have the best interests of charter schools and corporate America in mind, instilling the idea that public schools need a hostile takeover. An effective teacher can realize that growth mindset and grit, although important characteristics for students, situate the problem of learning in individual student motivation and ignore broader institutional and systemic inequities. If teachers are unable to deconstruct the deficit messages circulating in society about themselves, their students, or public education, they cannot successfully advocate for policies and practices that are research-based or ethically just.

The majority of professional development that PTs and practicing teachers receive from teacher education programs, their districts, and professional societies like NCTM do not focus on helping teachers understand or negotiate the politics they regularly face. Though we have made many advances in such things as how to appropriately use technology or how to build upon the linguistic and cultural resources that students bring to school, most programs in teacher education still work largely from the same set of assumptions about the kinds of knowledge bases teachers of mathematics need, which were developed in the late 1980s. Whether we call it pedagogical content knowledge (Shulman, 1986) or mathematical knowledge for teaching (Hill et al., 2008), teachers are expected to become fluent in content knowledge, pedagogical knowledge, and knowledge of students.

POLITICAL CONOCIMIENTO FOR TEACHING

I am arguing for a fourth kind of knowledge—political knowledge for teaching. I refer to this knowledge as *political conocimiento,* and I explain more thoroughly what that means in other papers (Gutiérrez, 2012, 2013b). What is important to understand here is that although the Spanish term *conocimiento* translates to "knowledge" in English, I am borrowing a version from Anzaldúa (1987) that acknowledges that all knowledge is relational. Things cannot be known objectively; they must be known subjectively. This is comparable in English to when we say, "Do you know that restaurant?" We are not expecting that knowledge to be a universal objective set of facts. Instead, the speaker is getting at your relationship with that restaurant: Are you familiar with it? What experiences do you have with it? Your knowledge of that restaurant may overlap with the knowledge that others have of it, but it will not be the same. For our purposes, key features of *conocimiento* are subjectivity, solidarity with others, and interdependence.

For mathematics teachers, political *conocimiento* is the kind of knowledge that helps you deconstruct and negotiate the world of high-stakes testing and standardization. It helps you connect and explain your mathematics to community members and district officials. It buffers you from a system or helps you reinvent or reinterpret systems so that you can be an advocate for your students. In essence, political *conocimiento* is the kind of knowledge that allows you to see how politics permeates everything we do, in education in general and mathematics in particular, and affects how we are connected to each other today and how we might envision a different, more humane connection for the future.

The key difference in this model versus other models is the idea that knowledge is *with* students and communities, not knowledge *of* them or *for* them (see Figure 2.1). We come to "know" students not in some kind of

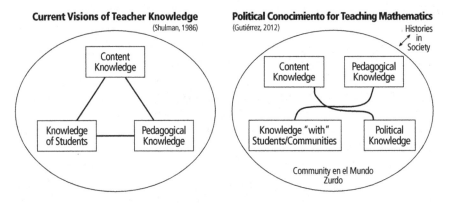

Figure 2.1 Teacher knowledge.

objectified way (Gutiérrez, 2009) but rather by standing alongside them, committed to being interdependent with them. All of this work is done not as individual teachers but in a supportive community with others. The term *el mundo zurdo* recognizes this community as the left-handed world of solidarity among people of color, people who are queer, historically looted, physically challenged, and resisting various forms of colonization (Moraga & Anzaldúa, 1981). The presence of the term *histories in society* recognizes that mathematics has been and is being practiced in different ways throughout the world. We are in a particular moment in time, not just in terms of modern mathematics, but in terms of what is happening with respect to mathematics and education today.

PTs who have developed political *conocimiento*—that useful knowledge that helps them deconstruct deficit narratives in society about students, teachers, or public education—are better prepared to question the world around them and to use their professional judgment when making decisions about the kinds of learning opportunities students need. They can see the benefits of using achievement data as a first step to identify who is not being served well by the school system, but they recognize the limitations of defining equity around such things as "closing the achievement gap." They understand that, more than just getting all kids to perform better or the same on tests of achievement, we should be invested in helping students become the kinds of people they want to be, fulfilling goals they have defined for themselves, which can mean different, not same outcomes (Gutiérrez, 2002b). Teachers with political *conocimiento* are able to question authority when outside entities come in and tell us that we need to focus on "bubble kids"or that we need to develop a "growth mindset" in our students. If we are telling students that it is really important for them to develop perseverance and grit or grow new dendrites to get smarter, but the system remains stacked against them, is that really a healthy perspective to promote? From the point of view of students of color and historically looted students,[5] does that just sound like a new version of "pull yourself up by your bootstraps"? When PTs and practicing teachers lack political *conocimiento*, they can unknowingly adopt simplistic reform packages and slogans that make them feel they are effectively addressing equity and social justice.

Creative Insubordination

When PTs are developing political *conocimiento*, they often feel a desire to do something to address the injustices they witness. This is where creative insubordination comes into play. *Creative insubordination* is a term grounded in the 1980s, a term I heard growing up in an activist family, a term used on a regular basis in my community. I later learned that creative insubordination was published in literature on principal leadership because some principals were found to stand up to the establishment to protect their teachers when decisions were being made that did not seem fair (Crowson & Morris, 1985). I find it extremely helpful for naming the work that community leaders and exceptional teachers do as a matter of their everyday practice (Gutiérrez, 2013a, 2015a, 2015b; Gutiérrez & Gregson, 2013; Gutiérrez, Irving, & Gerardo, 2013). Creative insubordination recognizes innovative work that individuals, in collaboration with others, do when they need to get a job done but when doing so will be met with resistance from those protecting the status quo. Teachers who are creatively insubordinate learn to bend rules and interpret things in ways that rely on a higher ethical standard. Rather than simply following what others around them are doing or telling them to do, they reflect deeply and base their decisions on professional judgment guided by doing right by students. I emphasize the creative part to highlight the fact that this work is not done foolishly or naively. It is done in a way that keeps teachers from being fired. In this sense, like any other professional knowledge, it requires skill and precision.

Teacher Education Programs Can Develop Political Knowledge

One set of issues in which mathematics teachers need to be able to reinterpret or bend rules is equity. When PTs enter classrooms for observation or student teaching, they receive strong messages that equity is about the achievement gap; equity is about growth mindset; equity is about grit and other things. So before they enter those sites, I try to help them grapple with a more sophisticated notion of equity. I present for them four dimensions of equity/learning (Figure 2.2) that they should consider: access,

Figure 2.2 Dimensions of equity/learning.

achievement, identity, and power (Gutiérrez, 2007, 2009) and get them to identify particular scenarios as being more or less about particular dimensions (Gutiérrez, 2006).

In doing so, they come to recognize the complexity and tensions that play out in our work as mathematics teachers (Gutiérrez, 2009, 2015a, 2015b). That is, our work can remain neat and tidy, aligned with most administrators and policymakers as well as the general public, including many parents, if we adhere to a mainstream definition of equity that concerns itself only with access (e.g., students having equal opportunities to learn, loaded terms like "quality" teachers and "rigorous" curriculum) and achievement (e.g., equal outcomes on standardized tests, equal numbers of mathematics courses taken, equal representation in the STEM pipeline). This is what I refer to collectively as the *dominant axis* of equity because it dominates the beliefs held by most educators, parents, and policymakers. But we might ask ourselves, is this definition of equity/learning adequate if we also care about the kinds of identities that students develop inside and outside of our classrooms? Does this definition of equity reflect justice if, in order to be seen as legitimate participants in mathematics, students can only follow the "standard algorithm" or speak English while doing mathematics? Does this definition of equity make sense if students never come to understand the historical and cultural aspects of mathematics as a human practice? Does this definition of equity encourage teachers to model how mathematics can be used as a lens to identify inequities in society and to then address those inequities in one's home community? Or is it simply concerned with students getting good grades and access to college?

What I aim for in my teacher education courses is that PTs will walk away asking themselves, "For any given definition of equity, who benefits?" When given the opportunity to think deeply about definitions of equity and learning that circulate in society and in coursework, most PTs are able to understand the importance of identity and power, which is the *critical axis* on the diagram. Here, I mean critical not as in fundamental or key, but as in a critique of the status quo. This axis considers what will be meaningful from students' perspectives. Whenever we think of equity, we always ask, "equity for what purpose and from whose point of view?"

The four dimensions of equity/learning are a useful taxonomy and mapping space. Rather than being a definition that PTs will adopt uncritically, the four dimensions provide language for discussing more nuanced situations that arise in the teaching of mathematics, something that terms like *mathematics for all, closing the achievement gap,* or simply *equity* do not easily capture. This language also helps PTs recognize that part of their job may involve helping students to "play the game" of mathematics, as in do well on standardized tests and develop proficiency in the eight Standards for Mathematical Practice (National Governors Association, 2010). Not to attend to such goals would put students in jeopardy of not having all options open in terms of career, college, or earning potential. The goal would be to attend to this axis at least enough for students to decide their own futures rather than having others dictate those futures. But another part of teaching may involve helping students to "change the game"—supporting students' identities and power, even when those are at odds with things like scoring well on standardized tests. Helping students to change the game may arise by using social justice mathematics curricula (Esmonde, 2014; Gregson, 2013; Gutstein, 2003, 2006; Turner & Strawhun, 2005) or assigning projects that draw upon students' experiences in home communities (Aguirre, Zavala, & Katanyoutant, 2012; Turner, Gutierrez, & Diez-Palomar, 2011). It could also involve changing the ways that we, as teachers, relate with mathematics and with our students and, again, it may require us to use our own sense of justice rather than that provided by our school or district. Changing the game is important because by not preparing students to do so, teachers are potentially keeping students from becoming the kinds of people they aim to become or from seeing a broader and more humane version of the activity we call mathematics. PTs may grapple with these cross-cutting goals, but those goals force them to think about their stance. What are they willing to stand for as a teacher? What definition of social justice will they use, and how will they know they are achieving it? A sophisticated definition of equity/learning would not allow a teacher to know she is achieving it without input from her students.

In the center of the diagram, there is the concept of *Nepantla*, a form of Nahua metaphysics. Nepantla is not only a space of tensions but from a kind of cosmological perspective is a way of interacting in the world that recognizes opposing forces and values and maintains those tensions rather than trying to shut them down (Anzaldúa, 1987; Anzaldúa & Keating, 2002). It is different from how we traditionally think of dealing with opposing views. Many PTs are familiar with cognitive dissonance, the psychological discomfort one experiences when recognizing two viable but seemingly irreconcilable perspectives (Festinger, 1957). They are able to value the idea that noticing competing views is an important component to motivating change in students. However, the goal in cognitive dissonance is to eliminate the dissonance, to choose one thing over another. With Nepantla, we want to maintain that dissonance for a while, to become comfortable with the tensions, because that is how we develop new knowledge.

The idea of Nepantla not only allows me to help PTs grapple with important tensions and ethics in teaching mathematics, it allows them to recognize that if their work as teachers will involve helping students to play the game and change the game, they, as teachers, will need to be able to do so as well. And by extrapolation, we as MTEs will need to learn how to play the game and change the game, a point I will return to later in this chapter.

As I have described, one way to help develop teachers' political *conocimiento* is to offer opportunities for them to interrogate mainstream definitions of equity, learning, and mathematics. By introducing a framework and a language for talking about mathematics teaching, MTEs help them learn to question the status quo in ways that set the stage for them to question the status quo in schools where they may work one day.

Conceptual Framework

Elsewhere, I have described how my research team and I created a program that allows us to support teachers to more fully develop political *conocimiento* with each other (Gutiérrez, 2015a; Gutiérrez et al., 2013). Figure 2.3 shows some of the key structural and conceptual components of our model.

This is our tapestry weave framework, and we use it to show that there are certain structural components—seminars, a teacher partnership, critical professional development sessions, an after-school mathematics club, and mentoring—that support the kinds of conceptual ideas we value. Broadening and challenging knowledge, developing an advocacy stance, noticing

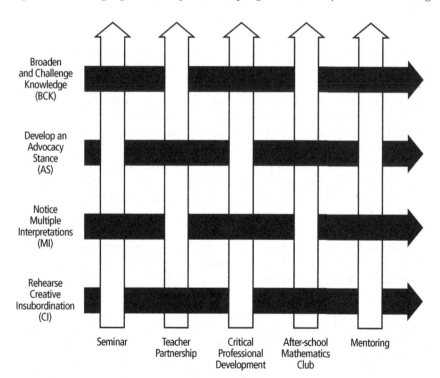

Figure 2.3. Conceptual framework for teacher education.

multiple interpretations, and rehearsing for creative insubordination are all things that we hope our PTs will develop as they move through the program. Our model involves more than just readings and reflections. The structural components provide the backbone for the conceptual components to take root, and all of these threads intertwine and provide support for each other to create the tapestry. At times, it involves becoming a *Nepantler@* (one who is comfortable living with tensions) and also becoming the "other" (one who is marginalized). It involves engaging with youth in nonschool settings and engaging with others who model creative insubordination, and it will help them rehearse for the kinds of political encounters that they will face as teachers. This model highlights that no single activity or aspect of the methods courses I teach prepares teachers to develop political *conocimiento.* Instead, they are exposed to a variety of situations and course assignments that reinforce that they should take seriously the idea that being a professional requires making professional judgments in connection with others; that such judgments require political clarity, a commitment to ethics; and that such judgments need to be defensible to parents, students, and other stakeholders. This is the basis upon which teachers will be able to carry out creative insubordination in their future work environments.

The Mirror Test

One key feature of the language and philosophy that undergirds my teaching is something I call the *Mirror Test.* The Mirror Test is a way of thinking about the profession that uses one's internal ethical compass as opposed to an external one, whether external is corporate America or one's district or professional society's ways of measuring whether one is a qualified professional. The Mirror Test suggests PTs look themselves in the mirror and ask: "Am I doing what I said I wanted to do in education when I set out to be in this profession? And, if I'm not, what am I going to do about that?" I do not mean to imply that a teacher candidate simply writes out why they want to enter teaching, what they plan to do, and then constantly returns to that list over the course of their career in order to decide if they are doing a good job. New teachers do not know everything they plan to stand for in teaching when they start out. What constitutes the Mirror Test is constantly being refined by the kinds of new knowledge bases, experiences, and solidarities that one creates with other groups. At the heart of the Mirror Test are one's core values and the willingness to act upon those values in order to advocate for students.

In My Shoes

In My Shoes is an activity I use with PTs during one of four methods courses. It provides PTs an opportunity to rehearse creative insubordination strategies and is explained in greater detail in the chapter by Gutiérrez,

Gerardo, and Vargas (see Chapter 10). The activity begins with a teacher leader (a practicing teacher, student teacher, or peer) describing a difficult situation in which they found themselves and were unsure what to do. It could be a new policy that is being enacted in the mathematics department or a new textbook adoption that does not seem to have students' best interests in mind. It could be hearing a colleague remark, "Students can't handle a more rigorous curriculum" or a student who says, "I can't believe she's not helping me just because I'm Black." In essence, the teacher leader is asking the group, "What would you do if you were in my shoes?"

After a scenario is raised with the group, several rounds of discussion and a role-play ensue. PTs discuss the situation and ask clarifying questions of the teacher leader who presented the scenario without the teacher leader revealing the response to that scenario. The facilitator then normally repeats back to the group the list of possible actions that have been offered. Next PTs identify, partly upon consensus in the group, which of the possible actions seems most worthy of taking up further through discussion, with the idea in mind that it will be the focus of the rehearsal or role-play. A member of the group volunteers or is chosen to become the main actor (protagonist) in the scenario. The role-play may take place immediately or, on the rare occasion, the protagonist may be asked to leave the room while others prepare for their roles. PTs role-play, with one member of the group acting as the protagonist who carries out the recommended action. Other members of the group take on different roles, including playing devil's advocates who aim to make it difficult for the protagonist to carry out their action in a productive manner. In this phase, the teacher leader takes on the role of the antagonist, or the one who caused the difficult situation in the first place. Next, PTs debrief how the role-play went, including how the protagonist felt about their actions and whether there was anything they might do differently. Other members offer feedback on which points of the role-play they thought the protagonist performed well and which points (including actions by the antagonist or devil's advocates) seemed to derail their efforts. The entire process can take upwards of an hour.

Unlike other kinds of rehearsals that may be scripted or are focused on a particular thing to learn about teaching, these role-plays are organic and are dependent on the perspectives and lived experiences of members of the group. The goal of this activity is not to prepare PTs for all of the important or even likely scenarios that they will face in their first few years of teaching. Rather, In My Shoes is an opportunity to reflect on all of the possible things they might do in a given situation and to feel what it will be like to engage in the complex process of negotiation inherent in situations where power dynamics are at play, where others will not necessarily agree with their points of view. More importantly, In My Shoes allows PTs to practice responding in a strategic manner and feeling all of the emotions that come with trying to articulate your view in the moment and standing up for something you feel is right.

From the point of view of the PTs, they know that the goal is not to come up with the *right* answer to the scenario. They also know that they are not simply trying to guess what the person who is offering the scenario actually did as a response. The beauty of In My Shoes is that one must learn to think like a marathoner, not a sprinter. Rather than going after every fight in a school, PTs need to pick their battles. They need time to ponder which things most warrant taking a stand. With The Mirror Test, one cannot stand for everything and all people and all things; one must have political clarity and pick the things that are really important in terms of advocating for students. These are the things that will run through one's mind when looking in the mirror each day. PTs, at this stage of their career, need to consider a variety of moves that they could use and the kind of language that would accompany those moves. Rather than assuming that PTs will figure that out on their own or see examples of it in their school placements, we structure opportunities for them to learn these moves. Negotiating the local politics in schooling is not a simple thing. Like other aspects of teaching that benefit from planning and rehearsals, helping PTs learn to deal with politics in a creative way that advocates for students while allowing them to keep their jobs is also an important skill worthy of rehearsing.

Teachers Learning Political Conocimiento

One way I identify if teachers have grown in their political *conocimiento* is when they participate in more sophisticated ways with others (e.g., peers, instructors, people in schools), more like professionals who have a clear stance on the field and less like students who are pleasing their professor or simply following what their cooperating teacher does (Brown & McNamara, 2005). Early in their program, few PTs see teaching as a political act; almost none would agree that mathematics is political. By the end of the program—partly through assignments where they are required to follow blogs or Twitter feeds; develop a working definition of mathematics; create critical dialogues with their cooperating teachers; and reflect on current events in teacher evaluations, national assessments, and learning standards— they are slowly exposed to the politics of teaching, and most demonstrate their understanding that teachers often need to take a stand one way or another if they are going to be able to look themselves in the mirror. It is unreasonable to expect PTs to have the level of sophistication that veteran teachers who possess political *conocimiento* would. Instead, I look for signs of growth. Those include the ability to deconstruct competing messages about concepts like equity, mathematics, and learning that circulate in society; consideration of not just one-to-one interactions but historical, systemic, and institutional aspects of schooling that affect particular students; the propensity to take a stand, even if PTs are unsure what that will look like in practice; and a well-developed evidence base that can back their claims that they are putting students' best interests first and supports their approach to teaching.

With respect to In My Shoes, in particular, I see growth when PTs are able to move from immediately suggesting actions to be taken—giving advice—to recognizing the importance of first gathering additional information and weighing the different kinds of approaches one could take. I also see growth when PTs are not just willing to consider more deeply the context of the situation and are able to offer viable actions for the teacher in the situation, but are willing to become the protagonist in the role-play. Other signs of growth are making connections between a given political scenario and other scenarios they have faced once the group is debriefing a role-play. They might remark, "So, something similar happened to me last week in *my* school." If PTs can make these connections and begin to do a meta-analysis—recognizing that two seemingly different scenarios have the same underlying theme—that will allow them to begin to understand larger discourses that operate in schools. Moreover, when they are able to bring new scenarios to the group regardless of how pleased or confident they felt in their response to that scenario, I see that they are moving beyond the idea that political situations have easy or correct answers. In some respects, I liken our teachers who are good at identifying these political situations to teachers who get good at identifying *group-worthy problems* (Horn, 2005). They develop a shared language and way of learning from and with each other that is meaningful and useful for teaching.

Partly as a result of various opportunities to develop political *conocimiento* and rehearsing for political situations through In My Shoes, the PTs with whom I have worked report doing a number of things in their schools to advocate for students. Some of these acts of creative insubordination occurred while they were student teachers and others once they had worked for a year or more. They include convincing others of the importance of taking a stand in a given situation and offering some viable ways to do it; organizing with other teachers in the building to challenge a principal who did not want to explicitly inform parents about their students' rights to opt out of high-stakes tests; challenging a cooperating teacher who blatantly disregarded students who put their heads down on their desks and completed no mathematical work in class; offering professional development workshops on Complex Instruction (Featherstone, 2011) and other useful ways to position students as experts; and creating spaces (e.g., Google Hangouts) for other teachers in the region to share their specific struggles, values, and approaches to deconstructing school policies and practices when they are not in the best interest of their students.

CONCLUSION

In doing this work, I have learned that teaching is about so much more than just planning for and carrying out instructional activities. In the same way that those teaching strategies that support emergent bilinguals and multilinguals tend to work well for other students, political *conocimiento* for teaching mathematics is critical for students who have been historically excluded or marginalized in mathematics, but such *conocimiento* is also helpful for all students. This kind of work is collaborative and intergenerational, meaning that the knowledge we create needs to be collaborative and in partnership with those who have come before us and will come after us. In our research group and in our interactions with PTs, we like to say, "We act ourselves into new ways of thinking, not vice-versa." That is, it is not just work we think about and create philosophy statements about; this work is action-oriented.

The language and frameworks to which we expose PTs go a long way towards helping them make sense of the profession. In the same way they would plan for and deal with students' conceptions of particular mathematical topics, they learn to recognize the politics and be able to plan for and deal with it when it arises. Moreover, it is not just *what* one learns in a teacher education program but actually *how* one learns that matters. If the forms of knowledge that we expect teachers to develop arise from habits of mind and actions that value tensions, rely on ethics, acknowledge politics, and are largely guided by what is in the best interest of students, this is modeling for them how they will do this work as lifelong learners. In other words, as I move forward in my career as a teacher, if I am learning through rehearsals and out-of-school spaces, if I am attending conferences and movies with veteran teachers and novices, if I am debriefing with others, it means that I am not going to expect to do this work on my own as a teacher. It also means that I am going to want to attend events with other people and debrief with other people. It means that I am not just going to look to textbooks or professors or peers for official knowledge, that I will continue to do this work in community with a diverse group of people, face-to-face and through blogs, social media, and using whatever means necessary.

The climate of high-stakes testing, new teacher evaluations, corporate America's growing interest in education, and the dismantling of schools like Railside and Union High show us that teaching mathematics requires much more than learning how to develop inquiry-based lessons and assessments, or cultivating relationships with students, or having goals of ambitious teaching in ways that have been traditionally defined. We must prepare teachers to take a stand and to reclaim the profession of mathematics education. Prospective (and practicing) teachers need opportunities to understand the broader education landscape as it relates to capital, identity, and power. They need to be able to deconstruct the messages that society sends to us about what is important in learning, teaching, and justice.

As we move forward as a field, I ask us to think hard about what is our Mirror Test as MTEs: By what standards will we judge ourselves to be excellent? Who or what will we look to in order to decide if we are really doing a good job in our teaching? Will we look externally to promotion and tenure reviews to decide we are excellent? Or will we have developed an internal compass that can tell us we are consistent with our ethics and what is best for students in the public schools, even if that means standing up to administrators, colleagues, or PTs who feel more comfortable maintaining the status quo? Can we honestly say we are preparing beginning teachers for the realities they will face if we ignore politics? For a long time, the mathematics education community has talked about equity, about a kind of "mathematics for all." From the point of view of students who are Black, Latin@/x, American Indian, recent immigrants, emergent bilinguals or multilinguals, or historically looted, if we do not prepare our teachers for the political nature of mathematics teaching, then we are doing little more than rearranging the chairs on the deck of the *Titanic*.

NOTES

1. This research was funded by the National Science Foundation, Grant # 0934901. Any opinions, findings, and conclusions or recommendations expressed in this material are those of the author and do not necessarily reflect the views of the National Science Foundation. Thank you to the teachers who so graciously shared their teaching struggles and accomplishments with me. Research assistants include Sonya E. Irving, Juan Manuel Gerardo, and Gabriela E. Vargas.
2. PARCC is not the only test with problems. Smarter Balanced tests also have been found to have egregious flaws (Heitin, 2015).
3. I use the term Latin@/x to indicate solidarity with individuals who identify as lesbian, gay, bisexual, transgender, questioning, and queer (LGBTQ). Both Latin@ and Latinx represent a decentering of the patriarchal nature of the Spanish language whereby groups of males and females are normally referred to with the "o" (male) ending as well as a rejection of the gender binary. For some, the circular line radiating outward represents gender fluidity; for others, the "x" represents a variable whereby any gender form could be represented. My choice to use this term reflects my respect for how people choose to name themselves.
4. I cite this article as 2010/2013 because it was published online through JRME in 2010 and some researchers began citing it as such then. It was not released

in print in JRME until 2013, and some researchers have cited it as such since. Because the focus of the article is on a particular point in history, the work should reflect the earlier date.

5. I use the term *historically looted* to emphasize the fact that certain students and their families are not just "low income." They have not been able to accrue wealth because others have stolen that wealth from them. See, for example, Madrigal (2014) for the inconsistent ways in which the Federal Housing Administration loans were distributed to citizens who were Black or White. See Weinberg (2003) for a brief history on how American Indians, Blacks, and poor Whites have been exploited for their labor.

REFERENCES

Aguirre, J., Zavala, M. R., & Katanyoutant, T. (2012). Developing robust forms of pre-service teachers' pedagogical content knowledge through culturally responsive mathematics teaching analysis. *Mathematics Teacher Education and Development, 14*(2), 113–136.

Alper, L, Fendel, D., Fraser, S., & Resek, D. (1997). *Interactive mathematics program, year 1.* Emeryville, CA: Key Curriculum.

Anzaldúa, G. (1987). *Borderlands/La Frontera: The new mestiza.* San Francisco, CA: Aunt Lute Books.

Anzaldúa, G., & Keating, A. L. (2002). *This bridge we call home: Radical visions for transformation.* New York, NY: Routledge.

Association of Mathematics Teacher Educators. (2017). *AMTE standards for mathematics teacher preparation.* Raleigh, NC: Author.

Balingit, M., & St. George, D. (2016, July 5). Is it becoming too hard to fail? Schools are shifting toward no-zero grading policies. *The Washington Post.* Retrieved from http://www.washingtonpost.com

Barnes, D. (director, writer), & Hill, B. (writer). (2012). *Won't back down.* USA: 20th Century Fox Film Corporation.

Blume, H. (2015, September 30). Backers want half of LAUSD students in charter schools in eight years report says. *Los Angeles Times.* Retrieved from http://www.latimes.com

Boaler, J. (2006). How a detracked mathematics approach promoted respect, responsibility, and high achievement. *Theory Into Practice, 45,* 40–46.

Boaler, J., & Staples, M. (2008). Creating mathematical future through an equitable teaching approach. *Teachers' College Record, 110,* 608–645.

Brown, T., & McNamara, O. (2005). *New teacher identity and regulative government: The discursive formation of primary mathematics teacher education.* New York, NY: Springer.

Chronaki, A. (1999) Contrasting the "socio-cultural" and "socio-political" perspectives in maths education and exploring their implications for teacher education, *European Educational Researcher, 5,* 13–20.

Crowson, R. L., & Morris, V. C. (1985). Administrative control in large-city school systems: An investigation in Chicago. *Educational Administration Quarterly, 21,* 51–70.

Esmonde, I. (2014). "Nobody's rich and nobody's poor.... It sounds good, but it's actually not": Affluent students learning mathematics and social justice. *Journal of the Learning Sciences, 23*(3), 348–391.

Featherstone, H. (2011). *Smarter together: Collaboration and equity in elementary classrooms.* Reston, VA: National Council of Teachers of Mathematics.

Festinger, L. (1957). *A theory of cognitive dissonance.* Stanford, CA: Stanford University Press.

Gaines, L. (2015). Evanston Township High School District 202 administrator weighs in on PARCC. *Chicago Tribune.* Retrieved from http://www.chicagotribune.com

Gojak, L. M. (2012, June 5). Let's keep equity in the equation. *National Council of Teachers of Mathematics Summing Up.* Retrieved from https://www.nctm.org/News-and-Calendar/Messages-from-the-President/Archive/Linda-M_-Gojak/Let_s-Keep-Equity-in-the-Equation/

Goosetree. (2015, March 2). To PARCC or not to PARCC? Retrieved from http://mrsgoostree.weebly.com/blog/to-parcc-or-not-to-parcc

Gregson, S. A. (2013). Negotiating social justice teaching: One full-time teacher's practice viewed from the trenches. *Journal for Research in Mathematics Education, 44*(1), 164–198.

Guggenheim, D. (writer, director), & Kimball, B. (writer). (2010, released October 29). *Waiting for Superman.* USA: Paramount.

Gutiérrez, R. (1999). Advancing Urban Latina/o youth in mathematics: Lessons from an effective high school mathematics department. *The Urban Review, 31*(3), 263–281.

Gutiérrez, R. (2002a). Beyond essentialism: The complexity of language in teaching Latina/o students mathematics. *American Educational Research Journal, 39*(4), 1047–1088.

Gutiérrez, R. (2002b). Enabling the practice of mathematics teachers in context: Towards a new equity research agenda. *Mathematical Thinking and Learning, 4*(2&3), 145–187.

Gutiérrez, R. (2006). *How would you classify it?* Unpublished professional development activity used in the secondary mathematics teacher education program at the University of Illinois at Urbana-Champaign.

Gutiérrez, R. (2007). Context matters: Equity, success, and the future of mathematics education. In Lamberg, T., & Wiest, L. R. (Eds.), *Proceedings of the 29th annual meeting of the North American Chapter of the International Group for the Psychology of Mathematics Education* (pp. 1–18). Stateline, NV: University of Nevada, Reno.

Gutiérrez, R. (2009). Embracing the inherent tensions in teaching mathematics from an equity stance. *Democracy and Education, 18*(3), 9–16.

Gutiérrez, R. (2010/2013). The sociopolitical turn in mathematics education. *Journal for Research in Mathematics Education, 44*(1), 37–68.

Gutiérrez, R. (2012). Embracing "Nepantla": Rethinking knowledge and its use in teaching. *REDIMAT- Journal of Research in Mathematics Education, 1*(1), 29–56.

Gutiérrez, R. (2013a). Mathematics teachers using creative insubordination to advocate for student understanding and robust mathematical identities. In M. Martinez & A. Castro Superfine (Eds.), *Proceedings of the 35th annual meeting of the*

North American Chapter of the International Group for the Psychology of Mathematics Education. (pp. 1248–1251) Chicago, IL: University of Illinois at Chicago.

Gutiérrez, R. (2013b). Why (urban) mathematics teachers need political knowledge. *Journal of Urban Mathematics Education, 6*(2), 7–19.

Gutiérrez, R. (2014, October). *When professional development is not enough: Secondary mathematics teaching in an era of high stakes education.* Selected presentation given at the annual meeting of the Society for Advancement of Chicanos and Native Americans in Science. Los Angeles, CA.

Gutiérrez, R. (2015a). Nesting in Nepantla: The importance of maintaining tensions in our work. In N. M. Joseph, C. Haynes, & F. Cobb (Eds.), *Interrogating Whiteness and relinquishing power: White faculty's commitment to racial consciousness in STEM classrooms* (pp. 253–282). New York, NY: Peter Lang.

Gutiérrez, R. (2015b). Risky business: Mathematics teachers using creative insubordination. In T. G. Bartell, K. N. Bieda, R. T. Putnam, K. Bradfield, & H. Dominguez (Eds.), *Proceedings of the 37th annual meeting of the North American Chapter of the International Group for the Psychology of Mathematics Education* (pp. 679–686). East Lansing, MI: Michigan State University.

Gutiérrez, R., & Gregson, S. (2013, April). *Mathematics teachers and creative insubordination: Taking a stand in high-poverty schools.* Paper presented at the Annual Meeting of the American Educational Research Association. San Francisco, California.

Gutiérrez, R., Irving, S., & Gerardo, J. M. (2013, April). *Mathematics, marginalized youth, and creative insubordination: A model for preparing teachers to reclaim the profession.* Paper presented at the annual meeting of the American Educational Research Association, San Francisco, California.

Gutstein, E. (2003). Teaching and learning mathematics for social justice in an urban, Latino school. *Journal for Research in Mathematics Education, 34.* 37–73.

Gutstein, E. (2006). *Reading and writing the world with mathematics: Toward a pedagogy for social justice.* New York, NY: Routledge.

Heitin, L. (2015, March 10). Math consultant: Smarter Balanced math tests have 'egregious flaws.' *Education Week.* Retrieved from http://blogs.edweek.org/edweek/curriculum/2015/03/math_consultant_smarter_balanc.html

Hill, H. C., Blunk, M. L., Charalambous, C. Y., Lewis, J. M., Phelps, G. C., Sleep, L., & Ball, D. L. (2008). Mathematical knowledge for teaching and the mathematical quality of instruction: An exploratory study. *Cognition and Instruction, 26*(4), 430–511.

Horn, I. (2004). Why do students drop advanced mathematics courses? *Educational Leadership, 62*(3), 61–64.

Horn, I. (2005). Learning on the job: A situated account of teacher learning in high school mathematics departments. *Cognition and Instruction, 23*(2), 207–236.

Jilk, L. (2010). Becoming a liberal math learner: Expanding secondary school mathematics to support cultural connections, multiple mathematical identities and engagement. In R. S. Kitchen & M. Civil (Eds.), *Transnational and borderland studies in mathematics education* (pp. 69–94). New York, NY: Routledge.

Kamenetz, A. (2016, March). Pearson's quest to cover the planet in company-run schools. *Wired.* Retrieved from https://www.wired.com/2016/04/apec-schools/

Klein, R. (2014, March 20). Winner of $1 million teaching prize has some depressing thoughts about the state of education. *The Huffington Post.* Retrieved from http://www.huffingtonpost.com/2015/03/20/nancie-atwell-prize_n_6910948.html

Lampert, M., Franke, M. L., Kazemi, E., Ghousseini, H., Turrou, A. C., Beasley, H., Cunard, A., & Crowe, K. (2013). Keeping it complex: Using rehearsals to support novice teacher learning of ambitious teaching. *Journal of Teacher Education, 64*(3), 226–243.

LaReviere, T. (2015, March 31). An open letter to Illinois State Board of Education Chairman, Reverand James Meeks. Retrieved from https://troylaraviere.net/2015/03/31/isbes-ungodly-stand-against-illinois-children/

Lifshitz, J. (2015). An open letter to my students: I am sorry for what I am about to do to you. *The Huffington Post.* Retrieved from http://www.huffingtonpost.com/jessica-lifshitz/an-open-letter-to-my-students-parcc_b_6808060.html

Look-Ainsworth, M. (2015, February 9). An open letter to Governor Walker. *The Marquette Educator.* Retrieved from https://marquetteeducator.wordpress.com/2015/02/09/an-open-letter-to-governor-walker/

Madrigal, A. C. (2014, May 22). The racist housing policy that made your neighborhood. *The Atlantic.* Retrieved from http://www.theatlantic.com

Martin, B. (1997). Mathematics and social interests. In A. B. Powell & M. Frankenstein (Eds.), *Ethnomathematics: Challenging Eurocentrism in mathematics education* (pp. 155–172). Albany: State University of New York Press.

Mighton, J. (2004). *The myth of ability: Nurturing mathematical talent in every child.* Toronto, ON: Walker Books.

Moraga, C., & Anzaldúa, G. (1981). *This bridge called my back: Writings by radical women of color.* Watertown, MA: Persephone Press.

Nasir, N. S., Cabana, C., Shreve, B., Woodbury, E., & Louie, N. (2014). *Mathematics for equity: A framework for successful practice.* New York, NY: Teachers College Press.

Natale, E. A. (2014, January 17). Why I want to give up teaching. *The Hartford Courant.* Retrieved from http://www.courant.com

National Governors Association Center for Best Practices & Council of Chief State School Officers. (2010). *Common Core State Standards for Mathematics.* Washington, DC: Authors.

National Research Council. (2001). *Adding it up: Helping children learn mathematics.* Washington, DC: National Academy Press.

National Council of Teachers of Mathematics. (2000). *Principles and standards for school mathematics.* Reston, VA: Author.

National Council of Teachers of Mathematics. (2008). *Equity in Mathematics Education.* Reston, VA: Author.

Otterman, S. (2011, August 13). In $32 million contract, state lays out some rules for its standardized tests. *The New York Times.* Retrieved from http://www.nytimes.com

Persson, J. (2015, September 22). CMD publishes list of closed charter schools (with interactive map). Retrieved from http://www.prwatch.org/news/2015/09/12936/cmd-publishes-full-list-2500-closed-charter-schools

Phillips, E. (2014). We need to talk about the test: A problem with the Common Core. *The New York Times.* Retrieved from https://www.nytimes.com

Powell, A. B., & Frankenstein, M. (Eds.). (1997). *Ethnomathematics: Challenging Eurocentrism in mathematics education.* Albany: State University of New York Press.

Rado, D. (2016, August 26). Math scores improve, English marks drop on state PARCC exams. *Chicago Tribune.* Retrieved from http://www.chicagotribune.com

Rosemond, J. (2004, February 1). Are public school teachers slackers or dedicated educators? *The Southern Illinoisan.* Retrieved from http://thesouthern.com

Seattle Education. (2015, November 13). Charter schools: A map of failure and a money vortex. Retrieved from https://seattleeducation2010.wordpress.com/2015/11/13/charter-schools-a-map-of-failure-and-a-money-vortex/

Shulman, L. S. (1986). Those who understand: Knowledge growth in teaching. *Educational Researcher, 15*(2), 4–14.

Singer, A. (2014, December 15). Pearson education can run, but it cannot hide. *The Huffington Post.* Retrieved from http://www.huffingtonpost.com/alan-singer/pearson-education-can-run_b_6327566.html

Stinson, D. W., & Bullock, E. C. (2015). Critical postmodern methodology in mathematics education research: Promoting another way of thinking and looking. *Philosophy of Mathematics Education Journal* [25th Anniversary Issue], *29,* 1–18.

Strauss, V. (2014, October 8). Pearson's wrong answer—and why it matters in the high-stakes testing era. *The Washington Post.* Retrieved from https://www.washingtonpost.com

Strauss, V. (2015, March 14). Pearson monitoring social media for security breaches during PARCC testing. *The Washington Post.* Retrieved from https://www.washingtonpost.com

Tough, P. (2016, June). How kids learn resilience. *The Atlantic.* Retrieved from http://www.theatlantic.com/

Turner, E. E., Gutierrez, M. V., & Diez-Palomar, J. (2011). Latina/o Bilingual Elementary Students Pose and Investigate Problems Grounded in Community Settings. In K. Tellez, J. N. Moschkovich, & M. Civil (Eds.), *Latino/as and mathematics education: Research on learning and teaching in classrooms and communities* (pp. 149–174). New York, NY: Information Age.

Turner, E. E., & Strawhun, B. T. F. (2005). "With math, it's like you have more defense": Students investigate overcrowding at their school. In E. Gutstein & B. Peterson, (Eds), *Rethinking mathematics: Teaching social justice by the numbers* (pp. 81–89). Milwaukee, WI: Rethinking Schools.

Valero, P., & Zevenbergen, R. (2004). *Researching the socio-political dimensions of mathematics education: Issues of power in theory and methodology.* Norwell, MA: Kluwer.

Vilson, J. (2012, November 28). An open letter to Chancellor Dennis Walcott and others on the idea of assessment. Retrieved from http://thejosevilson.com/open-letter-to-chancellor-dennis-walcott-and-others-on-the-idea-of-assessment/

Walkerdine, V. (1988). *The mastery of reason: Cognitive development and production of rationality.* London, England: Routledge.

Walshaw, M. (2001). A Foucauldian gaze on gender research: What do you do when confronted with the tunnel at the end of the light? *Journal for Research in Mathematics Education, 32*(5), 471–492.

Weinberg, M. (2003). A short history of American capitalism. New History Press. Retrieved from http://www.newhistory.org/AmCap.pdf

XKCD. (n.d.). *Purity*. Retrieved from https://xkcd.com/435/

CHAPTER 3

CHALLENGES IN MATHEMATICS TEACHER EDUCATION FROM A (MOSTLY) CONSTRUCTIVIST PERSPECTIVE[1]

Martin A. Simon
New York University

In this chapter, I am going to share some ideas that derive primarily from my research. In doing so, I hope to disrupt some thinking about mathematics teacher preparation. The cited articles go into greater depth on these ideas. A clarification before I begin: Although I was billed as speaking from a constructivist perspective, let me say I am not a constructivist. That is, I don't belong to the "church" of constructivism. I don't think educators can afford to belong to a theoretical church. Large, orienting theories of learning are useful tools, and in education we need to use as many of them as prove useful. We need to figure out what work each tool can do and put these tools to work for us.

Building Support for Scholarly Practices in Mathematics Methods, pages 39–48
Copyright © 2018 by Information Age Publishing

IS OUR GOAL TEACHER CHANGE OR TEACHER INDUCTION?

A fundamental distinction that needs to be made in thinking about the education of prospective teachers is whether our goal is education of teachers to change the current system or whether our goal is induction into the system as it is. My work and this talk are focused on teacher preparation for change. However, a lot of the solutions that are currently being proposed, which focus almost exclusively on experience in classrooms, are really solutions that foster induction—not change.

LIMITED BY WHAT I (WE?) DON'T KNOW

I don't know how to teach mathematics methods. I have done it about 25 times. However, I am never satisfied with the impact, so I change it every time. One of the reasons is the time I have with my students is not commensurate with the work that needs to be done. It is hard to get people in positions of power to understand the scope of the change that needs to take place and what is required to engender that change. The other reason I do not know how to teach methods is that I don't know enough about the development of prospective teachers. I don't understand enough about how to foster particular objectives, and I don't know enough about how to sequence those objectives. I believe these are also inadequacies in the knowledge base for mathematics teacher education.

EXPLORING THE PROBLEMS AND CONTRIBUTIONS TO SOLUTIONS

Time Efficiency

What can teacher education programs do to at least partially resolve the problem of time? First, let us consider the role of content courses. If we are going to change prospective teachers' understanding and image of what it means to teach mathematics conceptually, then we have to give them the opportunity to *learn* mathematics in compatible ways. Sending prospective teachers to the mathematics department to engage in *traditional* mathematics courses is part of the problem—not part of the solution. On the other hand, rigorous mathematics content courses taught conceptually in a way that builds on prospective teachers' knowledge have the potential not only to have a greater effect on teachers' content knowledge but also to engender more powerful images of mathematics learning and teaching—a solid foundation for mathematics education courses. With a foundation of this type, the work in a methods course can be more efficient and more advanced.

A second issue that works against time efficiency is the lack of appropriate mentor teachers. Generally, there is not a cadre of classroom teachers who can support the change in mathematics teaching that we are trying to promote in methods courses. So, that which prospective teachers see and are expected to fit into is very often not the kind of teaching that we would like them to learn. In most cases, prospective teachers are loyal to their mentor teachers and develop a good rapport with them. This creates a tension for the prospective teacher between what the mentor teacher does and what they are taught in their methods courses. This not only makes teacher education inefficient but reduces the impact as well. The lack of appropriate mentor teachers is part of a greater challenge of building an infrastructure for mathematics teacher education, an infrastructure that includes appropriately prepared people teaching content courses, teaching methods courses, serving as mentor teachers, and serving as practicum supervisors. Funding organizations, such as the National Science Foundation, are interested in how many teachers will be affected by a particular project. However, they generally do not ask, "How are you going to build the infrastructure to serve all these teachers without a very severe dilution effect?"

Insufficient Knowledge Base for Teacher Education

Earlier, I indicated that we do not know enough about teacher development. This is really a need for more and better research. However, in order to understand teacher development, we first have to understand the goal of development. We know that in teaching mathematics, specifying the mathematical concepts we want students to learn is crucial. I argue that specifying the pedagogical concepts that are the goal of teacher education is just as crucial. In particular, we need to specify pedagogical concepts related to how to help students learn a mathematical concept that they don't already know or understand.

We are teaching methods courses and running teacher development programs without sufficiently problematizing the goals. We have some goals related to attitudes and practices. We want teachers to listen to students, value different answers, and use collaborative groups, whole-class discussion, multiple representations, calculators, and nonroutine problems. However, none of these goals deals directly with how to foster the learning of a new concept. When students are not getting the concept, what does the teacher do?—listen to students, value different solutions, use manipulatives? These are all good tools—good practices. However, it is not enough to have faith that good practices will result in learning. As a field, we need to understand development and promote understanding in prospective teachers of how teachers can foster particular concepts.

In a traditional model of teaching, in which teaching was equated with imparting information, learning was assumed to be passive receiving and retention was based on repetitive practice. This was a clear, albeit ineffective, model. That model has been discredited, though we have not articulated clear models to take its place. By a model of teaching, I mean a framework for promoting the learning of particular concepts that is grounded in conceptualization of student learning. (In Simon, 2013a, and Simon, Placa, & Avitzur, 2016, we describe our emerging model of mathematics pedagogy based on a particular interpretation of reflective abstraction.)

I am arguing, teaching practices do not equal teaching. It is currently very popular to talk about practices—"ambitious practices" or "high-leverage practices." All the best practices together do not provide a conceptualization of teaching. We are leaving practicing and prospective teachers with a void. They are developing practices without a way to understand learning and fostering learning. They are carpenters with good tools without the knowledge of the principles of building a house.

I am arguing that mathematics teacher educators cannot be agnostic about teaching. We need to identify a conceptualization of teaching and learning and work to develop it in teachers. There is no generic teacher education program that prepares teachers for all conceptualizations of teaching. Based on particular conceptualizations of teaching that we are endeavoring to foster (we don't need a consensus—different conceptualizations can be the goal of different programs), we can engage in research on how prospective teachers come to conceptualize teaching in a compatible manner, and how we can foster such a conceptualization.

MAJOR ASSIMILATORY STRUCTURES: ONE REASON CHANGE IS SO DIFFICULT

Elsewhere (Simon, 2013b), I have postulated *major assimilatory structures* to explain the difficulty in engendering significant teacher change. What it means is that teachers, when they come to our teacher education programs, already have deeply ingrained images of what it means to teach mathematics. They are not coming in saying, "Please help me understand what it means to teach math." Rather, they expect us to prepare them to do what they already think mathematics teaching is. Major assimilatory structures are important because they affect everything teachers do as a teacher and because they have significant impact on the sense teachers make of the experiences we promote in their teacher education programs.

Why do I call it a major assimilatory structure? One, is it's made up of lots of different components. It includes images of students, images of the teacher's role, ideas of how students learn, and images of how teachers teach. It is also major because it has a major assimilatory role. That is, we can work to promote change in particular aspects of their conceptions of mathematics teaching, and they can assimilate the changes into their preexisting overall conception. For example, a popular and useful aspect of mathematics teacher education is helping prospective teachers learn to listen to students' ideas. However, because of their major assimilatory structure, prospective teachers often think they are learning to listen to students' thinking so they can figure out what students know and what students don't know—so they know what to teach. That is, they are not thinking that understanding students' thinking can provide a basis for building on the knowledge they already have. These two perspectives are very different and come from different assimilatory structures.

I will now discuss a set of major assimilatory structures based on our research (Simon, Tzur, Heinz, Kinzel, & Smith, 2000). We consider one assimilatory structure to be a *traditional perspective*. I define a traditional perspective as a conceptualization (perhaps unarticulated) that students learn by taking in that which is told to them and shown to them. So teaching is a transfer of teacher ideas to the students. I use this structure (surely oversimplified) only in contrast to the two structures that I now discuss.

A second structure that is commonly developed in our teacher education programs is a *perception-based perspective*. From this perspective, mathematics is transparently available in certain situations, certain experiences, certain tasks, and in certain representations. So, for example, base-10 concepts are readily available to students if you have them work with base-10 blocks. This perspective seems to have some advantages over a traditional perspective. Generally, a richer set of tools, representations, and tasks are used. However, the perspective is problematic. If you have used base-10 blocks with students, you may have observed that base-10 concepts are not visible to all in the blocks. The blocks represent base-10 concepts only if students bring certain understanding for making sense of the blocks. Otherwise they are just blocks of different sizes and shapes. The perception-based perspective is natural. Tasks and representations that, for the teacher, transparently represent the concept, perhaps the ones that in methods courses were so "illuminating," should be illuminating for the students as well. This expectation that mathematics is transparently portrayed in particular representations and tasks is the basis of this second, significant assimilatory structure, a perception-based perspective.

The third one, and for my research team a goal of teacher education, is a *conception-based perspective*. This perspective includes the following ideas. Students' knowledge affects what they perceive and the sense they make of what they perceive. What they know affects what they attend to. So, students' experiences are structured by what they already know and are not determined solely by the representation or the task. Further, students learn by building on prior knowledge. They don't take in new ideas from materials or from somebody else. Students don't perceive mathematical relationships; rather, they build concepts from their prior knowledge in the context of their experience.

Let me give a couple of examples. Nobody believes that students can develop a concept of number (cardinality) through traditional teaching. That is, children cannot learn cardinality by being told about numbers. Children cannot learn cardinality by being shown examples of particular numbers, for example four trucks and four elephants. Because cardinality is not something they can attend to, they cannot see it as what is common among sets of the same number of elements. It is like trying to demonstrate the colors blue and purple to a colorblind person. When students are counting initially, we know that they don't have a sense of cardinality. They learn to play the game. They have a song. The words are "one, two, three, four, five, six, seven" (in an English-speaking situation). They learn, after a while, to touch one object every time they say a word from their song. They learn to respond that way when asked, "How many?"

"How many are there?"
"One, two, three, four, five"
"How many?"
"One, two, three, four, five."
"So, how many are there?"
"One, two, three, four, five."{\UNL}

They have learned what to do. They have no understanding of cardinality. They could have just as well used a different song, "A, B, C, D, E." But they do eventually develop the concept of cardinality. They develop it from their activity. After a while they start to understand that the farther they get in their song, the more objects there are. Eventually, each number word comes to represent a different amount of stuff. These learners are building on their experience and what they already know. The concept of number was not in the objects (and not even in the "counting"), or they would have gotten it much more quickly. It also was not the result of somebody telling them. It was something they got from their activity of counting. For a detailed description of our emerging explanation of conceptual learning, see Simon et al. (2016).

In Heinz, Kinzel, Simon, and Tzur (2000), we discussed a practicing teacher, Ivy, who was very committed to helping her students understand mathematics. She was teaching a lesson on long division and was concerned that her students had previously learned it by rote. She had them do long-division tasks with base-10 blocks. In the classroom discussion that followed, she set up a juxtaposition for them to see how each step they did with the blocks corresponded to a step in the long division. It did not work. Why didn't it work? It didn't work for the same reason we have been discussing. Ivy had concepts that allowed her to see the juxtaposition, the parallels between those two systems. Students did not have those concepts, so they were not looking at the same stuff with the same knowledge and therefore were not seeing the same relationships. So this is an example of Ivy teaching from a perception-based perspective. For Ivy, the connection between the blocks solution and the algorithm was clearly visible, and therefore, she expected it would be so for her students.

SPECIFYING PEDAGOGICAL CONCEPTS

It is important that we know the particular understandings we want our prospective teachers to come away with. Pedagogical understandings are not the only goal of teacher education, but they are a big part of what we need to achieve. Providing experiences is not enough. What do we want prospective teachers to learn from those experiences? If we want to give prospective teachers experience leading class discussions—great; but what do we want them to learn about leading class discussions?

I argue here that aspects of the knowledge base in mathematics education, derived from empirical and theoretical research, can be critical content and goals for mathematics teacher education. However, we need to specify and clearly articulate the pedagogical concepts we wish to promote. I will give some examples of concepts that I have identified. I have chosen some challenging concepts that are not commonly part of the discussion of goals of mathematics teacher education.

The first example of a pedagogical concept pertains to negotiation of classroom norms. This is a social construct deriving from the work of Cobb and his colleagues (Yackel & Cobb, 1996) using an emergent perspective. I want prospective teachers to understand the following. A norm is what is expected in and what it means to successfully function in a classroom environment. Norms can be identified in every classroom, whether or not the teacher is consciously negotiating norms. Norms are negotiated, not imposed. Negotiation of norms is a back-and-forth interaction between teachers and students and between students and students. It is a negotiation in that each subsequent action or verbalization is a response to the prior one. For example, the teacher asks for a justification; the students offer a justification. The teacher asks a further question in light of the justification the students produced, and so on. All teachers negotiate norms through the way they interact with their students. If teachers are conscious about negotiating norms, they can make choices about the norms they intend to establish and the way they participate in the negotiation.

My second example of a pedagogical concept involves what it means to develop a new arithmetical operation (e.g., addition, subtraction, multiplication, and division of whole numbers and fractions). I want prospective teachers to understand the following. Every operation is a new way of viewing the world. When children are born, they do not separate the world into examples and nonexamples of division. At some point they develop a concept of division through which they can see different situations. Consider this contrast between two students. Student A is given the following problem. "You have 12 cookies and 3 friends. You want to give each friend the same number of cookies. How many cookies do you give each friend?" Student A takes 12 blocks to represent the 12 cookies and starts dealing them out into 3 piles. He concludes that each friend gets 4 cookies. Then the same student is given a second problem, "I have a toy train with three boxcars and I have 12 balls that I want the train to carry with the same number in each boxcar. How many balls should I put in each boxcar?" Student A once again takes 12 blocks and deals them out into 3 piles.

In contrast, Student B is given the same first problem. She uses the same blocks strategy as Student A. Then, Student B is given the second problem involving the train. Instead of using blocks, Student B looks at the teacher and says, as if it is a stupid question, "Four." What's different about these two students?

Student B can anticipate that she would undertake the same activity as in the prior task. When that anticipation is stable (i.e., she can recognize examples and nonexamples of the activity without going through the activity), she has a concept that can be symbolized and labeled, in this case partitive division. If a symbol or label is given prior to the development of that concept (anticipation), there is no meaning for the symbol or label; it does not refer to something that the learner can attend to. This is what I want prospective teachers to understand about students developing an arithmetical operation.

My third example of a pedagogical concept is a distinction I have discussed previously (Simon, 2006). It is another example of how contributions to the knowledge base on mathematics education can provide goals and content for mathematics teacher education. The distinction is between *reflective abstraction* and an *empirical learning process*. I will start with an example. A goal in the middle grades might be for students to understand what happens when two odd numbers are multiplied. In a first case, students take calculators and multiply pairs of odd numbers. They observe that the products are odd and conclude that the product of two odd numbers is odd. In this case, the students have *not* developed a mathematical concept. All they know is *that* the product of two odd numbers seems to be odd. They have no reason to *expect* that result. This is an example of an empirical learning process, drawing a conclusion by seeing a pattern of inputs and outputs. The process is a black box. A mathematical concept is never the result of empirical learning. Whereas there is sometimes a role for empirical work in a mathematics classroom, it never, in itself, produces a mathematical concept.

In contrast, I share a second case of students working on the same concept. In this class, an even number means that all elements can be paired. An odd number means that, after creating all possible pairs, one element remains unpaired. Students start with 5 times 3. They take blocks and make 3 groups of 5. They start by pairing within each group of 5. Because 5 is odd, each group has one unpaired block. They move the unpaired blocks together and realize quickly that they will not be able to pair them because 3 (one from each group) is odd. Through several such enactments of this activity with different pairs of odd numbers, the students come to be able to anticipate that if I have n odd numbers I'm going to have n leftovers, and if n is odd I'm not going to be able to pair them, so the product has to be odd. This is reflective abstraction. This is the development of a mathematical concept.

I define mathematical concept as *a researcher's articulation of student knowledge of the logical necessity involved in a particular mathematical relationship* (Simon, 2017). Knowledge of the logical necessity derives from reflective abstraction, not from empirical learning. I want teachers to understand the difference between empirical learning and reflective abstraction. I want them to understand that a mathematical concept involves knowledge of the logical necessity.

CONCLUSION

I began by asserting that after many years of experience, I do not know how to teach mathematics methods. Let me focus on what I think I do know. First, we need to change the image that prospective teachers have of mathematics teaching, and this is not an easy feat. However, anything less than this is not going to work. Therefore, we need to think about how to change their major assimilatory structures. I think that notion is important, because it gets us to conceptualize the challenge and not assume that because we change some practices we have changed their fundamental idea about teaching. Second, we have to problematize and specify the teaching that is the desired outcome of teacher education—not just the practices, but the underlying perspective on how to promote particular mathematical concepts. How can large groups of teacher educators get together to talk about the challenges, opportunities, and strategies for teaching prospective teachers and not problematize the goals, not problematize the teaching that is intended? Third, we must have specific conceptual goals—pedagogical concepts. This is necessary, but not sufficient. We need empirical studies on the development of these pedagogical concepts and conjectured trajectories for promoting the learning of these concepts.

NOTE

1. This chapter is an edited transcript of Simon's keynote address at the 2015 Scholarly Inquiry and Practices Conference (Sanchez, Kastberg, Tyminski, & Lischka, 2015).

REFERENCES

Heinz, K., Kinzel, M., Simon, M. A., & Tzur, R. (2000). Moving students through steps of mathematical knowing: An account of the practice of an elementary mathematics teacher in transition. *Journal of Mathematical Behavior, 19,* 83–107.

Sanchez, W., Kastberg, S., Tyminski, A., & Lischka, A. (2015). *Scholarly inquiry and practices (SIP) conference for mathematics education methods.* Atlanta, GA: National Science Foundation.

Simon, M. A. (2006). Key developmental understandings in mathematics: A direction for investigating and establishing learning goals. *Mathematical Thinking and Learning, 8*(4), 359–371.

Simon, M. A. (2013a). Issues in theorizing mathematics learning and teaching: A contrast between learning through activity and DNR research programs. *The Journal of Mathematical Behavior, 32,* 281–294.

Simon, M. A. (2013b). Promoting fundamental change in mathematics teaching: A theoretical, methodological, and empirical approach to the problem. *ZDM: The International Journal on Mathematics Education, 45,* 573–582.

Simon, M. A. (2017). Explicating mathematical concept and mathematical conception as theoretical constructs. *Educational Studies in Mathematics, 94*(2), 117–137.

Simon, M. A., Placa, N., & Avitzur, A. (2016). Participatory and anticipatory stages of mathematical concept learning: Further empirical and theoretical development. *Journal for Research in Mathematics Education, 47,* 63–93.

Simon, M., Tzur, R., Heinz, K., Kinzel, M., & Smith, M. (2000). Characterizing a perspective underlying the practice of mathematics teachers in transition. *Journal for Research in Mathematics Education, 31,* 579–601.

Yackel, E., & Cobb, P. (1996). Sociomathematical norms, argumentation, and autonomy in mathematics. *Journal for Research in Mathematics Education, 27,* 458–477.

CHAPTER 4

TEACHING A MATHEMATICS METHODS COURSE

Understanding Learning From a Situative Perspective[1]

Elham Kazemi
University of Washington

For the last 20 years I have been concerned with how teachers and teacher educators learn. I have been concerned with at least three big questions:

1. How do you make school a worthwhile place to be? For both teachers and students?
2. What kinds of learning environments get you inside practice, with others, to pay careful attention to content and to students as learners and as people?
3. How can you design and carry out powerful ways for adults to learn together about what really good teaching looks like?

Building Support for Scholarly Practices in Mathematics Methods, pages 49–65
Copyright © 2018 by Information Age Publishing

These questions are motivated by the knowledge that schools can be sites of both oppression as well as liberation. Schools play a critical role in creating what Mike Rose (1995) has called *possible lives*. As many scholars have aptly shown, schools are structured to put in place major barriers to our ability to pursue the goals we want to achieve for ourselves and for our families. My thinking is informed by many scholars in education who have called for us to be much more attentive not just to the quality of subject matter teaching but to the way we position students, whom schools do and do not serve, and the continual ways that those in power shape the life experiences of our students (e.g., Diversity in Mathematics Education, 2007). I am not a critical race scholar or a historian, and I want to be humble about my own understandings of the scholarly conversations on race, privilege, and power. That said, I am reading and listening and trying to expand my own role and responsibility as a teacher educator in helping prospective teachers (PTs) gain some insight into racialized, gendered, and class relations that shape our schools (e.g., Gutiérrez, 2013; Lareau, 2003; Wilkerson, 2010). I teach a course that is situated within a teacher education program that aims to help PTs become social justice educators (e.g., Murrell, 2001; Zeichner, 2012). We try to structure our courses and experiences so that teachers develop allies in schools and in the community to understand the lived experiences of students and families we serve. We want our students to be self-aware and critical of how schools function and how we ideally would like them to function in a more democratic and socially just society. From my perspective, in my course, I want students to be critically self-reflective of how the day-to-day interactions in our mathematics teaching contributes to or undermines the equity goals we have (see also Ramirez & Celedon-Pattichis, 2012; Turner et al., 2012).

At this conference, Scholarly Inquiry and Practices for Mathematics Education Methods, I was asked to describe my work in mathematics methods teaching and how situative perspectives on learning inform my work and help me understand what learning is happening in my classes. I want to acknowledge that hardly anything I talk about here is just my own thinking. I am thankful to my colleagues in the academy and in the community whom I have had the privilege to learn from. And in most recent years, the team of researchers from the Learning In, From and For Teaching (LTP) project (Lampert et al., 2013).

A SITUATIVE PERSPECTIVE

I'm going to draw heavily from two pieces by Jim Greeno and colleagues because they provide helpful summaries of what has been called a situated or situative perspective on learning (see also Herrenkohl & Mertl, 2010; Lave

& Wenger, 1991). Greeno (2006) has explained why he prefers the term *situative* to others because he thinks that when we use the terms *situated learning* or *situated cognition,* the use of *situated* can invite misconceptions "that some instances of action, cognition, or learning are situated and others are not" (p. 79). In what follows, as I state some of the basic premises of this perspective, I will use the term *situative.*

Greeno (2006) has shown how a situative perspective helps us analyze learning. I will in turn describe how it informs the way I think about the design of learning environments, specifically my mathematics methods course.

> In a situative study, the main focus of analysis is on performance and learning by an activity system: a collection of people and other systems. In a situative study, individual cognition is considered in relation to more general patterns of interaction.... The goal is to understand cognition as the interaction among participants and tools in the context of an activity. (pp. 83–84)

Because our understandings, goals, intentions, and expectations are shaped by our joint action, Greeno summarized that in order to study learning we must analyze

1. collaborative discourse
2. how we are positioned and position ourselves in participant structures of interaction because that positioning generates knowledge and information structures
3. the representations and representational practices people use in the activity system because they can make learning and knowledge visible (p. 86)

Before we dive into the activity system that constitutes my methods course experience, I want to remind us that situative perspectives are not prescriptive. Turning to another article by Greeno and colleagues (1998),

> As a scientific perspective, situativity does not say what educational practice should be adopted. However, it does say that the activities of different learning practices are important, not only for differences in their effectiveness or efficiency, but also because participation in those practices is fundamental in what student learn. (p. 14)

They go on to underscore that "all teaching and learning are situated; the question is what their situated character is" (p. 19). This idea is really important. I'm not suggesting that a situative perspective determines the particular design of my course, but rather that a situative perspective leads me to think about what the situated character of my course is and how that shapes the teaching and learning that occurs.

MATHEMATICS METHODS COURSE PRINCIPLES

To take you inside my methods class, it's important to learn about the principles that guide the course. These principles have grown over the last 5 years as I have interacted with my LTP colleagues and with Elizabeth Dutro at the University of Colorado, Boulder. I imagine they will continue to be refined. The principles guide me in thinking about what to make visible and explicit as we work together in the class.

We are guided by a set of principles for teaching:

1. Teachers must position students as sense-makers and knowledge-generators who desire to invest and succeed in school. This involves noticing children; building relationships with them, their families, and communities; valuing their perspectives; and attending to their thinking, curiosities, and capabilities.
2. Teaching is both intellectual work and a craft. Deep knowledge of content and pedagogy, creativity, and passion fuel both learning and teaching.
3. Teachers must design equitable learning environments in which all children are engaged in robust and consequential learning.
4. Teachers' instruction and student learning are always conducted within the context of larger social systems, structures, and hierarchies.
5. What we do and say matters and must be analyzed. Our language and action construct and constrain opportunities for children to build meaningful, positive, and sustained relationships to learning and one another.

The design of our activities and the coursework in general is guided by the following principles for learning to teach:

1. Teaching is intellectual work and requires specialized knowledge.
2. Teaching is something that can be learned.
3. Learning to do something requires repeated opportunities to practice.
4. There is value in making teaching public.
5. We all bring our histories forward. Our own learning experiences and identities shape what we know and do. Our developing identities as mathematics teachers matter to our work with children.

THE DESIGN OF THE MATHEMATICS METHODS COURSE

I'm going to share a set of images that help to provide an overview of the weekly cycle of investigation, enactment, and reflection on learning that we engage in our methods class so you get a sense of the activity system that we have created. And then I'll take you inside some of the conversations we have in these different ways of organizing learning.

The first of two quarters of our methods course takes place at a partner elementary school, and the examples in this chapter are drawn from the 2014–2015 school year, during which we worked closely with a third-grade class. The school population is comprised of a richly diverse community. That year, 32% of the population was Latina/o; 25% Asian American; 23% African American or Black; 3% Native Hawaiian or other Pacific Islander, 0.5% Native American or Alaskan Native; 11% White; and 5% identified as biracial or multiracial. Eighty-four percent of the school population in 2014–2015 qualified for free or reduced lunch. Seventeen percent qualified for special education services, and 43% were considered transitional bilingual.

Getting to know the teacher and the class is central to our work. We are attentive to the emotional and personal connections that matter in student–teacher relationships. We are playful with one another. We do team-building activities with our student buddies that get us to laugh and to see each other outside the context of academic discussions (see images in Figure 4.1). We have informal time with one another to learn about our interests and experiences. My course always begins with a human connection—getting to know each other, what matters to us, and how we narrate ourselves as people, teachers, and students.

As our class gets underway, our central activity in learning to teach is co-planning a set of intentionally selected instructional activities that enable us to learn about student thinking, central ideas in school mathematics, and teaching that is responsive to students' intellectual and social experiences (see tedd.org for a sample of the library of materials we use in the course). Choral Counting is one such activity (see Figure 4.2 for filled-out planning template for counting by fours starting at 4). We have written about the choice of these activities elsewhere (see Lampert, Beasley, Ghousseini, Kazemi, & Franke, 2010). A few important features of these activities that I want

Selfie from first day

Team building activities

Spontaneous
playful activity

Figure 4.1 Images of building playful relationships.

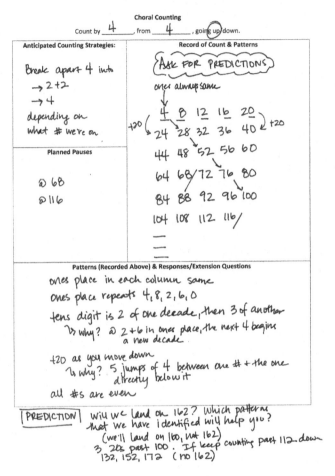

Figure 4.2 Planning template for the instructional activity of choral counting.

to repeat here are that we have selected activities that PTs can use across grade levels because their structures are accessible to novice teachers, and they include participation structures that encourage successful participation by students. One of the most pernicious ideas that we try to counter in our methods class is the ubiquitous use of "low," "medium," and "high" to describe mathematics learners. We use the particular instructional activities because they have helped us develop more productive narratives about the capabilities of students and the abilities they have to think mathematically and make public their brilliant ideas (see Leonard & Martin, 2013; Ramirez & Celedon-Pattichis, 2012). We have designed a basic protocol or lesson-plan sketch for each instructional activity from which teachers can improvise to be responsive to students and that open up the possibility of student-centered learning. We want to disrupt the traditional IRE patterns of classroom discourse that center disciplinary authority on the teacher and subscribe competence to a limited group of students (Cazden, 2001). To develop a productive learning community, students need to be seen and heard, and teachers need to communicate that their ideas are valued.

We engage in a cycle of learning that begins with learning about the structure of the instructional activities by watching videotaped excerpts or experiencing them ourselves. Through the study of each instructional activity's structure and how teachers improvise within that structure, our PTs consider what our course principles mean and how to create classroom learning environments where students learn that their ideas matter, that they are capable of doing mathematics, and that mathematics is worthy of study. We use both detailed protocols and planning guides (go to tedd.org for examples and see Figure 4.2 for an example of a choral-counting planning tool) to help PTs prepare for specific enactments. Other course readings on student thinking, classroom discourse, and culturally responsive mathematics teaching help PTs contextualize the work we are doing and engage in their own identity-building work as mathematics teachers.

Our 4-hour class follows a similar structure each week. The class is divided into four teams of PTs. Our partner class of third graders is also divided into four teams of students. The teams are paired together and work together throughout the quarter. In addition each child in the third-grade class is partnered with one PT who is referred to as a "Husky Buddy." This term was coined by our very first partner classroom and has stuck ever since. The Husky is the University of Washington mascot, and once the quarter is over, we host a field trip for the class at the University campus where they experience a mock lecture (e.g., a lecture on globalization and the development of cities), meet with undergraduates who describe their pathways to college, tour the UW campus, and visit classes in session (e.g., the landscape architecture studio). We cultivate a personal connection to our third-grade buddies, and all of our classroom visits include informal time to chat with one another. We also typically engage in team-building activities within our paired teams or join the third graders for recess.

Prospective teachers come to class with their lesson plans, prepared to teach a particular instructional activity. They spend a few minutes finalizing plans and may briefly review their goals with a teacher educator who will support the team for the day. Because we have four teams, we like to have four teacher educators: the classroom teacher (the program pays for a half-day substitute), the lead university instructor, and two doctoral-student teaching assistants (who also coach/supervise the teachers as part of their broader responsibilities in the program).

1. Teams rehearse and refine lesson plan

4. Teams debrief students' thinking
and instructional decision-making

2. University and classroom teacher
co-teach whole class

3. Teams teach students with support from teacher educator

Figure 4.3 Cycle of learning activities in a typical course session.

The images in Figure 4.3 provide a glimpse into the activities that constitute our learning cycles.

1. PTs in their teams rehearse and refine their instructional plan with their colleagues and with feedback and support from a teacher educator.

2. We visit our partner classroom and often begin with the lead instructor and classroom teacher coleading an activity that typically foreshadows the next week's focus.
3. PTs break up into their teams to coteach students in the partner classroom.
4. PTs leave the partner classroom and debrief the plan with support from a teacher educator.
5. Teams summarize what they have learned, and the lead university teacher-educator uses the last hour of class to prepare PTs for the following week.

When we debrief our learning experiences, we consider the following questions about each student in our small group (see Figure 4.4):

1. What did each student do or say? What do they seem to be understanding?

Date:	Activity:	Recorder:
Focus Questions: What did each student do or say? What do they seem to be understanding? What questions do you have about the students' ideas?		

Name	Student Notes	Goal/thoughts for next week

Figure 4.4 Structure for taking notes during debriefing session.

2. What questions do we have about the students' ideas?
3. What did we learn about our group? What do we want to keep in mind for next time?

We pay attention to our students' social, emotional, and intellectual participation, paying close attention to what we noticed, what we're curious about, and how to make instructional decisions that will bring out student voice and investment in our disciplinary work together. We share questions and curiosities about the mathematical ideas themselves, getting clearer on our objectives and how to interpret what students are doing and understanding.

INTERACTIONS INSIDE METHODS COURSE

Now that I've described how the typical course session is structured, I'd like to zoom in to one particular session and use a situative lens to understand what the structure and the nature of participation that we are trying to cultivate in the classroom affords for learning.

Returning to Greeno (2006), he explained that from a situative perspective, our understandings, goals, intentions, and expectations are shaped by our joint action, so to study learning, we must analyze

1. collaborative discourse
2. how we are positioned and position ourselves in participant structures of interaction because that positioning generates knowledge and information structures
3. the representations and representational practices that people use in the activity system because they can make learning and knowledge visible (p. 86)

Let's consider one course session as a way of developing some insight about the activity system—cycle of learning—that structures our interactions. I will draw on the three aspects of joint action that Greeno stated we need to analyze in order to help explicate the kinds of understandings that I think are being shaped in the methods course. In the description of the course, I have already shared the range of representations of practice that guide our interactions and convey that teaching is about being responsive to students' disciplinary ideas.

One important part of being with the students in both formal and informal interactions is that we see, albeit in a limited way, that children are different in different situations. One striking contrast my team of PTs starts to voice is the observations they have made about children's modes of communication in these different settings. As mentioned above, one of the norms of our class is to avoid labeling children with categories like "low" and "high" or "shy" or "difficult" and instead to describe and ask questions about what we notice. So we pay attention to how we position our students as people and as learners. Consider these observations that one team of PTs made about three students in the group of third graders they worked with every week.

- Isra loves to talk but in formal lessons, her voice almost becomes a whisper.
- Taavi is all smiles and easily approachable but in formal lessons his thinking can be really hard to follow as a listener because he often talks without pausing.
- Aisha says she is terrible at math and is very reticent in small-group activities.

We learn and we worry about these things. We want to position students competently, and we face uncertainties in our lessons about how to support our team of students. In typical school discourse, these students could easily be labeled and described as being "low" students. How do our observations about their participation and their self-descriptions play out in the way we design, enact, and debrief our lessons together? These are questions that we take seriously, and each week as we observe and interact with our students, we try to complicate any simple pronouncements about what they are like as students. Doing so helps us pay attention to the way we are positioning students as people and as learners.

The protocols or lesson sketches, planning sheets, and debriefing questions are representations of practices that begin to make visible how we approach learning and how we build a narrative arch over the course of the quarter about what it means to be a good math teacher. On one particular day, the instructional activity we take on in methods class was launching word problems (Jackson, Garrison, Wilson, Gibbons, & Shahan, 2013) and selecting an intentional goal for comparing and connecting strategies (Kazemi & Hintz, 2014). We were learning about how to pose problems so that students understand the problem context and have a way of starting without leading students to solve the problem in a particular way. We began our first exploration into these teaching and learning issues by having one of the university teacher educators (Andrea, one of the two doctoral teaching assistants) launch the problem with the whole class of third graders, and then students worked independently to solve the problem. During the classroom visit, we huddled to think about what to focus on.

The problem of the day was:

Problem: *14 husky buddies are sharing snack. They each eat 3 apple slices. How many apple slices did they eat?*

Prior to going into the classroom, we conducted a rehearsal. Because Andrea was going to launch the problem with the third graders, she was the one who rehearsed. I think this is important here—it was an authentic rehearsal for us because all of us needed opportunities to think through our decision-making and our goals. Rehearsal is not experienced as only something that novices do. Rather it is a participation structure, which teachers with any level of experience can use to think through a lesson with colleagues.

Before the rehearsal began, Andrea framed the rehearsal by sharing what kinds of issues she considered in making her plan for launching the problem. She decided to take a particular approach to launching the problem that involved acting the problem situation out with different quantities. For the launch, she decided to remove the number 14 from the problem but leave in the 3 in order for students to be able to think about the problem situation without being able to actually solve the problem. She also wanted students to think about "what the answer to the problem will sound like" in order to see if the students would be able to contextualize the numerical answer in relation to this particular problem situation.

During the rehearsal, there were several pauses where the group discussed different instructional choices, ways to phrase questions, and how to effectively use the acting-out strategy. Remember that Andrea was hoping to help the students consider what their answer was going to sound like. I asked her how she would pose that question to the students after the problem had been acted out:

> **Elham:** If they are ready to go, what will that sound like? So we just pretended two or four [husky buddies got apple slices], what will you say when you put the real number in there?
>
> **Andrea:** (*trying out her questioning*) Now that you've thought about this problem with 2 husky buddies or 4 husky buddies, I'm going to tell you my number in a second, the actual number of husky buddies. But first I want you to think about what our answer is going to sound like once we solve it? We're going to have some number, what are the words going to be after that number?
>
> **Elham:** Ohhhh. I know what. Why don't we connect that back to the 12 when people said, 12.
>
> **PTs:** Ooohhhh.

Elham: Would that be better?

PTs: Yeah!!

Andrea: So when we got 12, when there were 4 husky buddies, 12 what?

PTs: (*different voices saying different things*) apple slices, apples!

Andrea: 12 apples? 12 apple slices? What does the 12 mean?

Elham: (*commenting on the teacher move*) Is it 12 apples? 12 apple slices? 12 husky buddies? Yeah, that's a good question.

PT: (*interjecting to ask a question*) I'm thinking about a student in our group who is an English language learner, I'm thinking that he's hearing there's 3 and 6 and 12. I'm wondering how he's going to process this? What would you do in that situation? And then there's going to be 14 and 3. Would you just check in with that student and make sure he's understanding that before he works independently? I'm just imagining I'm him, and I've got so many numbers in my head. And I'm so confused.

Andrea: That's a good question.

I will stop the exchange here although it did go on for another 2 minutes as the group considered how to connect acting out the strategy back to the sentences in the problem and whether or not to use realia (actual apple slices) during the acting out. The classroom teacher also weighed in on what he thought would be supportive of the students. The class laughed together as someone offered up her own apple slice snacks for the good of the cause. What I would like for you to notice from this interaction is that the discourse during rehearsals conveys that teaching is both intellectual work and a craft. We don't portray teaching as following a script, but we try to make visible what we are thinking about and weigh different decisions based on what we hope to achieve and how they impact our learners. The PTs can then begin to wonder about what decisions to make based on their developing knowledge about the specific students they are coming to know.

When we went into the classroom, the teacher educator launched the problem and then each third grader had a chance to work on the problem. In the team I led, all students successfully arrived at the answer. Interestingly, the four girls in the group used mathematical models that match the problem situation and solved it as 14 groups of 3. Both boys in the group solved it as 3 groups of 14.

We huddled as a team and chose to focus on comparing how two of the girls, Isra and Aisha, counted the total differently. And we decided to recreate their work on larger posters to aid the group conversation that we would have when we called the students back together. These students were also chosen because of our goal of helping them become more confident sharers by positioning them as students with legitimate strategies and whose representations were worthy for the rest of the group to figure out (see Figure 4.5). Aisha had created a representation that modeled each of the groups, and she had counted the total by skip counting by sixes. Isra had represented each group with numerals and kept a running total as she skip counted by threes. She also wrote an equation 14 × 3 = 42. During the huddle before we called the students back in, I discussed with the two PTs coleading the lesson how we could focus the discussion on how each representation accounted for the 14 groups of 3 and to identify which were the people and which were the apple slices. We also decided to ask the students to think about the differences in the ways the two girls figured out the total.

The group discussion proved worthy of our choices. The group worked hard to understand what the numbers were on each poster and how they corresponded back to the problem situation and how they were connected to one another. We ran into some trouble because it seems like Taavi had some trouble understanding why we were counting by 3s, since he counted by 14s.

A number of mathematical, social, and instructional issues came up for us in the debrief session following the classroom visit. Following Greeno's (2006) frame for what to analyze when taking a situative perspective that I summarized earlier, our discourse, which I will summarize here, conveys that the work of teaching is complex, often ambiguous, and worthy of continual inquiry. It also conveys that there are always important ways in which we have to consider and question our own mathematical knowledge. Using our debriefing sheet (see Figure 4.4), an important artifact that also conveys that considering each student is important for future lesson planning, we discussed that Riad and Taavi, our two boys, recognized this problem as being modeled by the expression 14 × 3 and decontextualized 14 × 3

Aisha   Isra

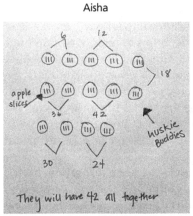
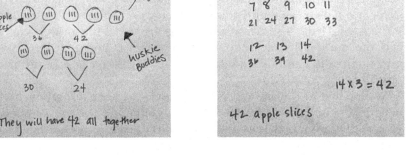

Figure 4.5 Student work for discussion.

from the specific problem situation while computing the sum of three 14s. As a team of PTs and a teacher educator, we tried to figure out what is important about this observation and if Riad and Taavi's strategies of doing three 14s affected their ability to understand the other students' representations. We had a breakthrough with Aisha being proud to share her thinking but noticed that Isra still showed some reticence in projecting her voice so others could hear.

What's important about this activity system that we have created? What do our discourse, our representational practices, and our positioning convey about what we are doing in this space? I offer the following observations as an insider to this world:

- We care about our K–5 student partners and how they talk about themselves.
- We want to be successful in teaching mathematics for understanding. We make intentional use of lesson structures, planning, and debriefing protocols that support teachers and students to engage in meaningful disciplinary work.
- We face all kinds of uncertainties in teaching and building relationships with each other and with students.
- We learn that taking risks in front of one another is invigorating.
- We open up space to talk to each other authentically about practice.
- We wrestle with what it means to teach for equity.

As the teacher educator, I am concerned that the PTs authentically experience that children are brilliant in all kinds of ways that might not be evident at first. I particularly do not want PTs to fall into viewing children as low and high. The PTs also want to feel competent—and they'll tell me sometimes in exit cards, "We need less support now. Can you sit back more and let us try things?" The fact that they can say this to me I think is a good sign that we are establishing an environment of mutual respect.

Each year that we enact this particular design, we change things, we learn from our PTs, and we learn from our classroom partners. We try harder to link our methods experiences to other courses on management, on supporting English language learners, on working with special education students, on developing meanings of being a social justice educator. We are not complacent or feel that we have reached the answer. We experiment with better ways of engaging our mentor teachers in partner schools. We consider how to develop deeper, longer relationships with community leaders and with schools. The enterprise of learning to teach is a journey.

NOTES

1. This chapter was informed by Kazemi's keynote address at the 2015 Scholarly Inquiry and Practices Conference (Sanchez, Kastberg, Tyminski, & Lischka, 2015).

REFERENCES

Cazden, C. (2001). *Classroom discourse: The language of teaching and learning.* Portsmouth, NH: Heinemann.

Diversity in Mathematics Education (DiME). (2007). Culture, race, power, and mathematics education. In F. Lester (Ed.), *Handbook of research on mathematics teaching and learning* (pp. 405–433). Charlotte, NC: Information Age.

Greeno, J. (2006). Learning in activity. In R. K. Sawyer (Ed.), *The Cambridge handbook of the learning sciences* (pp. 79–96). New York, NY: Cambridge University Press.

Greeno, J. G., & Middle School Mathematics Through Applications Project Group. (1998). The situativity of knowing, learning, and research. *American Psychologist, 53,* 5–26.

Gutiérrez, R. (2013). The sociopolitical turn in mathematics education. *Journal for Research in Mathematics Education, 44,* 37–68.

Herrenkohl, L. R., & Mertl, V. (2010). *How students come to be, know, and do: A case for a broad view of learning.* New York, NY: Cambridge University Press.

Jackson, K., Garrison, A., Wilson, J., Gibbons, L., & Shahan, E. (2013). Exploring relationships between setting up complex tasks and opportunities to learn in concluding whole-class discussions in middle-grades mathematics instruction. *Journal for Research in Mathematics Education, 44,* 646–682.

Kazemi, E., & Hintz. A. (2014). *Intentional talk: How to structure and lead more productive mathematical discussions.* Portland, ME: Stenhouse.

Lampert, M., Beasley, H., Ghousseini, H., Kazemi, E., & Franke, M. (2010). Using designed instructional activities to enable novices to manage ambitious mathematics teaching. In M. K. Stein & L. Kucan (Eds.), *Instructional explanations in the discipline* (pp. 129–141). New York, NY: Springer.

Lampert, M., Franke, M., Kazemi, E., Ghousseini, H., Turrou, A. C., Beasley, H., Cunard, A., & Crowe, K. (2013). Keeping it complex: Using rehearsals to support novice teacher learning of ambitious teaching in elementary mathematics. *Journal of Teacher Education, 64,* 226–243.

Lareau, A. (2003). *Unequal childhoods: Class, race, and family life.* Oakland, CA: University of California Press.

Lave, J., & Wenger, E. (1991). *Situated learning: Legitimate peripheral participation.* Cambridge, England: Cambridge University Press.

Leonard, J., & Martin, D. B. (Eds.). (2013). *The brilliance of Black children in mathematics: Beyond the numbers and toward new discourse.* Charlotte, NC: Information Age.

Murrell, P. (2001) *The community teacher.* New York, NY: Teachers College Press.

Ramirez, N., & Celedon-Pattichis, S. (2012). *Beyond good teaching: Advancing the mathematics education of ELLs.* Reston, VA: National Council of Teachers of Mathematics.

Rose, M. (1995). *Possible lives: The promise of public education in America.* New York, NY: Penguin.

Sanchez, W., Kastberg, S., Tyminski, A., & Lischka, A. (2015). *Scholarly inquiry and practices (SIP) conference for mathematics education methods.* Atlanta, GA: National Science Foundation.

Turner, E. E., Drake, C., Roth McDuffie, A., Aguirre, J., Bartell, T. G., & Foote, M. Q. (2012). Promoting equity in mathematics teacher preparation: A framework for advancing teacher learning of children's multiple mathematics knowledge bases. *Journal of Mathematics Teacher Education, 15,* 67–82.

Wilkerson, I. (2010). *The warmth of other suns: The epic story of America's great migration.* New York, NY: Vintage.

Zeichner, K. (2012). The turn once again toward practice-based teacher education. *Journal of Teacher Education, 63*(5), 376–382.

SECTION II

USING PERSPECTIVES TO INFORM SCHOLARLY INQUIRY AND PRACTICE

CHAPTER 5

USING THE *KNOWLEDGE QUARTET* TO SUPPORT PROSPECTIVE TEACHER DEVELOPMENT DURING METHODS COURSEWORK

Tracy L. Weston
Middlebury College

Mathematics teacher educators (MTEs) can use frameworks in mathematics methods courses as organizational structures and to inform the design of activities. Examples of mathematics teaching frameworks that could be used in this way include Mathematical Knowledge for Teaching (MKfT; Ball, Thames, & Phelps, 2008), Mathematics-for-Teaching (MfT; Davis & Simmt, 2006), Political Knowledge for Teaching (Gutiérrez, 2013), Mathematical Knowledge in Teaching (MKiT), and the Knowledge Quartet (KQ; Rowland, 2014). Each framework conceptualizes knowledge for teaching mathematics differently and therefore links practice and knowledge differently as well. This chapter will explain one empirical framework, the Knowledge

Building Support for Scholarly Practices in Mathematics Methods, pages 69–83

Quartet (Rowland, 2014), provide a rationale for its use, and describe how the author uses the KQ to organize and design activities in elementary methods coursework to support the development of prospective teachers' (PTs') MKiT and practices in regards to observation and lesson planning.

THE KNOWLEDGE QUARTET

Rowland, Huckstep, Thwaites, and Turner researched knowledge in teaching mathematics with the aim of offering a way to build professional knowledge for mathematics teaching (Rowland, 2008). Similar to Ball and her colleagues (2008), Rowland, Huckstep, and Thwaites (2005) focused on Shulman's (1986) three content-specific categories (subject matter knowledge, pedagogical content knowledge, and curricular knowledge) rather than the four generic categories of teacher knowledge. The links to Shulman's categories and comparisons to Ball et al.'s work are well rehearsed elsewhere, as is the approach that generated the Knowledge Quartet (Rowland, 2008). In this chapter I report on my use of the Knowledge Quartet (KQ) for PT instruction by integrating the framework into methods coursework including PT observation and lesson planning.

The Knowledge Quartet is an empirical framework for specific content knowledge that impacts mathematics teaching, which is demonstrated in both planning and teaching. It was developed by researchers Rowland, Turner, Thwaites, and Huckstep (2009) based on analysis of mathematics teaching. The aim of their project was to "identify different aspects of teacher knowledge that have an impact on teaching from observations of that teaching" (p. 26). Their ultimate goal was to help beginning teachers develop their knowledge for teaching mathematics, a goal that aligns with the work of MTEs in methods courses.

To develop a conception of mathematical content knowledge based in teaching, Rowland et al. (2009) observed PTs' teaching, identified portions of the lessons informed by PTs' mathematical content knowledge or mathematical pedagogical knowledge, and analyzed these moments through grounded theory. This resulted in 18 initial codes, which were grouped into four broader dimensions of MKiT called the *Knowledge Quartet* (see Table 5.1). Since its origination, the KQ has been modified based on additional research of classroom teaching, with some codes being eliminated and others added. At the time of this writing, the Knowledge Quartet framework includes 21 codes, with each of the four dimensions containing four to seven aspects of MKiT. In the balance of this section, I provide an overview of those four dimensions and indicate how I view connections to cognitive, situative, and sociopolitical perspectives.

TABLE 5.1 Knowledge Quartet Framework		
		Focal Knowledge Quartet Code With Brief Definition
Dimension	**Contributory Codes**	**Areas of PT focus during planning and reflection**
Foundation: Knowledge and understanding of mathematics per se and of mathematics-specific pedagogy, beliefs concerning the nature of mathematics, the purposes of mathematics education, and the conditions under which students will best learn mathematics.	• Awareness of purpose	
	• Adherence to textbook	
	• Identifying errors	
	• Concentration on procedures	**Concentration on understanding** The teacher uses instructional practices to develop *relational* rather than *instrumental* understanding.
	• Overt display of subject knowledge	
	• Theoretical underpinnings of pedagogy	
	• Use of mathematical terminology	**Use of terminology** The teacher uses mathematically correct language and written notation is accurate.
Transformation: The presentation of ideas to learners in the form of analogies, illustrations, examples, explanations, and demonstrations.	• Choice of representation	**Choice of representation** The teacher critically selects a representation (e.g., number line, place-value grid, manipulatives) that reflects the concepts or procedures. The teacher uses the representation correctly to explain the concept/procedure and focus students on the mathematical concept.
	• Teacher demonstration (to explain a procedure)	
	• Use of instructional materials	
	• Choice of examples	**Choice of examples** The teacher critically selects or intentionally plans tasks, problems, activities, or examples for use with students that are high cognitive demand, purposefully uses numbers and operations to focus student attention on the desired concept or strategy, and addresses (or at least does not add to) misconceptions.

(continued)

TABLE 5.1 Knowledge Quartet Framework (continued)

Dimension	Contributory Codes	Focal Knowledge Quartet Code With Brief Definition / Areas of PT focus during planning and reflection
Connection: The sequencing of material for instruction and an awareness of the relative cognitive demands of different topics and tasks.	• **Anticipation of complexity**	**Anticipation of complexity** The teacher is aware of different levels of difficulty that exist within a topic, concept, or strategy and use this knowledge to break down the steps so they can be understood by students.
	• **Decisions about sequencing**	**Decisions about sequencing** Ideas and strategies need to be introduced in a progressive order. Sequencing can occur across a unit, across a lesson (ordering the sections of the lesson and topics), and across examples (using tasks/ activities/ examples in a progressive and intentional way).
	• Making connections between concepts	
	• Recognition of conceptual appropriateness	
	• Making connections between procedures	
Contingency: The ability to make cogent, reasoned, and well-informed responses to unanticipated and unplanned events.	• Deviation from agenda	
	• Teacher insight during instruction	
	• Responding to the (un)availability of tools and resources	
	• **Responding to students' ideas** (use of opportunities)	**(Reflection Only)** **Responding to students' ideas** The teacher needs to respond to student ideas as evident by their verbal comments, questions, (in)correct answers, and interactions with the task.

Source: Adapted From Rowland, 2014, p. 25

The first dimension of the KQ is *foundation*, which comprises a teacher's mathematical content knowledge and theoretical knowledge of mathematics teaching and learning, including beliefs about how students best learn mathematics. As the name indicates, the foundation dimension supports and is drawn upon in each of the other three dimensions. Codes in the foundation dimension include a teacher's content knowledge, correct use of mathematical terminology and notation, and identification of student errors.

The foundation dimension is a category of mathematical knowledge, knowledge of mathematics pedagogy, and beliefs. Based on my interpretation of the KQ and the three perspectives investigated at the 2015 Scholarly Inquiry and Practices Conference (Sanchez, Kastberg, Tyminski, & Lischka, 2015), I see the strongest alignment between the foundation dimension and the cognitive perspective. Within the cognitive perspective, a "central organizing metaphor is that of knowledge as an entity that is acquired in one task setting and conveyed to other task settings" (Cobb & Bowers, 1999, p. 5). Transfer of an individual's knowledge from one setting to another (Beach, 1995) is similar to Turner and Rowland's (2011) description of the foundation dimension of the KQ:

> [Foundation] differs from the other three [dimensions] in the sense that is about knowledge 'possessed' [footnote omitted], irrespective of whether it is being put to purposeful use. For example, we could claim to have knowledge about division by zero, or about some probability misconceptions—or indeed to know where we could ask advice on these topics—irrespective of whether we had had to call upon them in our work as teachers. (p. 200)

In addition to the general harmony between the cognitive perspective and foundation dimension, the contributory codes within the foundation dimension further indicate areas of overlap. For example, codes in the foundation dimension include *subject matter knowledge* (SMK) and *identification of student errors* (IE). Similarly, an area investigated by researchers who operate within a cognitive perspective includes a teacher's understanding of student mathematical conceptions (Heinz, Kinzel, Simon, & Tzur, 2000). The focus on "internal cognitive activity" (Cobb & Bowers, 1999, p. 6) by first-wave cognitive theorists seems to have strong overlap with the foundation dimension on the KQ. See Turner and Rowland (2011, p. 200) for a more thorough description including distributed knowledge.

The other three KQ dimensions—transformation, connection, and contingency—are categories concerned with situations that may arise in the course of a mathematics lesson, as opposed to types of knowledge. *Transformation* is the dimension most similar to Shulman's (1986) conceptualization of pedagogical content knowledge—that is, how a teacher transforms their content knowledge into forms that are accessible and pedagogically powerful to pupils. This category pays special attention to the teacher's use of representations, examples, and explanations. A third dimension is *connection*, which is whether a teacher makes instructional decisions with an awareness of connections across the domain of mathematics and that mathematics is not, after all, a subject that contains discrete topics. It also encompasses the need to sequence experiences for students, anticipate what students will likely find "hard" or "easy," and understand typical student misconceptions within a given mathematical topic. Because not all aspects of a lesson can be planned for ahead of time, *contingency* is the dimension concerned with a teacher's need to think on their feet in unplanned and unexpected moments, such as when responding to students and when making decisions to deviate from the lesson plan for the sake of student learning.

Based on my interpretation of the KQ and cognitive, situative, and sociopolitical perspectives, I perceive alignment in the transformation, connection, and contingency dimensions with the situative perspective. As Cobb and Bowers (1999) synthesized, the "primary metaphor of the situated learning perspective is that of knowing as an activity that is situated with regard to an individual's position in the world of social affairs" (p. 5). From a situative perspective, learning happens as a result of participation in social practice. As supported by Putnam and Borko (2000), it is important for PTs to develop their professional knowledge of teaching through authentic activities contextualized in classrooms. Similarly, Rowland et al.'s (2009) work is grounded in the belief that "mathematical content knowledge for teaching will be most clearly seen in the action of teaching" (p. 25).

The contingency dimension *responding to student ideas* (RSI) is, by definition, something that cannot be fully planned for ahead of time. Working within a situative framework acknowledges the need for this kind of decision making. For example, McDonald, Kazemi, and Kavanagh (2013) explained PTs need experiences in which they are prepared for "the constant in-the-moment decision making that the profession requires" (p. 378). As Turner and Rowland (2011) noted, "Teaching requires knowledge in several different domains, and a number of taxonomies reflect this multidimensional perspective" (p. 196).

My interpretation is that within the configuration of the KQ framework there is strong integration of two perspectives: cognitive and situative. Because the KQ intentionally focuses on mathematics-related issues, it deliberately sets aside issues related to teaching all subjects, such as sociopolitical (Gutiérrez, 2013). At the same time, the foundation dimension incorporates knowledge of learners and instructional decisions in other areas (for example, *choice of representation*) and should be enacted in a way that attends to broader needs and concerns in classrooms and society.

The work of teaching is a complicated endeavor, and within the taxonomy of the KQ (see Table 5.1) elements that relate to the cognitive and situative perspectives are evident, connected, and drawn upon by the teacher. As Turner and Rowland (2011) explained, "The distinction between different kinds of mathematical knowledge is of lesser significance than the classification of the situations in which mathematical knowledge surfaces in teaching" (p. 196).

MOTIVATION FOR USING A CONCEPTUAL FRAMEWORK

The KQ has been used in a variety of research studies and projects (e.g., Rowland, 2008; Weston, 2013; Weston, Kleve, & Rowland, 2012). I first encountered the KQ as a researcher, when conducting a yearlong study of PTs' teaching in their final year of a university-based teacher education program that included a fall methods course and field placement and spring full-time student teaching (Weston, 2011). I used the KQ to analyze PTs' teaching and found that MKiT codes that were most often demonstrated by PTs at a minimal level (e.g., *choice of examples*: at least one example was procedures with connections [Stein, Smith, Henningsen, & Silver, 2000], although other examples may have been low cognitive demand), only about half of the codes were consistently demonstrated, and improvement on a code over the course of student teaching was rare. Having aspects of MKiT absent or enacted at minimal levels without evidence of improvement resulted in my commitment to focus methods coursework on identifying and developing the knowledge and practices required to teach mathematics well.

Focus group and individual interviews indicated PTs were unclear of the existence of MKiT aspects that they were trying to develop. For example, a PT named Holly said, "Right before I started my student teaching... I had watched my teacher do enough whole class math that I thought, 'Okay, well at least I can mimic her,' which I could do fine. And so I did, I mimicked her" (Weston, 2011, p. 312).

Despite being one of the top students in the program, earning an A in her mathematics methods course, and having high mathematics SAT scores, Holly was not able to describe what the work of mathematics teaching entailed or what knowledge and practices she should work to develop in her field placements. Instead she was left to mimic her cooperating teacher. Similarly, another PT from the same cohort and study, McKenzie, explained, "I might look at a whole lesson, and think, 'That was good.' But maybe there was one thing that wasn't effective, but since I thought the whole lesson was good I wouldn't notice" (Weston, 2011, p. 281). Statements such as Holly's and McKenzie's along with their relatively low enactment of MKiT in their teaching, despite being top students and clearly dedicated to their studies, professional development, and field placements, compelled me to find a way for PTs (not just MTEs) to name and develop important aspects in mathematics teaching. Therefore, as an MTE I became invested in finding an empirical framework to explicitly use with PTs to better develop their MKiT.

I selected the KQ as a promising framework to use with PTs for three reasons. First, the KQ identifies classroom situations that have mathematical potential and thus offers a valuable window into mathematics teaching. My principle motivation for using the KQ was that MKiT is most visible through observation of classroom situations during actual teaching, and the KQ is organized around such situations, as it is largely in the situative domain. In contrast, a framework organized around types of knowledge in the purely cognitive domain would function as a heuristic for MKiT rather than the way MKiT would be revealed in a teaching situation, which could make it more difficult for PTs to use because they would first need to operationalize the framework. A second reason for selecting the KQ was that the contingency domain is not found in other frameworks. The real-time decisions teachers make is an area of known difficulty for PTs and a significant aspect of the work of teaching (McDonald et al., 2013; Rowland et al., 2009). Finally, the KQ was selected for its applicability across grade levels and curricula, as confirmed by a recent international project that spanned multiple settings, K–16+ grade levels, and curricula (Weston et al., 2012). Given that as an MTE I am preparing my students to teach across a range of grades (K–6), curricula, and contexts, using a framework that will continue to apply not only during the methods course but also in new and varied future settings was an important consideration.

USE OF THE KNOWLEDGE QUARTET: FOCAL CODES

Once I selected the KQ to use in my elementary methods course, I used it as the organizational framework for the course rather than mathematical topics or practices. I made an early decision to focus on a core subset of the KQ codes rather than trying to cover all 21. I did this for two reasons. First, I realized it would be difficult for PTs (or any teacher) to simultaneously work on 21 aspects of teaching and certainly not possible to do so with any amount of depth. Second, some codes are, it seems, more important than others. As their definition indicates, MKiT is mathematically related knowledge that teachers can draw upon to support student learning (Rowland & Ruthven, 2011). To be sure, all of the elements on the KQ impact student learning, but not all are an equally advantageous first step into mathematics teaching. This interpretation seems consistent with the thinking of the original KQ research team, who suggested focusing attention on six codes in lesson planning and seven in reflection.

The seven codes that I selected to focus on in the methods course (and subsequently in student teaching) were the following: concentration on understanding, use of terminology (foundation), choice of examples, choice and use of representations (transformation), anticipation of complexity, decisions about sequencing (connection), and responding to student ideas (contingency). Six of these codes were identified by Rowland, Thwaites, and Turner (personal communication, November, 2011) as the areas for PTs to focus on during planning and reflection. One adjustment I made was to name *concentration on understanding* as one of the focal areas, whereas the Cambridge team recommended using a different code from the foundation dimension, *theoretical underpinning of pedagogy*. My use of concentration on understanding is a reversal of the KQ foundation code *concentration on procedures*, which identifies a teacher's disadvantageous emphasis on procedures rather than on developing a conceptual understanding of the underlying concept. Concentration on understanding labels the idea with a positive directive, paralleling other codes.

A focus on these seven aspects of MKiT helps me facilitate explicit dialogue and design activities to provide consistent practice and target these crucial areas during coursework and field placements. A brief definition of each of these seven areas is provided in the right column of Table 5.1. There are many ways to design experiences around these seven areas to develop PTs' knowledge and practices. In my course, I use these to focus PTs' observations of teaching episodes that occur in class through video, at a school site through live teaching demonstrations, or at their field placement through observation of their cooperating teacher. I also use these seven areas as the focus for PT lesson planning, their self-reflection on their teaching, and my observation and feedback on their teaching. In this way I have deeply integrated the KQ framework into the course so that PTs have many experiences discussing, observing, planning, and reflecting on these specific aspects of MKiT, with the goal of developing their own mental habits to attend to these elements of teaching mathematics.

It is not possible to cover the extent to which I integrate the KQ through readings, assignments, observations, lesson planning, reflection, and feedback. Rather, the purpose for the remainder of the chapter is to provide a description of two ways I use the KQ as examples of how MTEs can use conceptual frameworks to inform the generation of activities to support PT development. The two areas on which I will focus are PT observations of teaching and PT lesson planning, both of which are aspects that all MTEs are likely to address in their work with PTs.

USE OF THE KNOWLEDGE QUARTET: OBSERVATIONS

At the beginning of the course, I introduce the KQ to the PTs by providing an overview of the four dimensions and then introducing the seven focal codes. I use *Developing Primary Mathematics Teaching* (Rowland et al., 2009) as the main text in the class, which describes each of the four dimensions and provides many scenarios from elementary classrooms that illuminate the framework. Along with other readings and assignments, this text helps PTs understand the importance of using a conceptual framework for teaching mathematics, become familiar with the specific dimensions and codes on the KQ framework, and gain insight into teaching mathematics.

After introducing the KQ and while PTs continue to read throughout the course to further develop their knowledge of the framework, I have PTs focus on the seven selected codes (see Table 5.1) while observing excerpts of mathematics lessons. Research on novice teacher noticing indicates that it is challenging for PTs to "see" dimensions of MKiT as they observe a teaching episode (Jacobs, Lamb, & Philipp, 2010). Therefore, live teaching demonstrations during class, analysis of video, and observations during field experiences provide opportunities to teach PTs to attend to specific aspects of MKiT, rather than focusing on superficial or nonmathematical concerns such as classroom management. Naturally there are different ways to approach using the KQ to support PT observations. For example, whichever of the four KQ dimensions or seven codes is the focus in the methods course at a given time can be the PTs' primary focus as they observe teaching during class or at a field placement. Even with the reduction of codes from 21 to 7, it is helpful to first focus on one aspect at a time. One way to do this is to find short videos that are strong exemplars of a given code and have PTs watch for evidence of that code and debrief about that aspect of MKiT. At a later time in the course, after they have experience reading about, discussing, and observing the constituent elements, PTs can be asked to observe more holistically to attend to moments when the four dimensions or seven codes are evident.

I will provide an example of PT coursework before and after introducing the KQ to show the difference that occurs in their attention to specific aspects of MKiT when observing mathematics teaching as a result of using the KQ. On the first day of class, all PTs watched a 6-minute video segment and wrote what they noticed as they watched and what feedback they would give the teacher. They repeated this exercise with the same video just 1 week later, and in between read about *concentration on understanding* (CU). For the second viewing, PTs were told to pay attention to moments where CU occurs or moments that seem like missed opportunities for CU. The clip is particularly useful because the teacher seems to be poised and organized; however, the lesson is not mathematically strong. PTs watch the same video a third time at the end of the course, this time without a specific prompt other than to pay attention to what they notice (the same prompt they had on Day 1).

Excerpts from the written activity following the first viewing include mostly positive statements about a rhyme that the teacher has students repeatedly say to remember when to round a number up or down, such as "great use of the rhyme to help students remember" and "keeps all students engaged by asking about their rounding saying." There are also many comments about generic aspects of teaching including behavior management, such as "good teacher presence, authoritative, loud, moving around" as well as many comments about her "control" (their wording) and voice. As part of the second viewing, PTs are given back a copy of what they wrote the week before and asked to cross out or write more to reflect their current thinking. All of the aforementioned comments were crossed out and replaced with quotes such as "it would have been helpful to explore other methods," that students "could get caught up in knowing (the rhyme) but not understanding it," and that the rhyme was "too procedural. (Students) need to understand concepts and use own thinking." By the third viewing, PTs have had many experiences to read, discuss, observe, plan, and reflect on the focal KQ aspects, and these show up in their writing without explicit prompting. PTs included analysis of the use of terminology, suggestions for choosing a representation that would better illuminate the concept, and responses to student ideas in analyzing the class discussion. The language and concepts of the KQ were evident in PT lesson analysis after repeated practice. This is one example of the development of PT articulation of MKiT through consistent use of the KQ framework and methods course experiences that seek to embed the framework through regular use.

USING THE KNOWLEDGE QUARTET TO SUPPORT PT LESSON PLANNING

To help PTs attend to the six focal areas during planning (*responding to student ideas* is omitted from planning because contingent aspects are not possible to plan for ahead of time), I developed a preplanning template that PTs work through in advance of writing their full lesson plan on a program-wide generic lesson plan template. Without this step, I find that PTs are less likely to attend to mathematically important considerations during lesson planning. Instead, they tend to start lesson planning by finding an activity, rather than by analyzing the mathematics and related information and using these as the basis from which to select an activity (Weston & Henderson, 2015).

I organized the template in a sequential order so that subsequent categories build from thinking done in previous portions of the template. In the first section, PTs identify what *concentration on understanding* (CU) entails within the mathematical domain they are teaching, after which they identify corresponding Common Core State Standards. Next, PTs identify and define terminology that is likely for them or students to use during the lesson, as called for by *use of terminology* (UT). Third, PTs work on *anticipation of complexity* (AC), in which they indicate mathematical aspects students will likely find challenging and possible errors or (mis)conceptions. PTs do this based on information from class, course readings, and preassessment data. Next, they think about *decisions about sequencing* (DS). Based on the information in AC, PTs outline general features of the mathematics they will include over the course of the lesson. Next they select a task for *choice of examples* (CUE). Here it is important to note that "examples" means the task, activity, or problems done by teachers or students in the lesson, not only those done by the teacher to demonstrate a procedure. It is worth underscoring that task selection comes at this point, following consideration of all of the previous aspects, again to shift away from "activity mania" and instead encourage thoughtful consideration for selecting or designing a task. Lastly, PTs identify the *choice of representations* (CUR) that they or students will use while completing the tasks in the lesson.

I introduce this template to PTs during the second or third week of class, after the KQ introduction and while they are beginning to read about the framework and use it for observations during class and field placements. I share examples of templates my former students completed and some I completed (see Figure 5.1). These model the content and level of specificity

CONNECTION	
Anticipation of complexity	Ordering fractions can be difficult. Students will not know about common denominators. They can use a visual model (fraction bar kit) as well as reasoning strategies that build understanding of fractions (attention to the parts and whole). **Common errors/misconceptions:** Incorrect terminology You can't compare fractions Drawing is a helpful strategy (it does not tend to be—too imprecise) The bigger the denominator, the bigger the fraction The bigger the numerator, the bigger the fraction You can ignore the number on the top You can ignore the number on the bottom Lack of familiarity with using a number line
Decisions about sequencing	When asking students to put fractions in order, students will first do fractions that are easier to compare (unit fractions). Then they will also work with benchmark fractions (1/4, 1/2) and like denominators. Last they will encounter like numerators and fractions that are 1/d from being whole.
TRANSFORMATION	
Use of examples	Easy: 1/6, 1/10, 1/5, 1/3, 1/2 Medium: 5/12, 1/2, 1/7, 2/3, 3/4 Hard: 31/32, 8/9, 2/3, 7/11, 7/13 **Comparison worksheet:** use mainly fractions with denominators 2, 3, 4, 5, 6, 8, 10, 12, 100. Also have students write to explain their thinking and develop some examples on their own.
Use of representations	Drawings are a poor strategy for fractions. Pre-made fraction blocks are available in the room, but in order to develop students' understanding of fractions and the relationship between parts and whole, students will construct their own fraction bar kit. The number line will also be used to put fractions in order.

Figure 5.1 Excerpt from Knowledge Quartet lesson plan template.

expected to help PTs approach their lesson planning. After they complete the KQ template, PTs complete the full lesson plan in which they outline more specifics including directions, explanations, and key questions. PT lesson plans built using the preplanning template informed by the KQ are much more focused on core mathematical ideas and more specific than plans PTs wrote prior to my integration of the KQ.

In Figure 5.1, four of the focal KQ codes are identified in the left column. The right column contains preplanning work for a lesson about ordering fractions that fall between zero and one. The excerpt illustrates work following identification of the mathematical focus and related standards as well as relevant mathematical terminology and definitions. Using that information and reflecting on *anticipation of complexity* (AC) resulted in the identification of some likely areas of difficulty in the mathematical terrain, including a list of potential errors or (mis)conceptions. The second row of Figure 5.1 concerns *decision about sequencing* (DS). Having done this foundational work, in *choice of examples* (CE) some numbers were generated that correspond to the work done in AC and DS. After making some initial decisions about the content and numbers to use, *choice of representations* (CR) provides a place to match the representation to the mathematics and provide a rationale. The template is meant to be straightforward and provide the fewest prompts necessary to focus attention on the KQ focal codes. One way to encourage and develop skills and practices around intentional planning focused on mathematics is to have PTs explicitly think through these areas before writing a full lesson plan.

Given the daunting task of teaching PTs how to teach in very little time, methods courses can address many topics but fail to provide PTs with a unifying framework by which to organize all of the information and experiences they have during the course and field placements. I am much more satisfied after shifting away from a course organized around a series of topics to one that is organized around a conceptual framework. I have found the KQ to be both robust and novice-teacher friendly. PTs develop a language that is grounded in the framework and their discourse in class and when processing classroom episodes is much more focused on these core aspects. Integrating the KQ into a methods course provides PTs with useful footholds and a generative way to orient themselves to the work of teaching mathematics that is applicable across grade levels, curricula, and settings. Multiple intentional experiences learning about and using the framework layer together to develop dimensions of MKiT through observing and beginning to teach so that a conceptual basis for teaching mathematics is better established in advance of student teaching and program completion. PTs are not left in Holly's or McKenzie's position of imitating a teacher's behaviors or not knowing what areas of mathematics they should reflect on.

As Simon (2008) explained, oftentimes MTEs' learning goals for their PTs are underspecified. In teacher education, as in any kind of instruction, "knowledge to be learned must be clearly identified" (Simon, 2008, p. 19). Using a conceptual framework, such as the KQ, is one way for MTEs to specify the knowledge they are trying to help PTs develop and can guide the intentional design of methods course experiences. I have found the use of the KQ provides clarity to both the MTE and PTs in their work together.

REFERENCES

Ball, D. L., Thames, M. H., & Phelps, G. (2008). Content knowledge for teaching: What makes it special? *Journal of Teacher Education, 59*, 389–407. doi:10.1177/0022487108324554

Beach, K. (1995). Activity as a mediator of sociocultural change and individual development: The case of school-work transition in Nepal. *Mind, Culture, and Activity, 2*, 285–302.

Cobb, P., & Bowers, J. (1999). Cognitive and situated learning perspectives in theory and practice. *Educational Researcher, 28*(2), 4–15.

Davis, B., & Simmt, E. (2006). Mathematics-for-teaching: An ongoing investigation of the mathematics that teachers (need to) know. *Educational Studies in Mathematics, 61*(3), 293–319.

Gutiérrez, R. (2013). The sociopolitical turn in mathematics education. *Journal for Research in Mathematics Education, 44*(1), 37–68.

Heinz, K., Kinzel, M., Simon, M. A., & Tzur, R. (2000). Moving students through steps of mathematical knowing: An account of the practice of an elementary mathematics teacher in transition. *Journal of Mathematical Behavior, 19*, 83–107.

Jacobs, V. R., Lamb, L. L., & Philipp, R. A. (2010). Professional noticing of children's mathematical thinking. *Journal for Research in Mathematics Education, 41*(2), 169–202.

McDonald, M., Kazemi, E., & Kavanagh, S. S. (2013). Core practices and pedagogies of teacher education: A call for a common language and collective activity. *Journal of Teacher Education, 64*(5), 378–386. DOI: I0.1177/0022487113493807

Putnam, R., & Borko, H. (2000). What do new views of knowledge and thinking have to say about research on teacher learning? *Educational Researcher, 29*(1), 4–15.

Rowland, T. (2008). Researching teachers' mathematics disciplinary knowledge. In P. Sullivan & T. Wood (Eds.), *International handbook of mathematics teacher education: Vol. 1. Knowledge and beliefs in mathematics teaching and teaching development* (pp. 273–298). Rotterdam, The Netherlands: Sense.

Rowland, T. (2014). The Knowledge Quartet: The genesis and application of a framework for analysing mathematics teaching and deepening teachers' mathematics knowledge. *SISYPHUS Journal of Education, 1*(3), 15–43.

Rowland, T., Huckstep, P., & Thwaites, A. (2005). Elementary teachers' mathematics subject knowledge: The Knowledge Quartet and the case of Naomi. *Journal of Mathematics Teacher Education, 8,* 255–281. doi:10.1007/s10857-005-0853-5

Rowland, T., & Ruthven, K. (Eds.). (2011). *Mathematical knowledge in teaching.* London, England: Springer.

Rowland, T., Turner, F., Thwaites, A., & Huckstep, P. (2009). *Developing primary mathematics teaching: Reflecting on practice with the Knowledge Quartet.* London, England: SAGE.

Sanchez, W., Kastberg, S., Tyminski, A., & Lischka, A. (2015). *Scholarly inquiry and practices (SIP) conference for mathematics education methods.* Atlanta, GA: National Science Foundation.

Shulman, L. S. (1986). Those who understand: Knowledge growth in teaching. *Educational Researcher, 15*(2), 4–14. doi:10.3102/0013189X015002004

Simon, M. (2008). The challenge of mathematics teacher education in an era of mathematics education reform. In B. Jaworski & T. Wood (Eds.), *International handbook of mathematics teacher education: Vol. 4. The mathematics teacher educator as a developing professional* (pp. 17–29). Rotterdam, The Netherlands: Sense.

Stein, M. K., Smith, M. S., Henningsen, M. A., & Silver E. A. (2000). *Implementing standards-based mathematics instruction: A casebook for professional development.* New York, NY: Teachers College Press.

Turner, F., & Rowland, T. (2011). The Knowledge Quartet as an organising framework for developing and deepening teachers' mathematics knowledge. In T. Rowland & K. Ruthven (Eds), *Mathematical knowledge in teaching* (pp. 195–212). London, England: Springer.

Weston, T. L. (2011). *Elementary preservice teachers' mathematical knowledge for teaching: Using situated case studies and educative experiences to examine and improve the development of MKT in teacher education* (Doctoral dissertation). Available from ProQuest Dissertations and Theses database. (UMI No. 3477561)

Weston, T. L. (2013). Using the *Knowledge Quartet* to quantify mathematical knowledge in teaching: The development of a protocol for initial teacher education. *Research in Mathematics Education, 15*(3), 286–302.

Weston, T. L., & Henderson, S. C. (2015). Coherent experiences: The new missing paradigm in teacher education. *The Educational Forum, 79*(3), 321–335.

Weston, T. L., Kleve, B., & Rowland, T. (2012). Developing an online coding manual for the knowledge quartet: An international project. *Proceedings of the British Society for Research into Learning Mathematics, 32,* 179–184.

CHAPTER 6

THREE LEARNING PERSPECTIVES FOR TRANSLATING CURRICULUM INTO INSTRUCTION

Darrell Earnest
University of Massachusetts, Amherst

Julie M. Amador
University of Idaho

When I become a teacher, I want a curriculum that is easy to understand. I don't want to be wasting time just trying to figure out what it is I'm trying to teach.

—An elementary prospective teacher (PT)
in her second month of a 9-month program

In this chapter, we consider instruction in mathematics methods course-work from each of the three theoretical perspectives: cognitive, situative, and sociopolitical. Learning—whether the focus is PTs or children—is a complex and nuanced characteristic of human ontogeny. Theories to explain the context of learning highlight particular properties of the human condition—whether intellectual and/or social—to explain critical aspects of how learning transpires. Any of the three perspectives is a lens

Building Support for Scholarly Practices in Mathematics Methods, pages 85–97
Copyright © 2018 by Information Age Publishing
All rights of reproduction in any form reserved.

illuminating the work of teaching and preparing teachers by highlighting particular aspects of classroom culture and interactions. We contend that the three perspectives are in fact complementary. To consider the connections across perspectives and highlight their interrelatedness in preparing new teachers, we examine PTs' analysis and translation of curriculum materials, with each perspective enabling different insights related to instruction in mathematics methods coursework.

Consider the quote beginning this chapter. A PT in one of our elementary methods courses viewed curriculum as a reality of classroom teaching that she hoped would not impede her from implementing effective instruction. In her view, an ideal curriculum was more like a theatrical script complete with stage directions. Such a perspective is not unreasonable, as PTs have little to no experience with classroom teaching, let alone curriculum. However, as we have argued elsewhere (Amador & Earnest, 2016; Earnest & Amador, 2017), translating curriculum for a classroom of students is an intellectual and dynamic process that is consequential to classroom learning. As a result, curriculum is a unique context in which to consider the three perspectives. Rather than treating the perspectives as mutually exclusive, we contend that each offers a powerful lens on reading curriculum in mathematics methods coursework to make visible the dynamics of elementary instruction.

In this chapter, we focus on the case of Maya and Vanessa (classmates to the PT quoted at the beginning of this chapter) as they translate curriculum into a vision of enactment (Remillard & Heck, 2014). PTs' analysis and translation of curriculum materials is a context that serves two purposes. First, it allows PTs opportunities to grapple with intellectual aspects of thinking and learning for both the individual and the collective. Second, it allows mathematics teacher educators to identify and address areas in which PTs need further support. In this chapter, we consider the role of PTs' own curriculum analysis to support their insights from each perspective. We conclude with considerations of how in turn this process supports the work of mathematics teacher educators as well.

CURRICULUM

Before moving into our case study, we first provide further detail on the importance of curriculum. Research has shown that almost every mathematics classroom will have some form of curriculum (Banilower et al., 2013) and that curriculum influences teacher planning as well as student learning (Matsumura et al., 2006; Stein, Remillard, & Smith, 2007; Tarr, Chávez, Reys, & Reys, 2006). Recent research has also found that PTs interpret curriculum through varied lenses that may reflect a concern towards the learning of mathematics or, at another extreme, that students in their classes are enjoying themselves (Nicol & Crespo, 2006). Given these studies, we argue that curriculum—a tool that we almost certainly can expect our new teachers to encounter—is a unique context with which to consider how students learn mathematics and that such learning is situated in the context of socially organized participation structures (see Borko, 2004) and reflects entrenched historical narratives of identity and power (Gutiérrez, 2013). Cognitive, situative, and sociopolitical perspectives not only *can* be coordinated but *must* be coordinated in order to support the mathematical understandings of all learners.

For the purposes of this chapter, we consider how the focal PTs translated the introductory lesson on fractions from the curriculum *Investigations in Number, Data, and Space* (Russell et al., 2008). As the introductory lesson, the curriculum first details how third graders are tasked with splitting a brownie fairly between two people as a way to introduce equal parts of a whole. As detailed below, Maya and Vanessa's efforts to translate curriculum into instruction revolve around a focus on constructing equal pieces of brownies, an activity depicted at the launch of the lesson in the curriculum materials.

THE ACTIVITY

The activity featured in the case study, which occurred as part of an elementary mathematics methods course for licensure, was designed both to engage PTs in the work of translating curriculum materials and at the same time to research the processes through which teachers notice the context of curriculum materials (see Amador, Males, Earnest, & Dietiker, 2017 and Males, Earnest, Dietiker, & Amador, 2015 for elaboration on curricular noticing). The intent of the activity was to support PTs in considering their curricular decision making and encourage them to reflect on how they think about materials. Further, the purpose was to support them in substantiating their decisions about how they could use resources for effective instructional outcomes.

The activity was administered as part of coursework in elementary mathematics methods courses at two universities. Assigned to work in pairs, PTs in both courses completed the activity using assigned curricula. We selected a third-grade lesson, the introductory lesson on fractions, and provided PT pairs with all accompanying resources within the materials related to that lesson. All data collection and analysis addressed PTs' reading and use of this one lesson from the materials.

The activity was intentionally designed to position PTs in a context parallel to that of practicing teachers in which they would have to read materials, plan a lesson, and then enact the lesson. Grossman et al. (2009) have argued for the importance of decompositions, representations, and approximations of practice. The activity included three components that we considered helpful in supporting these types of pedagogies of practice for PTs: (a) a tool for analyzing (decomposing) curriculum as well their generating of (b) a lesson plan, and (c) an animation (considered an approximation). The following describes these three in greater detail.

The Curriculum Spaces Analysis Tool (Drake et al., 2015) is a template to support the critical analysis of curriculum. The template asks about the central mathematical goals or ideas in a lesson and the standards to which the content relates. It includes several prompts addressing such elements as individual tasks; opportunities for activating or connecting to family, culture, or community; and potential spaces for discussion, for example. We considered this tool to offer an opportunity for PTs to decompose (Grossman et al., 2009) the materials they were reading on the basis of their envisioned lesson. As we argue below, we regard their reading as considering each of the three perspectives both individually and collectively. Although the tool prompted PTs to analyze curriculum, we stress that such prompts functioned as a way to start the conversation rather than yielding expert examinations of the materials.

Following completion of the Curriculum Spaces Analysis Tool (Drake et al., 2015), PTs were asked to write a lesson plan using a template provided at one of the two universities. The template included space for PTs to write about the subject, standards, big mathematical ideas, objectives, and materials. There were then sections for the three-part lesson format (Van de Walle, Karp, & Bay-Williams, 2012), specifically launch, investigate, and summarize. Within each of these sections, the PTs were asked to be specific about the role of the teacher and the role of the students and also to provide rationales for their decisions. We claim that this process also exemplified decomposition of the curricular materials (Grossman et al., 2009) because PTs could select components from materials to include, could include components beyond those mentioned in the materials, or could adapt the materials—all while considering the nuances of each of the components of the lesson.

Finally, PTs approximated practice by generating an animation of the first 5 minutes of their lesson, which constituted the opening part of the launch. For this component, they used online cloud software from goanimate.com (GoAnimate for Schools, 2015) to generate animations of the enacted curriculum. Figure 6.1 shows a screenshot of an animation from the pair described later in the chapter. (For information on the role of this technology in teacher education, please see Amador & Earnest, 2016).

Following this three-part activity, PTs were asked to reflect on the Lesson Planimation process as a whole. Specifically, they were asked about the process and why they included specific components in their lesson plan and animation. They were asked to revisit their responses on the Curriculum Spaces Analysis Tool and consider the adaptations they had originally identified when they had decomposed the curricular materials. Following this,

Figure 6.1 Screenshot of the focal pair's animation.

they were prompted about the extent to which they made these adaptations in their lesson plan, along with reasons for decisions. They were then asked similar questions about adaptations in their animation, along with ratio-nales. Finally, they were asked about other resources they may have used in designing their lesson plan or animation.

DATA SELECTION AND ANALYSIS

For the purposes of this study, we selected the complete Lesson Planimation activity from one pair of PTs for analysis. The pair was intentionally selected because their plan, animation, and reflection provided opportunities for analysis of their considerations of the cognitive, situative, and sociopoliti-cal perspectives. We focus on Maya and Vanessa in order to illuminate how the analysis, and translation of curriculum in particular, enables insights related to germane properties of each perspective to emerge.

To analyze data, we considered the work of these PTs to be a case (Yin, 2009). Specifically, we considered the Curriculum Spaces Analysis Tool, les-son plan, animation, and reflection to all provide insight to the perspec-tives of the PTs. We reviewed literature relevant to cognitive, situative, and sociopolitical perspectives. Then, both researchers individually analyzed each assignment component three times, assuming one of the three per-spectives for each round of analysis. For one round of analysis, we assumed a cognitive perspective to compare the data components. To do so, we identified those elements that related to the cognitive perspective, specifically text related to the learning and understanding of mathematical content and what students may or may not know or have experienced (e.g., "Students may think of other things in the world that can be folded in half, like a butterfly or the letter 'W.'"). We marked such elements as *cognitive*. Second, we conducted a similar analysis process from a situative perspective, again working through analysis by assuming a situative lens and considering data elements as they related to participation and task structures in instruction (e.g., "Students check with a partner to see if they divided the brownie in the same or different ways"). Finally, we repeated the same process for the sociopolitical perspective as related to power and identity (e.g., "English language learner extension suggests that teachers preteach the American idea that a fair share is an equal share"). At the conclusion of these three rounds of analysis, we open coded using the constant comparative method (Corbin & Strauss, 2008) and wrote memos for each perspective. These memos served as an overall descriptor of that perspective. We then compared codes and memos to come to an agreement on how we viewed PTs' perceptions of the curriculum materials using cognitive, situative, and sociopolitical perspectives. Note that a particular sentence or chunk of text may have been marked as more than one perspective.

CASE STUDY: GRADE 3 INTRODUCTORY LESSON ON FRACTIONS

We now consider the pair of PTs and their translation of curriculum materials as reflecting concerns within and across each of the three perspectives. In order to present the case, we first consider each of the strands alone: cognitive, situative, and sociopolitical. As we found in our analysis of 33 PTs in our study, PTs adapt considerably as they interpret curriculum and design lessons rather than giving full authority to the original materials (Amador & Earnest, 2016; Earnest & Amador, 2017). The pair we present here is no exception to this. As mentioned above, the Lesson Planimation activity targeted the introductory lesson for fractions in Grade 3. To constrain the focus of this chapter, we focus in particular on a key idea underlying fractional quantities—the role of equal parts of the whole—and how this became explicit across the pair's materials.

Cognitive Perspective

Introducing children to fractions involves a new way of conceptualizing quantity. Although children typically have everyday experiences of fair sharing, introducing fractional notation involves a new mathematical operation—which Simon (2008) identified as a key pedagogical concept—and effectively supporting PTs requires consideration of how children experience fractional quantities. In particular, the CSAT (Drake et al., 2015) provoked PTs' reflection on both development and mathematics.

The PTs began their analysis of curriculum materials by reflecting on theory in development as they considered what a typical third grader likely understands. Specifically, they focused on *conservation*, stating that third graders "should have developed the ability to understand conservation, suggesting that they should be able to place together multiple slices and conceptualize that they constitute equal parts." In the materials, a worksheet featuring a two-by-four array is provided, with the curriculum indicating students should use the rectangles to divide a brownie into two equal pieces. Two questions are provided in the materials for teachers to observe their students at work: "Do students make equal pieces?" and "Can students provide that their shares are equal by cutting, measuring, folding, or reasoning?" (Russell et al., 2008, p. 26).

In their lesson plan as well as their animation, Maya and Vanessa provided an example of how they visualized this discussion about equal shares taking place. In their written plan and their animation, a student volunteers information about sharing a cookie with a sibling; the teacher responds by saying, "You shared your cookie with your brother? That was very nice of you. How did you share it with your brother? Did you give him a little crumb?" In this case, PTs' questions—while consistent with the tone of the curriculum—are not featured in the original materials but represent the PTs' thinking about how to make connections between content targeted in the curriculum and students' own experiences.

In the process of translating the original materials into a visualization of practice, Maya and Vanessa began to consider the epistemology of fractions: What does it mean to understand equal pieces? When PTs ask children, "Did you give him a little crumb?", they reveal a sophisticated mechanism to authentically support children's understanding of equal parts through experiences they have with fair shares. Children would very likely have a strong understanding of food crumbs as compared to a whole piece, and the PTs leverage this idea in their introduction of the lesson.

Situative Perspective

Mathematics instruction is a unique context featuring socially organized participation structures (Borko, 2004). Teachers integrate mathematical ideas with children's own experiences in the context of instruction, thereby bringing meaning to notation and new operations. The PTs interpreted the curriculum materials in terms of particular activity structures that might support student learning. Like most curricula, materials provided to PTs included an activity structure, including materials needed, potential dialogue, and duration for each section. The curriculum states in the beginning of the lesson, "Suppose that this is a brownie and that two people want to share this brownie. Think about how you can cut the brownie so that two people can have equal shares. You are going to use only straight lines or straight cuts, as if you were cutting your brownies with a knife" (Russell et al., 2008, p. 25). As they designed their lesson, the focal pair drew upon ideas of fair sharing as a key element to partition the brownies.

The PTs wrote in their lesson plan, "OK, let's pretend this rectangle... is one whole brownie. You want to share this brownie with your friend. Because you are friends, you want to make sure that your friend gets an equal or the same size piece that you do." Their animation is consistent with this, with the teacher beginning the lesson by stating, "Class, today we are going to talk about sharing. Who can give me an example of a time when they had to share something with a friend, a group of friends, or maybe with your siblings or members of your family?" The pair adapted the materials with an eye towards a cultural narrative of fair sharing.

We see this pair's efforts as an example of the situative perspective, as their adaptation of the activity reflected concerns about how to engage their learners in mathematical ideas of fractional quantities through a connection to the people with whom they interact. Although the original materials did include references to fair sharing in a section for supporting English language learners, we consider PTs' efforts as adapting with a situative lens for two reasons. First, the PTs adapted the body of the lesson through this particular lens for all students, not just one particular group within the class. Second, and considering the cognitive lens above, the text within the section for English language learners establishes sharing and fair sharing through equal parts, stating, "'A *fair share* means we each get the same amount.' Have partners demonstrate sharing items and repeat the sentences: I share my _____ with _____. This is my share, and this is her/his share. We each get a fair share." Although a final discussion in the lesson does bring in the idea of unequal parts, the PTs' adaptations reflect concerns for the structure of the activity as related to connecting students' everyday lives and experiences with sharing together with the target mathematical ideas of the lesson.

Sociopolitical Perspective

Doing mathematics and making mathematics accessible to children involves asking questions about identity and power in learning mathematics. Gutiérrez (2013) considered the sociopolitical turn of mathematics education (from a poststructuralist view) as questioning historical patterns of power and how discourse patterns may replicate or disrupt such patterns. To her, "knowledge and power are inextricably linked" (p. 8). In one interpretation of this, despite the fact that a curriculum is written for *any* classroom in *any* town or city across the United States, implementation involves the consideration of local resources and funds of knowledge to enable access to the mathematical ideas valued by the field (Turner et al., 2011). Student-centered mathematics instruction seeks to build on and mathematize students' own experiences. We focus here on how PTs identified the context of the lesson—sharing brownies—as a potential barrier that may deny access to some children and, as a result, adaptations they made.

The pair stated in their lesson plan, "Many American children like brownies. With that said, many American students don't eat 'typically American' foods. The teacher may opt to use an item that children have seen in the classroom. Graham crackers may work, if they're a common classroom snack." From the perspective of children learning about fractions for the first time, the pair emphasized the importance of children's own familiarity with the context and slight modifications they could make to provide more access. Because not all students may be familiar with brownies at home, Maya and Vanessa drew attention to the fact that many teachers have classroom snacks available with which children would be familiar. In doing so, they acknowledged the range of backgrounds children may have and at the same time alluded to historical themes of power ("typically American"). Their reading of curriculum materials involved interrogating a question of "Mathematics for whom?" Their lesson plan reflected goals of providing access to all students, including those who may be from marginalized populations by emphasizing that not all students have had the same experiences before entering the classroom.

The pair also displayed concern for differentiating instruction for different learners. From a sociopolitical perspective, we see their concern—which elaborates on support provided in the material—as addressing a broader question regarding who has or should have access to powerful mathematical ideas. Furthermore, providing access to different learners may involve differentiated instructional approaches. In their lesson plan and later in their animation, the pair demonstrated a concern for the everyday experiences of their students and potential connections to the mathematical ideas symbolized in fraction notation. In their plan, Maya and Vanessa described showing pictures of brownies in particular to support English language learners' understanding of the context: "We are going to pretend that these rectangles (teacher holds up paper and points to rectangles) are brownies. Each rectangle is one whole brownie (teacher moves finger around the outside of the rectangle and write [sic] the word 'whole' on the board)." The

PTs considered how they could involve and make connections to the lives of all students in the class, to recognize their varying backgrounds.

Connections Across Strands

The process of translating curriculum materials to lesson enactment is one that involves each of the three perspectives detailed throughout this book. Although we attempted above to shine light on each perspective individually, we underscore that working with curriculum materials involves a constant coordination and negotiation across perspectives.

As the PTs generated their plan and animation from the curriculum, they considered how third graders might engage in the big ideas of fractional quantities. From our viewpoint, this is an example of their utilizing a cognitive lens to read curriculum. Questions such as juxtaposing a crumb with the whole brownie sought to support students' authentic grappling with the mathematical idea of equal parts. Yet doing this involved careful consideration of the activity structure, thereby exemplifying a situative perspective. In their interpretation of the materials, they adapted the lesson to relate to children's own lives, specifically the experiences they have sharing with friends. Furthermore, generating their lesson involved the question of mathematics *for whom*, reflecting sociopolitical considerations. Although differentiating instruction is common in curriculum materials, the pair's attention reflected clear concern for appealing to the everyday experiences that a range of children may bring to the classroom and how to design instruction to support all students, not just those who are English-dominant. Reading curriculum materials involved consideration across the three perspectives.

To close the case of Maya and Vanessa, we wish to underscore their PT status. As such, they approached the process of translating curriculum materials into lesson enactment through naïve eyes that, with experience in the field and in relation to their own classroom of children, will involve increasing coordination across the three perspectives.

CONCLUDING THOUGHTS: A PERSPECTIVE ON PERSPECTIVES

Teaching involves constant orchestration of concerns related to each of these three perspectives. Yet teaching is messy and challenging and, in line with this, not reducible to any one perspective over another. In the preparation of PTs, a goal is to support their building of a toolbox that allows each of these perspectives to be applied when appropriate. But beyond this, we hope that our preparation of new teachers honors the complex character of teaching mathematics. As demonstrated in the case above, the reading of curriculum materials for the purposes of visualizing a lesson involves each perspective and connections across them. We see the focus on curriculum in methods coursework as critical in order to support PTs in the use of a tool that we can be quite certain they will encounter.

Yet most importantly, engaging PTs in curriculum analysis as they prepare lessons engages them in the complicated work of teaching captured in the interconnectedness of perspectives, making the three lenses greater than any one of them independently. Looking at curriculum may in fact serve as a productive mechanism for teacher educators to assess the ideas their PTs are taking up in coursework and, if applicable, areas that merit further attention in coursework. In the project mentioned above, we made subsequent decisions in methods coursework to focus in particular ways, for example, on the role of equal pieces. Most PTs did not consider the instructional role that "a crumb" could play; in fact, some pairs replaced the suggested brownies with candy or food prepartitioned into equal parts, such as chocolate bars, thereby altering participation structures and decreasing the cognitive demand. Our focal pair above demonstrated how learning, as framed through each of the perspectives, leads to strategic translation of curriculum for the purposes of instruction. Yet, as reflected in the quote beginning this chapter from Maya's and Vanessa's classmate, mathematics teacher educators cannot expect PTs to have such an orientation to the work involved in using curriculum. We encourage mathematics teacher educators to provide such opportunities to translate curriculum materials and reflect on that process through the three perspectives in order for our new teachers to develop the conceptual tools to support all learners in mathematics.

REFERENCES

Amador, J., & Earnest, D. (2016). Lesson Plan-imation: Animation as a transformational lesson planning extension for preservice elementary mathematics educators. In M. Niess, S. Driskell, & K. Hollerbrands (Eds.), *Handbook of research on transforming mathematics teacher education in the digital age* (pp. 241–271). Hershey, PA: IGI Global.

Amador, J., Males, L., Earnest, D., & Dietiker, L. (2017). Curricular noticing: Theory on and practive of teachers' curricular use. In E. Schack, M. Fisher, & J. Wilhelm (Eds.), *Building Perspectives of Teacher Noticing* (pp. 427–444). New York, NY: Springer.

Banilower, E. R., Smith, P. S., Weiss, I. R., Malzahn, K. A., Campbell, K. M., & Weis, A. M. (2013). *Report of the 2012 national survey of science and mathematics education.* Chapel Hill, NC: Horizon Research.

Borko, H. (2004). Professional development and teacher learning: Mapping the terrain. *Educational Researcher, 33*(8), 3–15.

Corbin, J., & Strauss, A. (2008). *Basics of qualitative research* (3rd ed.). Los Angeles, CA: SAGE.

Drake, C., Land, T. J., Bartell, T. G., Aguirre, J. M., Foote, M. Q., Roth McDuffie, A., & Turner, E. E. (2015). Three strategies for opening curriculum spaces. *Teaching Children Mathematics, 21*, 346–353.

Earnest, D., & Amador, J. (2017, online first). Lesson planimation: Preservice elementary teachers' interactions with mathematics curricula. *Journal of Mathematics Teacher Education.*

GoAnimate for Schools. (2015). *Make animated videos in the classroom* [Computer software]. Available from http://www.goanimate4schools.com/

Grossman, P. L., Compton, C., Igra, D., Ronfeldt, M., Shahan, E., & Williamson, P. W. (2009). Teaching practice: A cross-professional perspective. *Teachers College Record, 111*, 2055–2100.

Gutiérrez, R. (2013). The sociopolitical turn in mathematics education. *Journal for Research in Mathematics Education, 44*, 37–68.

Males, L. M., Earnest, D., Dietiker, L., & Amador, J. M. (2015). Examining K–12 prospective teachers' curricular noticing. In T. G. Bartel, K. N. Bieda, R. T. Putnam, K. Bradfield, & H. Dominguez (Eds.), *Proceedings of the thirty-seventh annual meeting of the North American Chapter of the International Group for the Psychology of Mathematics Education* (pp. 88–95). Lansing, MI: Michigan State University.

Matsumura, L. C., Slater, S. C., Junker, B., Peterson, M., Boston, M., Steele, M., & Resnick, L. (2006). *Measuring Reading Comprehension and Mathematics Instruction in Urban Middle Schools: A Pilot Study of the Instructional Quality Assessment. CSE Technical Report 681*(CSE Technical Report No. 681). Los Angeles, CA: National Center for Research on Evaluation, Standards, and Student Testing (CRESST). Retrieved from http://www.eric.ed.gov/ERICWebPortal/detail?accno=ED492885

Nicol, C., & Crespo, S. (2006). Learning to teach with mathematics textbooks: How preservice teachers interpret and use curriculum materials. *Educational Studies in Mathematics, 62*, 331–355.

Remillard, J., & Heck, D. (2014). Conceptualizing the curriculum enactment process in mathematics education. *ZDM Mathematics Education, 46*, 705–718.

Russell, S. J., Economopolous, K., Wittenberg, L., Bastable, V., Bloomfield, K. H., Cochran, K., . . . & Sillman, K. (2008). *Investigations in number, data, and space* (2nd ed.). Glenview, IL: Pearson Scott Foresman.

Simon, M. (2008). Mathematics teacher education in an era of reform. In B. Jaworski & T. Wood (Eds.), *The international handbook of mathematics teacher education, Volume 4: The mathematics teacher educator as a developing professional* (pp. 17–29). Rotterdam, The Netherlands: Sense.

Stein, M. K., Remillard, J., & Smith, M. S. (2007). How curriculum influences student learning. In F. K. Lester, Jr. (Ed.), *Second handbook of research on mathematics teaching and learning* (pp. 319–369). Charlotte, NC: Information Age.

Tarr, J. E., Chávez, Ó., Reys, R. E., & Reys, B. J. (2006). From the written to the enacted curricula: The intermediary role of middle school mathematics teachers in shaping students' opportunity to learn. *School Science and Mathematics, 106,* 191–201.

Turner, E., Drake, C., Roth McDuffie, A., Aguirre, J., Bartell, T., & Foote, M. (2011). Promoting equity in mathematics teacher preparation: A framework for advancing teacher learning of children's multiple mathematics knowledge bases. *Journal of Mathematics Teacher Education, 15,* 67–82.

Van de Walle, J., Karp, K. S., & Bay-Williams, J. M. (2012). *Elementary and middle school mathematics methods: Teaching developmentally* (8th ed.). New York, NY: Allyn and Bacon.

Yin, R. K. (2009). *Case study research: Design and methods* (4th ed.). Los Angeles, CA: SAGE.

CHAPTER 7

DIVERSE PERSPECTIVES ON SOCIOPOLITICAL FRAMINGS FOR MATHEMATICS METHODS

Frances K. Harper
University of Tennessee

Beth Herbel-Eisenmann
Michigan State University

Andrea McCloskey
Pennsylvania State University

During the Scholarly Inquiry and Practices Conference (Sanchez, Kastberg, Tyminski, & Lischka, 2015), mathematics teacher educators (MTEs) grouped by theoretical perspective (cognitive, situated, or sociopolitical) worked to develop goals and activities for mathematics teacher education. Our sociopolitical group resisted naming objectives and activities to align with our perspective. Some felt that an emphasis on objectives suggested a cognitive

Building Support for Scholarly Practices in Mathematics Methods, pages 99–114
Copyright © 2018 by Information Age Publishing
99

underpinning that conflicted with our perspective. Instead, we discussed creating experiences for prospective mathematics teachers (PTs) to reflect on the social, cultural, and political dimensions of mathematics teaching and learning. We agreed that fostering equitable teaching requires MTEs to challenge the status quo across courses rather than through isolated activities. Moreover, we recognized that unique experiences with privilege and oppression inform sociopolitical perspectives in deeply personal ways, and consequently, MTE practices informed by such perspectives cannot be generalized. While honoring this complexity, we recognized the need for feasible (for MTEs) and accessible (for PTs) approaches to focusing on the sociopolitical nature of mathematics within already packed methods courses.

We each share a commitment to challenging hegemonic practices from within and outside the classroom that impact mathematics teaching and learning. For us, this commitment lies at the heart of the sociopolitical perspective, but we weave different experiences, ideas, and theories together to translate this commitment into our MTE practice. In this chapter, we invite readers to find their own way into a sociopolitical framing for mathematics methods by describing our three complementary approaches. In each section, we highlight a key sociopolitical idea underlying our work: equity, positioning, and ritual, respectively. Each narrative describes (a) the key idea informing our work (equity, positioning, ritual), (b) how we make the idea central, (c) how the idea informs our practice, and (d) what we have noticed about engaging PTs with sociopolitical perspectives. We integrate these components of our narratives in different ways. We hope these snapshots inspire others to find ways of drawing on their experiences and perspectives to challenge the status quo in their methods courses.

ORIENTING EQUITY IN MATHEMATICS EDUCATION FROM ALL STUDENTS TO EVERY STUDENT

I (Frances) am intentional in the way I talk about equity in mathematics education. Namely, I have shifted from saying "*all* students" to saying "*every* student" because I feel the latter better captures my orientation towards equity. My word choice reflects an effort to challenge how the *mathematics for all* rhetoric maintains the status quo in mathematics education (see Martin, 2003). A blanketed focus on *all* students ignores "the complexities and particularities of race, minority/marginalized status, [and] differential treatment" (Martin, 2003, p. 10). Framing equity as mathematics for *all* students oversimplifies teaching and learning and maintains a focus on those students who are already well served by the status quo in mathematics education. Shifting the emphasis to *every* student demands attention to unique mathematics experiences. Equitable teaching cannot be generalized for all but instead must be contextualized in order to respond to and support each individual's learning. Moreover, contextualizing an individual's experiences can draw greater attention to the larger structural inequities within schooling and mathematics that affect an individual student's ability to succeed (see Chubbuck, 2010). The focus on *every* student, combined with considerations of larger structural inequities, encourages challenging the status quo and emphasizes the need for responsive mathematics instruction for students who are marginalized in mathematics.

Making "All Students" to "Every Student" Orientations to Equity Central

The goal of shifting from an "*all* students" orientation to equity to an "*every* student" orientation has guided my work with PTs in secondary mathematics methods. PTs easily adopt an orientation that frames equity as the mathematics success of *all* students. PTs naturally want their future students to succeed, but most have limited experience attending to the unique needs of individual students, particularly those from traditionally and systematically marginalized groups. Instead, they have the most experience being high-achieving mathematics students themselves who have navigated and benefited from the current status quo within mathematics education. My MTE practice guides PTs to question and disrupt a system that mostly worked for them.

Using backwards design (Wiggins & McTighe, 2005), I foster PTs' evolution from an "*all* students" to an "*every* student" orientation by designing the two-course mathematics methods sequence around *big ideas* about equitable mathematics teaching that I hope all PTs will carry throughout their careers. I adopted the four principles of equitable mathematics teaching from Horn (2012) as the big ideas because these principles sparked emerging ideas about mathematics for *all* students while also allowing understandings of equity to deepen over time. Moreover, I could envision connecting both traditional methods activities and overtly sociopolitical activities to these principles. I unpacked each big idea through *essential understandings* (Wiggins & McTighe, 2005) and adapted existing learning targets, which I connected to the essential understandings and course assignments (Table 7.1). Here, I discuss how I leverage one principle: *learning is not the same as achievement* (Horn, 2012, p. 13), to reorient PTs' equity focus.

Essential Understanding 1: An "All Students" Orientation to Equity

Horn (2012) emphasized the need for better mathematics learning across the achievement spectrum. When PTs read about the big idea initially, I prompt them to reflect on their mathematics experiences and to

TABLE 7.1 Course Design Using Big Ideas, Essential Understanding, and Learning Targets

Big Idea	Essential Understandings	Learning Targets	Assignments
Learning is not the same as achievement.	1: Promoting learning involves anticipating and *assessing students' thinking during various stages* of instruction.	Develops strategies of formative assessment to gain insights into student learning to inform teaching.	Student Thinking Interview
	2: *Leveraging multiple math competencies* allows you to promote learning by recognizing and positioning students' various math backgrounds and competencies as valuable.	Demonstrates knowledge to create environments that leverage multiple math competencies and resources to support individual and collaborative learning.	Groupworthy Task Adaptation & Community/ Social Justice Math Lesson

consider opportunities they had to explore the conceptual underpinnings of procedures. For example, asking PTs to explain why they multiply by the reciprocal when dividing by a fraction illuminates limitations in their opportunities to learn mathematics, despite their high achievement. This reflection helps PTs recognize limitations of traditional measures of achievement to assess learning and of traditional methods of instruction (i.e., focus solely on procedures) to give students opportunities to learn rich mathematics. PTs begin to see a need for a different way of teaching mathematics and for deeper learning for all students.

I build on PTs' need for a different way of teaching mathematics for all students by first focusing on formative assessment. Mathematics teaching that prioritizes learning over achievement requires insight into students' mathematical thinking beyond correctly using a procedure at the end of an instructional unit. Instead, instruction must respond to and assess student thinking at various stages of learning (Essential Understanding 1). PTs complete readings and assignments to explore students' mathematical thinking and develop strategies for formative assessment (Learning Target 1). Their work culminates in a student-thinking interview conducted in their field-placement classroom. The interview provides PTs an opportunity to investigate a student's thinking about a mathematics problem and to reflect on how the experience shaped their ideas about teaching that promotes mathematics learning for all students.

During the interviews, PTs look for evidence of mathematics thinking and learning (e.g., ability to explain and justify thinking about a conceptually rich mathematics problem). Situating the assignment within the big idea helps PTs begin to challenge the status quo by questioning the relationship between learning and achievement. This connection is an important entry point for the more explicit sociopolitical focus on *every* student that follows.

Essential Understanding 2: An "Every Student" Orientation to Equity

The shift to an explicit sociopolitical focus happens across two units, each lending a different perspective to Essential Understanding 2 (Table 7.1). Recognizing the numerous yet undervalued ways students can demonstrate competence challenges PTs to reimagine what counts as success in mathematics. This shift brings a more explicit sociopolitical focus, drawing attention to how limited abilities and competencies (e.g., using procedures quickly and accurately) are traditionally valued in mathematics and how those ways of knowing are situated within broader systems of privilege' (e.g., White, male, heteronormative). The focus is still on mathematics learning rather than achievement, but learning takes on a different meaning towards a more critical stance.

The first unit introduces *complex instruction*, an approach to mediating group work in diverse classrooms by fostering more equitable interactions and relations among students (Cohen, 1994; for full unit see http://francesharper.com). The unit culminates with PTs adapting an existing high cognitive demand task to be *groupworthy* (i.e., require a range of abilities and competencies such that an individual cannot possibly complete the task alone; Horn, 2012). PTs must first recognize how students might demonstrate a range of abilities and competencies based on various student backgrounds that are valuable for learning mathematics (Essential Understanding 2), and then they must situate those multiple abilities and competencies within the specific mathematical work of a task. For example, a groupworthy task might require asking good questions, explaining well, and knowing how to use different representations. PTs' task adaptation must create opportunities for different students to contribute their unique strengths, which move the mathematical work of the group along in valuable ways.

PTs enjoy the assignment because they develop practical strategies for leveraging multiple mathematical abilities to support collaborative learning (Learning Target 2), and they recognize complex instruction as an approach to teaching mathematics that could benefit *all* students. I then push PTs further to consider *every* student through the emphasis on students with low status within complex instruction. Differences in academic or social status influence how students interact around mathematics with low status restricting opportunities to learn (Cohen, 1994). Thus, teachers must pay attention to individual students, particularly those who are traditionally and systematically marginalized in mathematics. An important part of the groupworthy task adaptation is identifying abilities that give students with low status greater opportunities to learn. Over time, and through recognition by teachers and peers, the valuable mathematics contributions of students with low status position them as competent.

During the subsequent unit on critical mathematics education, I encourage PTs to think even more critically about broader structural influences and the mathematics curriculum by introducing other pedagogical approaches, such as culturally relevant pedagogy (e.g., Tate, 1995) and teaching mathematics for social justice (e.g., Gutstein & Peterson, 2013). I challenge PTs to explicitly name the role of race, gender, class, and so on in mathematics teaching and learning by sharing personal stories from my own teaching with historically marginalized youth (Harper, 2016) and by engaging them with example mathematics lessons. The unit culminates as PTs integrate community mathematics practices or a social justice issue (assignment adapted from Turner et al., 2015) into a mathematics lesson. I have been surprised by how enthusiastically PTs have welcomed these (even more radical) approaches to mathematics teaching and the orientation to equity that accompanies such approaches and their enthusiasm for a critical stance to teaching mathematics shows in the lessons they have developed (see http://francesharper.com).

By the end of methods, my goal is that PTs no longer imagine students' various mathematics backgrounds and competencies under the blanket of *all* students. Instead, I hope they recognize the importance of tailoring lessons for specific students, with consideration for how race, gender, class, and so on influence each student's mathematics background inside and outside the classroom. Viewing mathematical competency as rooted in issues or practices that matter in specific contexts (within and beyond school) fosters instruction that challenges the status quo in mathematics in order to promote the meaningful learning of *every* student.

POSITIONING AS ONE SOCIOPOLITICAL LENS TO INTERPRET MATHEMATICS CLASSROOM DISCOURSE

Recently I (Beth) have been collaborating with colleagues to design professional development (PD) materials for secondary mathematics teachers[1] on the basis of my previous collaboration with teacher-researchers in Iowa (see Herbel-Eisenmann & Cirillo, 2009). We designed PD materials that would introduce secondary mathematics teachers to discourse moves as alternatives to the pervasive Initiate-Respond-Feedback/Evaluate (IRF/E, Mehan, 1979) pattern. Across 4 years, we piloted and revised the materials and have used many of the activities in mathematics methods courses with PTs.

Making Positioning Central

We began with the *talk moves* (Chapin, O'Connor, & Anderson, 2009), which we modified into *teacher discourse moves* (TDMs; see Herbel-Eisenmann,

Steele, & Cirillo, 2013, for more information). We attended to "productive" discourse or how the TDMs helped teachers support students' "access to mathematical content and discourse practices" (Esmonde, 2009, p. 250) and found that the ways teachers thought about aspects of mathematics classroom discourse became more nuanced (Herbel-Eisenmann, Johnson, Otten, Cirillo, & Steele, 2015). We had not given the same attention to another important opportunity to learn: students' "(positional) identities as knowers and doers of mathematics" (Esmonde, 2009, p. 250) or what we came to call "powerful" discourse. Our first version of the materials focused on consideration of classroom norms and social goals. Yet, we noticed that discussions about these ideas often devolved into a focus on students' misbehavior or nonconformity rather than ideas about students' developing identities and relationships with mathematics. Although I used the idea of positioning to understand issues of authority, agency, and voice in my research (e.g., Wagner & Herbel-Eisenmann, 2009), I had not operationalized it in my work with teachers. We needed to give similar careful attention to how the TDMs could support powerful discourse, so we made positioning theory central to the materials (e.g., Harré & van Langenhove, 1999). This also emphasized important sociopolitical aspects of mathematics classroom discourse.

Operationalizing Positioning for Work With Teachers

Every time mathematics teachers and students interact, they communicate about mathematics content, negotiate relationships, and send implicit and explicit messages about students' positional identities. Positioning is "the ways in which people use action and speech to arrange social structures" (Wagner & Herbel-Eisenmann, 2009, p. 2). Positioning happens all the time and is often unintentional. We operationalized positioning by focusing on interactions between people (the *student–student* and *teacher–student* interactions) as well as interactions between people and mathematics.

When considering positioning in *student–student* interactions, issues of status (Cohen, 1994) and smartness (Featherstone et al., 2011; Horn, 2012) can be considered. For example, in a short synthesis of research (which we call a *touchstone* document), we highlight that students can position themselves and others in implicit and explicit ways:

> In the classroom, a student may communicate other students' positioning implicitly by ignoring them when they speak or always taking up the ideas of other students. A student may also be positioned implicitly when they are told, "That is a great idea!," "Let's try that," or "That will never work." Students may also communicate about themselves or others' positionings more explicitly. Statements such as, "I've never been good at math," "I just put that

answer because Najah did," or "Wow! Steven is so smart in algebra. He always gets As," position students directly. (Herbel-Eisenmann, Cirillo, Steele, Otten, & Johnson, in press)

These status issues play out in classroom interactions and are also made transparent through structural and institutional practices like tracking. When particular positionings are repeated over time, they can impact students' identities (Anderson, 2009) and dispositions (Gresalfi, 2009) in ways that may be positive or negative in terms of students' perceptions of themselves as people who can know, do, and understand mathematics.

Our focus on *teacher–student* interactions involves consideration of teacher authority and helps secondary mathematics teachers think about the ways in which they control classroom practice. It also helps them think about how they might disrupt positionings of students by assigning competence (Cohen, 1994; Featherstone et al., 2011; Horn, 2012). Often, these positionings are based on stereotypes and thus link some of the interactional positionings to broader structures that may be used to marginalize students in classrooms. Finally, considering the positioning of mathematics involves reflecting on what teachers engage students in and how students come to see those tasks, activities, practices, and so on as being what mathematics *is*. For example, if students primarily work in silence, practice manipulating equations, and take timed tests, they could come to believe that knowing and doing mathematics is all about individual work, perfecting procedures by following what others told them to do, and getting right answers quickly. Although people familiar with positioning theory would identify this as calling into question the storyline of traditional school mathematics, we made this a type of positioning to avoid introducing more new terminology.

Engaging Teachers in Activities and Discussions About Positioning

In the materials, we begin with explorations of the key ideas without actually naming them. We consulted with Lisa Jilk, a former Railside School teacher (see Boaler & Staples, 2008) and experienced mathematics teacher educator and researcher who supports teachers to use complex instruction. She gave us feedback on activities and reflections for teachers to consider these ideas. For example, we have teachers identify how they are smart and ask questions about how they participated in mathematics tasks they solved, whose ideas were taken up, and so on. We reflect on the messages various tasks and practices in video excerpts might send to students about what it means to know or do mathematics. When teachers look at a set of student solutions that are all technically correct, but in which students communicate their thinking differently, we talk about the tendency to make assumptions about the students on the basis of their language use. Sometimes when teachers look at these solutions, for example, they talk about them as being produced by "high" and "low" students (see, e.g., Suh, Theakston-Musselman, Herbel-Eisenmann, & Steele, 2013). We then follow up with discussion questions like: If all of these solutions are technically correct, why might someone talk about the solutions as if they thought a "high-level student" or "low-level student" had produced it? What implications might these hidden values have for assessing student learning?

After some exposure to these ideas, we introduce positioning through having teachers read a touchstone document focused on this idea. Teachers then trace a transcript of a classroom lesson in which the teacher used the Hidden Triangles task (for a description of a modified version of this task, see Harper, Sanchez, & Herbel-Eisenmann, this volume). Teachers first consider positioning more generally and then are assigned a particular student to focus on across the lesson. From this point on, some of the questions we consider whenever we watch videos, look at mathematics tasks, or look at student work include: Who is considered "smart" in this classroom? About what (e.g., procedures? concepts? representations?)? Who is considered a "struggling" learner? By whom? Who is talking (the teacher, which students specifically)? What do they talk about? What kind of do language they use to communicate about mathematical objects or processes? What might students say it means to "know" mathematics in this classroom? In what ways does the teacher use her authority in this classroom?

In terms of starting to consider these ideas in their own classrooms, for example, we ask teachers to gather information about students' conceptions of mathematics by conducting short interviews or having students take an available conceptions instrument (see Online Evaluation Resource Library, n.d.). We ask teachers to compare their responses to their students' and consider why differences might matter. For example, one group of teachers found that they strongly agreed that mathematics makes sense, yet many of their students disagreed with this statement. Additionally, they found that many students agreed with a statement that doing mathematics mainly involved following procedures given to them in the textbook or by their teacher. These and other differences were rather startling to the teachers and led them to start using high cognitive demand tasks and to adopt different and common classroom norms as a department (see Busby et al., in press). As the teachers incorporate TDMs in their own classrooms, we ask them to reflect on the purposes for using various TDMs through the lens of positioning. These investigations often continue into action research projects.

A Few Observations about Teachers' Engagement With the TDMs and Positioning

For the past 4 years, I have collaborated with a group of mathematics teachers. In the first year, when they were introduced to the idea of positioning, the teachers said they had thought a little bit about aspects of positioning but not to the extent we considered it. As they began to engage in action research projects, they incorporated the TDMs in ways that considered positioning. They have also incorporated new activities and reflections in which students share how they are thinking about issues of positioning. For example, they have asked students to write about times in class when another student impacted how they thought about something, how they thought about themselves as mathematics learners, or how they participated. They have also designed short surveys about positioning that they have students fill out bimonthly.

The idea of positioning has not only impacted their practice but also served an unanticipated role in our PD context. The word *position* has become a term that I have been able to use to call out negative or unproductive positionings of students in our conversations. When the teachers talked about their "low-level" students, for example, I can ask if that was the way they really wanted to position their students because of the implications for their practice. My sense was this term became something that allowed us to talk about positionings of students, while allowing the teachers to save face. Additionally, as time passed, they eventually started using this term to keep each other accountable for the ways they talked about students.

As we move forward in our work and further consider aspects of positioning, I am aware that the idea is limited as we examine sociopolitical aspects of mathematics teaching and learning. The idea of positioning focuses primarily on local contexts, while recognizing that people draw on storylines in their interactions when they position one another (Herbel-Eisenmann et al., 2015; Wagner & Herbel-Eisenmann, 2009). Yet our materials do not treat the institutions, structure, and societal views in ways that do justice to broader systems of privilege and oppression. Although these ideas can help teachers understand, to some extent, how status relates to participation, there is more work to be done to link the broader systems of privilege and oppression to these patterns of interaction.

RITUAL AND MY SOCIOPOLITICAL TURN IN MATHEMATICS METHODS INSTRUCTION

Although the notion of ritual might not immediately seem like a key sociopolitical idea, my scholarly interest in the construct of ritual was very much related to my personal experiences teaching mathematics methods courses. I (Andrea) have been teaching mathematics methods courses for elementary PTs on a consistent basis (i.e., an average of two courses per year) for the past decade. It is one of my favorite courses to teach, for a variety of reasons. The students are good at student-ing. They are motivated to learn because they intend to teach elementary school students in the very near future. They only have 1 or 2 more years of fulltime study ahead of them, so the clock is ticking. They are mostly kind and receptive. They like classrooms and schools and teachers. That's why they want to be teachers. We bond over a shared love for school supplies.

And then the content itself is fun. There is no shortage of interesting mathematical content or pedagogical content that we could explore, and long ago I let go of the charade that I was even trying to cover everything that might appear on someone else's "must-cover" list, so I choose to spend quality time on a few topics. We talk about how children learn about our place-value system, how teachers can foster environments in which sound mathematical reasoning is the currency of value, and what are good resources like books and websites that they can turn to when they have their own classroom in the very near future. It's fun. Every semester, it's fun. My teaching evaluations are reasonably high, and I sometimes hear back from former students about how much they learned.

Nevertheless, my responsibilities and commitments as a mathematics teacher educator are not only to be an effective instructor but also to participate in research and scholarly conversations about methods instruction. And it was at the intersection of these two domains of my work (teaching and researching teaching) that I encountered a tension. Several years ago I began to become frustrated in my ability as a mathematics education researcher to describe activity in mathematics classrooms as anything more than their teachers' and students' knowledge, beliefs, and behaviors. Furthermore, I was personally dismayed at the frequency with which I and others in the field of mathematics teacher education attended to the deficiencies in these areas. On the one hand, it felt too easy and too disrespectful to visit classrooms and then document only the ways that teachers or students fell short. On the other hand, documenting mainly success stories—studying only teachers and students who experience mathematics learning in schools in affirming and empowering ways—also felt like a disservice. Such a stance could misrepresent how bleak mathematics in schools has become, especially for groups of students who were already underrepresented from participation.

Perhaps more germane, I was dissatisfied in my work as a MTE, which is primarily composed of my instruction of elementary mathematics methods courses but also includes regular interactions with practicing teachers and school district personnel. I felt like I was engaging with my PTs in very limited and overly rational ways. I was talking to them (and even "with" them, as we are encouraged to do) about particular mathematics content and the teaching and learning thereof, but I was not engaging with them as whole human beings. I turned to Noddings' (1984) notion of *care*, and that helped me frame mathematics teaching and learning as fundamentally relational enterprises in which developing trust and care within a mathematics learning community is not at odds with developing mathematical and pedagogical understandings. Balancing the caring for PTs with the caring about mathematics as a discipline is not a zero-sum game. I extended the notion of care to a study on professional development so that work with practicing mathematics teachers could be similarly broadened (McCloskey, 2012).

This ethical turn helped me justify approaches to teaching and research that felt better to me, but I was left still unable to make sense of the mixed messages I was encountering when I talked with a variety of groups about *what* and *how* children *should* learn in school mathematics. Conversations with parents were particularly perplexing. I came to the conclusion that contemporary U.S. society at large has a deeply ambiguous relationship with mathematics, children, and schools and the intersection thereof. I have talked with parents who described their negative memories of their own time as a student in mathematics class, yet in the next breath they said they wanted their children's homework to look just like theirs did. I have heard from teachers who think good teaching means making limited use of a textbook, yet they rely on their own teacher's manual to choose what and how they should teach.

Nonetheless, the notion that mathematics teaching and learning is complicated, multivocal, and always contested and negotiated is not sufficiently explanatory in the face of the very real, material role that mathematics achievement plays as a gatekeeper. In particular, mathematics achievement is defined in narrow and unequal ways so that rhetoric about "achievement gaps" persists and drowns out other narratives within the public discourse about public schools and their so-called failings. And so I sought a framework that allowed me to view mathematics classrooms through both postmodern and critical lenses, and I found that framework through Quantz's (2011) scholarship of ritual. I have written elsewhere (McCloskey, 2014) about the adaptations I made in developing a definition of ritual for use in mathematics education contexts, but—in short—the definition I use is that *ritual* is that aspect of practice that is symbolic, traditionalized, formalized performance.

Making Ritual Central

Because I now view mathematics classrooms primarily as places through which cultural expectations about mathematics learning are confronted, embodied, negotiated, and sometimes contested, I understand the mathematics methods course as a central site in which I can engage PTs in developing awareness (knowledge) and appreciation (a less rational and more aesthetic, embodied response) for the power and ubiquity of rituals in our mathematics schooling tradition. And so, for example, in my methods instruction, I do require timed, paper-and-pencil quizzes and tests, and I do require that my students write 5-page essays. I am not only using these assignments as opportunities to assess their *knowledge* about mathematics content and mathematics teaching principles, although I do look for and expect a level of expertise. In this case, my framework of ritual helps me regard the practices of "testing," "grading," and "writing papers" as familiar rituals that my students have been engaging in as university students and in their many years learning mathematics in school. They are traditionalized performances. My students think they know what to do. And so, I change the form of these activities. I disrupt what it looks like to perform a mathematics test by allowing them to work with partners and by including items with concepts they have never seen before. We discuss the symbolism that is carried through these practices—such as taking a mathematics test—in ways that they might not have been aware of before they became unfamiliar. My intention is that by deliberately changing and then discussing familiar rituals such as grading or taking a test, PTs can develop a sensitivity to the powerful ways classroom activities can "embody memory in collective time" (Tippett, 2016, p. 58).

My point here is not that structuring tests in a mathematics methods course as partner tests is particularly innovative or that this activity is a reliable way to convey sociopolitical ideas to PTs. No one activity or even a set of activities can do this. The larger point is that ritual emphasizes the way that classroom structures—the ways activities (i.e., the "content" of the course itself) are presented, enacted, and referenced again—are a central part of what students learn in my course. *How* we do things is as important to me as *what* we do. We learn through what we do—anthropological research reveals just how much cultural value is learned through small, unobvious moments of ritual (Goffman, 1967).

The primary goal for my mathematics methods courses is to elicit stories from the PTs about their experiences in mathematics classrooms and then to piece these individual stories together as well as to bring in additional voices (this is one way I use research to inform my instruction—to broaden our class community's narrative of experiences of teachers and learners in mathematics classrooms). I no longer simply critique the heck out of my students' stories about their favorite mathematics teachers (I used to do this!), but I have a newly developed appreciation for the role the ritual of telling stories about favorite teachers plays for PTs, and I make an effort to honor these stories. My commitment to social justice also leads me not just to analyze classroom practices for ritual aspects but to critique the ways that these practices play out in patterned ways through time and place.

The question could be raised: *To what extent does the idea of ritual exclusively inform my instruction?* Other, similar subquestions that address this same idea are: *How much are the students in my class aware of my ritual perspective? How would my course be different if I was not drawing upon the framework of ritual?* The answer to the first subquestion is: not at all. I do not intentionally use the word *ritual* in my teaching. I do not explicitly use much of the vocabulary itself. The answer to the second question is: This is impossible to answer definitively. The best I can do is speculate; and in doing so, I have come to the conclusion that my study of ritual has fundamentally changed my mathematics methods instruction, from my in-the-moment decision making to my articulation of learning goals for the course.

SOME FINAL THOUGHTS

Our three narratives provide a glimpse into the way that scholarly inquiry has guided our approaches to framing mathematics methods from sociopolitical perspectives. We have drawn on a range of ideas and theories (e.g., orientations to equity, backwards design, complex instruction, positioning, care, ritual) to inform our MTE work. A variety of paths have led us to using sociopolitical perspectives in our methods courses, but we each share a willingness to take risks as we adapt our practices to highlight social, cultural, and political dimensions of mathematics teaching and learning and to challenge the status quo in mathematics education. Although our narratives and the practices they describe are complementary, they are each different because they are informed by who we are as individuals and the different contexts in which we work. In the words of bell hooks, "One of the things that we must do as teachers is twirl around and around, and find out what works with the situation that we're in. Our models might not work. And that twirling, changing, is part of the empowerment" (1999, p. 128). We hope our stories inspire other MTEs to take risks with their practice and to explore what sociopolitical models for mathematics methods work for them.

NOTES

1. This work was supported with funding from the National Science Foundation ([NSF], Award #0918117, Herbel-Eisenmann (PI), Cirillo, & Steele (co-PIs)). Any opinions, findings, and conclusions or recommendations expressed in this material are those of the authors and do not necessarily reflect the views of the NSF.

REFERENCES

Anderson, K. (2009). Applying positioning theory to the analysis of classroom interactions: Mediating micro-identities, macro-kinds, and ideologies of knowing. *Linguistics and Education, 20*, 291–310.

Boaler, J., & Staples, M. (2008). Creating mathematical futures through an equitable teaching approach: The case of Railside School. *Teachers College Record, 110*, 608–645.

Busby, L., Goff, C., Hanton, D., Herbel-Eisenmann, B., Jones, L., Loeffert, C., Pyne, E., & Wheeler, J. (in press). Supporting powerful discourse through collaboration and action research. In A. Fernandes, S. Crespo & M. Civil. (Eds.), *Access and Equity: Promoting high quality mathematics in grades 6-8*. Reston, VA: NCTM.

Chapin, S. H., O'Connor, M. C., & Anderson, N. C. (2009). *Classroom discussions: Using math talk to help students learn* (2nd ed.). Sausalito, CA: Math Solutions.

Chubbuck, S. M. (2010). Individual and structural orientations in socially just teaching: Conceptualization, implementation, and collaborative effort. *Journal of Teacher Education, 61*(3), 197–210. doi:10.1177/0022487109359777

Cohen, E. (1994). *Designing groupwork: Strategies for the heterogeneous classroom*. New York, NY: Teachers College Press.

Esmonde, E. (2009). Mathematics learning in groups: Analyzing equity in two cooperative activity structures. *Journal of the Learning Sciences, 18*, 247–284.

Featherstone, H., Crespo, S., Jilk, L., Oslund, J., Parks, A., & Wood, M. (2011). *Smarter together! Collaboration and equity in the elementary math classroom*. Reston, VA: National Council of Teachers of Mathematics.

Goffman, E. (1967). *Interaction ritual*. New York, NY: Anchor.

Gresalfi, M. S. (2009). Taking up opportunities to learn: Constructing dispositions in mathematics classrooms. *Journal of the Learning Sciences, 18*, 327–369.

Gutstein, E., & Peterson, B. (2013). *Rethinking mathematics: Teaching social justice by the numbers* (2nd ed.). Milwaukee, WI: Rethinking Schools.

Harper, F. K. (2016). Challenging patterns to change my world: Using my personal evolution of critical race consciousness in mathematics teacher education. In N. M. Joseph, C. Haynes, & F. Cobb (Eds.), *Interrogating whiteness and relinquishing power: White faculty's commitment to racial consciousness in STEM classrooms* (pp. 151–169). New York, NY: Peter Lang.

Harré, R., & van Langenhove, L. (Eds.). (1999). *Positioning theory: Moral contexts of intentional action*. Oxford, England: Blackwell.

Herbel-Eisenmann, B., & Cirillo, M. (Eds.). (2009). *Promoting purposeful discourse: Teacher research in mathematics classrooms*. Reston, VA: National Council of Teachers of Mathematics.

Herbel-Eisenmann, B., Cirillo, M., Steele, M. D., Otten, S., & Johnson, K. R. (in press). *Mathematics discourse in secondary classroom: A case-based professional development curriculum*. Sausolito, CA: Math Solutions.

Herbel-Eisenmann, B., Johnson, K. R., Otten, S., Cirillo, M., & Steele, M. (2015). Mapping talk about the mathematics register in a secondary mathematics teacher study group. *Journal of Mathematical Behavior, 40*, 29–42. doi:10.1016/j.jmathb.2014.09.003

Herbel-Eisenmann, B., Steele, M., & Cirillo, M. (2013). (Developing) teacher discourse moves: A framework for professional development. *Mathematics Teacher Educator, 1*(2), 181–196.

hooks, b. (1999). Embracing freedom: Spirituality and liberation. In S. Glazer (Ed.), *The heart of learning: Spirituality in education* (pp. 113–129). New York, NY: Jeremy Tarcher.

Horn. I. S. (2012). *Strength in numbers: Collaborative learning in secondary mathematics.* Reston, VA: National Council of Teachers of Mathematics.

Martin, D. B. (2003). Hidden assumptions and unaddressed questions in "Mathematics for All" rhetoric. *The Mathematics Educator, 13*(2), 7–21.

McCloskey, A. (2012). Caring in professional development projects for mathematics teachers: An example of stimulation and harmonizing. *For the Learning of Mathematics, 32*(3), 28–33.

McCloskey, A. (2014). The promise of ritual: A lens for understanding persistent practices in mathematics classrooms. *Educational Studies in Mathematics, 86,* 19–38.

Mehan, H. (1979). *Learning lessons.* Cambridge, MA: Harvard University Press.

Noddings, N. (1984). *Caring: A feminine approach to ethics and moral education.* Berkeley, CA: University of California Press.

Online Evaluation Resource Library. (n.d.). Teacher education instruments. Retrieved from http://oerl.sri.com/instruments/te/teachsurv/instr55.html

Quantz, R. (2011). *Rituals and student identity in education.* New York, NY: Palgrave Macmillan.

Sanchez, W., Kastberg, S., Tyminski, A., & Lischka, A. (2015). *Scholarly inquiry and practices (SIP) conference for mathematics education methods.* Atlanta, GA: National Science Foundation.

Suh, H., Theakston-Musselman, A., Herbel-Eisenmann, B., & Steele, M. (2013, October). *Teacher positioning and agency to act: Talking about "low-level" students.* Research paper presented at the 35th annual meeting of the North American Chapter of the International Group for the Psychology of Mathematics Education, Chicago, IL.

Tate, W. F. (1995). Returning to the root: A culturally relevant approach to mathematics pedagogy. *Theory Into Practice, 34*(3), 166–173.

Tippett, K. (2016). *Becoming wise: An inquiry into the mystery and art of living.* New York, NY: Penguin.

Turner, E., Aguirre, J., Drake, C., Bartell, T. G., Roth McDuffie, A., & Foote, M. Q. (2015). Community mathematics exploration module. In C. Drake et al. (Eds.), *TeachMath learning modules for K–8 mathematics methods courses.* Retrieved from http://teachmath.info

Wagner, D., & Herbel-Eisenmann, B. (2009). Re-mythologizing mathematics through attention to classroom positioning. *Educational Studies in Mathematics, 72*(1), 1–15.

Wiggins, G. P., & McTighe, J. (2005). *Understanding by design.* Alexandria, VA: Association for Supervision and Curriculum Development.

SECTION III

LEARNING GOALS AND ACTIVITIES
IN MATHEMATICS METHODS COURSES

CHAPTER 8

EXPERIENCES USING CLINICAL INTERVIEWS IN MATHEMATICS METHODS COURSES TO EMPOWER PROSPECTIVE TEACHERS

A Conversation Among Three Critical Mathematics Educators

Theodore Chao
The Ohio State University

Jessica Hale
Georgia State University

Stephanie Behm Cross
Georgia State University

We are three mathematics teacher educators (MTEs), operating at differing phases of our careers with differing levels of power and privilege, working closely with prospective teachers (PTs) across two universities and three

Building Support for Scholarly Practices in Mathematics Methods, pages 117–131
Copyright © 2018 by Information Age Publishing
All rights of reproduction in any form reserved.

departments. Through narrative vignettes and reflective conversation, we share our continual struggles in using the clinical interview as an activity in our mathematics methods courses. We grapple with what it means to prepare teachers who value and seek out students' mathematical funds of knowledge through the experience of the clinical interview, and we contemplate the sociopolitical implications that arise from this work.

To illustrate our current thinking and continued wonderings, we share vignettes from our methods courses to show the different ways we justify and adapt the clinical interview. Our use of vignettes, or "narrative snippets that crystalize illustrative issues" (Graue & Walsh, 1998, p. 213), is purposeful; writing for us has become a part of the interpretive process, illustrative of our struggles as teacher educators operating from a sociopolitical theoretical perspective.

Clinical Interviews

The use of clinical interviews as an activity in education is not new. In fact, one-on-one interviews about mathematical problem solving were central to Piaget and Inhelder's (1969) early work in attempting to understand children's thinking. Additionally, Vygotsky's (1978) work on how children move from an inner to a social speech was based heavily on one-on-one mathematical interviews with children. Even Freire's (1970) emancipatory work with Brazilian literacy teachers revolved around one-on-one interviews in which Freire engaged teachers to use whatever means they wished (e.g., drawings, symbols, gestures) to describe their thinking.

In mathematics methods courses, the clinical interview is most often used to help PTs practice listening to children's mathematical thinking. Ginsburg's *Entering the Child's Mind: The Clinical Interview in Psychological Research and Practice* (1997) serves as a guideline for how PTs can plan and implement a clinical interview with children. Ginsburg's model of the clinical interview focuses heavily on eliciting mathematical explanations from children. In elementary mathematics methods courses, a Cognitively Guided Instruction (CGI; Carpenter, Fennema, Franke, Levi, & Empson, 1999) approach is often used to situate the clinical interview as a place for PTs to develop the teaching skill of listening to children's mathematical thinking (Empson & Jacobs, 2008).

Sometimes, as detailed in Jessica's vignette later, CGI-style clinical interviews are focused only on one particular learning opportunity: PTs learning to diagnose where on the CGI framework a child is thinking for a particular content area (e.g., What strategies does a child use when solving a Part-Part-Whole Part Unknown task involving double-digit numbers?). This diagnostic approach to the clinical interview focuses on assessing a child's mathematical thinking, a positivist/post-positivist mindset towards mathematical learning that ignores the mathematics teacher's role in socioconstructed or critical aspects of how children create mathematical knowledge.

Sociopolitical Perspectives

We draw upon Gutiérrez's (2013) description of the sociopolitical turn in mathematics education in analyzing our own practice. We ask, "What forms of power and authority are enacted in determining what...students learn and from whose perspectives?" (p. 3) and "What is missing from the mathematics classroom because I am required to cover this concept?" (p. 1). As MTEs attempting to help our PTs consider the intersection of their sociopolitical and mathematics teacher identities (Felton-Koestler, 2015), we create instructional activities, we hope with lasting residue, to help PTs consider "how students are constructing themselves and being constructed with respect to mathematics" (p. 1).

SCENE 1: ADDRESSING PROSPECTIVE TEACHERS' DEFICIT PERSPECTIVES

Stephanie, an assistant professor teaching a middle-grades student teaching seminar, sits down to read her PTs dilemmas of practice. It's the middle of the semester, and she's been worried about her PTs' language around students and their families.

Stephanie draws in a quick breath as she starts reading the posts (All quotes and vignettes come directly from PTs' assignments in her methods courses.):

> Tam (all student names are pseudonyms): The majority of my students do not remember how to add, subtract, multiply or divide fractions.... They don't pay attention because they get distracted so easily.

Stephanie pauses, wonders how to respond to the written dilemma, and finally suggests that Tam shadow a student for a day after her full-time student teaching unit is complete. Another PT, Krystal, writes,

> My students, for the most part, do not do their work. They refuse to show their work when solving equations, and they do not turn in homework. My mentor teacher prepares notes for the students to write in their notebooks, but over half the students have lost the notebooks. The students end up taking notes on a sheet of paper, but they leave it on the floor at the end of the period! They never ask questions or ask for help. There are several students that do nothing!

Stephanie thinks back to Gutiérrez's (2013) article and copies a few sentences for her PTs:

> Our rush to move onto the next mathematical concept almost ensures we will not ask why this concept. Who benefits from students learning this concept? What is missing from the mathematics classroom because I am required to cover this concept? How are students' identities implicated in this focus? (pp. 1–2)

Stephanie reads Jennifer's dilemma of practice next:

> I have been in the classroom for almost three months. Seeing the day-to-day realities of the educational system have been disheartening, to say the least. I am constantly trying to instruct students that are more concerned about the latest athletic shoe than the fact that they cannot multiply and divide. The students also come from extremely difficult home environments. Rather than striving for a way out, they seem content to continue the cycle of their families.

Stephanie finishes reading with a sinking feeling in her heart. She feels overwhelmed by her PTs' negative views of their students and how they seem to think their job is to impart knowledge rather than facilitate student inquiry into the world. She knows that many teacher candidates come into certification programs with negative beliefs about students in "urban schools" but thought her students had gotten past that. Stephanie also knows that most of her PTs experienced direct instruction in schools—the sort of instruction they appear to be seeing and using in their student teaching. But in this student teaching seminar course, she talks with her PTs about inquiry-oriented approaches to mathematics instruction—about ways to draw on students' understandings and cultural funds of knowledge to meet them where they are, to develop instructional tasks designed to push students in their thinking. She pauses and wonders, "We did this, right? What else can I do?"

Scene 1 Conversation: How Do We Help PTs Overcome Deficit Perspectives?

In our first conversation, we focus on Stephanie's experiences that led her to question if she needed to change her approach to her courses. Her PTs' reflections helped us to unpack the complex task of teaching. Our dialogue digs into the impact of student teaching on reifying or dismantling what we teach in our methods courses, leaving us feeling like the collective hard work MTEs and PTs do together can unravel in the field. We wonder how Jessica and Theodore's work using clinical interviews might fill this gap.

Theodore: Stephanie, you mentioned that your PTs learn about cultural funds of knowledge and that you enact an inquiry-oriented approach to teaching during your seminar, focusing on what students are coming to school with, as opposed to what students are not coming to school with. Yet when you look at these reflections, your PTs seemed to ignore those ideas from their coursework, focusing instead on what they think they see and generalizing their students through limited experiences.

Jessica: Furthermore, how does this reversion to a deficit mindset get carried into practice when you become a classroom teacher on your own?

Stephanie: I think it's about the veteran teacher a PT is partnered with and the environment a PT is teaching in. How do we mess up the structure of student teaching while we still have these PTs, so they can actually see the readings and discussions from our methods courses as relevant? A lot of my PTs' frustration comes from being up in front of the class and trying to teach for the first time and not feeling prepared. There's something about jumping from classes into full-time teaching while you're student teaching that's just not working. We don't provide them space to sit down with a child one-on-one as the classroom teacher in order to understand their students as people first, to listen to what they're thinking about mathematics. My PTs dove right into having to teach the full class. And when they move to full-time teaching in the field, the same thing happens. They don't have the space and room to consider, to wonder, to come alongside a student, or to work with families or communities to disrupt the deficit narrative. As I reread my vignette, I keep thinking, what else can I do? What should I do differently? Is using clinical interviews before and during student teaching the answer? I'm not sure it's that simple.

SCENE 2: CREATIVE INSUBORDINATION

One week before the start of the semester, Jessica, a 3rd-year doctoral student assigned to teach her first elementary mathematics methods course, wipes away tears as she finishes reading the common syllabus she has just been handed.

The excitement of being asked to teach an elementary mathematics methods course, a course Jessica had wanted to teach long before she started the doctoral program, was wearing off. Jessica's plans for a course that pushed PTs to view mathematics teaching and learning as a sociopolitical act were replaced with clinical interviews meant to diagnose student thinking and multiple choice exams. Once the tears dried, Jessica set out to do what she asks of her PTs, to find a way to do what was best for those she taught, but in a way that would avoid getting her fired. She asked trusted colleagues how to adapt what she had been given. The syllabus focused on using clinical interviews with a CGI framework to classify students' mathematical thinking, culminating with a summative assessment in which PTs matched sample student responses with CGI classifications—a strictly diagnostic interpretation of the clinical interview.

So on the first day of class, Jessica walked in feeling much like she did the first time she walked into her middle school classroom: excited, anxious, and not quite sure how she was going to pull this off. She struggled to teach what felt right, while also staying within what was asked of her. She constantly questioned herself and wrote in her teaching journal:

> What if I'm not preparing them for their next methods class [note: at Jessica's institution, PTs take two mathematics methods courses in their teacher preparation sequence]? What if their supervisors push back against the ideas they're creating in my class? Do I have teaching observations in this position; is this going to hurt those? No matter what I push in class conversation, the work I'm asking them to do with kids doesn't go far enough. I know that these are the same things they're likely to struggle with as they begin their teaching careers, can I use this somehow?

Shaking as she spoke, Jessica opened herself to the vulnerability of sharing what she was grappling with. As the course progressed, Jessica witnessed PTs starting to connect what she shared about her teaching decisions, as they began to implement ethical and critical teaching practices within their field placements. It was not perfect. A small number of students still focused only on the assignments in the syllabus, simply conducting interviews of a dozen questions, classifying children's solutions, and using the data to calculate next steps. But the other students engaged on a more sociopolitical level. Many chose to expand beyond the assigned interview, theorizing questions they could ask to get to children's mathematical thinking outside of school. Others discussed visiting students' homes, wondering about engaging in mathematical discussions with families or groups of children over meals instead of through a one-on-one interview in the library. One PT focused on a child whose mathematical brilliance was not demonstrated in her school's current assessments, collecting multiple interviews of this child's sophisticated mathematical thinking and bringing these data to the child's individualized education plan (IEP) meeting. Another PT used the clinical interview as an opportunity to teach in Spanish to her Spanish-speaking children, an affordance denied by her school's language policy.

But there are always risks involved with creative insubordination. As Jessica's first group of PTs transitioned to their second mathematics methods class, their ideas were not warmly received. Jessica received panicked text messages from students: They were told that fully memorizing the CGI framework was a requisite for passing the edTPA end-of-program assessment. One PT told Jessica, "We've learned that outside of your class, we just shouldn't ask these questions when it comes to math." Jessica was left wondering if she had failed her PTs by trying to get them to think critically about clinical interviews, to get to know their children rather than diagnose them. What was the impact of modeling creative insubordination?

Scene 2 Conversation: How Do We Enact and Model Creative Insubordination?

As we continue to consider the enactment of a sociopolitical perspective on mathematics teaching and learning, the three of us dialogue around Jessica's experience teaching a course using CGI clinical interviews in a mandatory diagnostic way. Our conversation centers on how Jessica enacted (and modeled) creative insubordination in this space, the purpose of the clinical interview in teacher preparation, and the impact of Jessica's creative insubordination.

> **Jessica:** I tried really hard to model what creative insubordination looks like, to be transparent. And I loved seeing some of my PTs do the same thing with their assignments, taking things that felt problematic and adapting them during student teaching. That felt really good. But now, over a year later, I've sat down and had conversations with some of these same PTs and deficit views about the children they work with come pouring out from them, like mocking the way their children speak or mocking their children's mathematical abilities. But I've also had PTs go into their second methods class and convince their professor to consider the sociopolitical work they were doing. So it's this balance of feeling like some things went really, really well, and some not at all. It seemed to have some residue for some, but not for others.
>
> **Theodore:** Does this connect to what we were talking about earlier, that our PTs are just enacting what they see modeled? I like to think that your PTs were buying into what you were doing; they were authentically enacting it. And maybe three

semesters later, they're mirroring what they see someone else in a position of authority doing. I hope your creative insubordination approach is sowing seeds to see how to navigate their own careers. There's a mimicking hierarchy. Our PTs look to us, and we look to our mentors, the people who taught us to be teachers: We're products of the experiences we have the opportunity to participate in.

Stephanie: That's why it's important for our PTs to engage in collaborative practices. They need to push each other to think differently about mathematics, to think differently about children, and to identify deficit perspectives. Just like we are collaborating here as critical mathematics teacher educators, we need to help our PTs support one another and push each other in critical ways.

Theodore: I'm really taken by what sort of power you had, Jessica, and what sort of vulnerability you exposed yourself to. Yet, even after all that, to hear your PTs espouse deficit views a year and a half later—wow. It's troubling, but I'm not surprised. It gets to what Stephanie was talking about in her vignette: Our students are sponges and will enact different philosophies and pedagogies depending on where they are. We fall into the trap of conceptualizing teacher belief change as instantaneous and permanent, but we know it's not. Our activities don't always have the impact and residue we hope. Jessica, your story opens up the idea that even if you are in a space where you are not supported to do critical work— to be critical about the clinical interviews—the students will feed off your passion. There's a lot of hope here; your students will probably reflect back and say, "Remember that one semester I had Professor Hale? Now I get it. Now I understand what she was doing."

SCENE 3: LISTENING TO CHILDREN

Theodore, an assistant professor in his 5th year of teaching elementary mathematics methods courses, reads through his PTs' letters. After engaging in 5 weeks of one-on-one clinical interviews with the same child, his PTs have written letters to the child's family to showcase what they learned.

Theodore opens up a letter, noticing the third paragraph:

Savannah: Levi asked me if he could help write the problem. He told me he wanted to solve a problem in which "a dude" was buying an Xbox and some games for it and "he had $150." I followed Levi's lead, using his ideas to develop a multi-step word problem: "Our dude has $150 to buy an Xbox and some games. An Xbox costs $100 and each game costs $10. How many games can the dude buy?" While the problem could come off as humorous, it is actually a very complex problem involving both subtraction and division, and Levi was able to solve it using an inventive strategy. He used Unifix cubes to represent the games, and then used base ten blocks to help represent the groupings of tens and hundreds in the problem. He started with a hundreds "flat," and counted out five groups of tens to represent $150. He moved the hundreds "[flat]" to show that he was taking away $100 for the Xbox, and then dragged a Unifix cube to each rod of ten since "all the games cost $10" and the Unifix cube "is a game." He was able to solve a problem that involved two operations and developed a unique way of keeping track of the groupings of tens to divide.

Theodore smiles and opens up another letter. Tina is struggling in her 2nd-grade placement at a school that primarily serves low-income, Black students. Tina often uses deficit language to describe her students and their families.

Tina: I have looked through [Jamie's] math worksheets and she does know how to subtract when presented with a minus symbol or written conventionally as a number sentence; yet, as I have discovered through my interviews, unless explicitly told to subtract, she will add numbers in a story problem. I would continue to work with her by having her find and define key words in story problems that would denote addition or subtraction.

Theodore sighs. Tina focuses on keywords and categorizing tasks based upon a specific operation or procedure, rather than allowing Jaime to create her own strategy or model of the task. Theodore reads on:

Tina: While I have attained knowledge about my student through these interviews and my observations, I would like to learn more about my student's home life in order to incorporate more of it into lessons for her. Since she has a strong connection with her mother, I would like to know about her other relationships with her brothers and father and use this to allow her to feel ownership in her math.

"There it is. We're starting to get somewhere, Tina," thinks Theodore. Tina, who still espouses a mechanistic view of mathematics learning, has started internalizing the importance of listening and getting to know her students in order to understand how the worlds they live in connect to their mathematical knowledge. The experience from the clinical interview activity helped Tina realize she needs to know her student better in order to be able to reach her as a mathematics teacher.

Scene 3 Conversation: Helping PTs Learn to Listen and the "Ideal" Clinical Interview

In our third conversation, we discuss the different philosophies surrounding CGI, particularly the troubling ways in which clinical interviews can be framed as a diagnostic tool. Even when a tool is "good," Foucault (1997) reminded us that it is still dangerous. We discuss how the clinical interview helps our PTs learn to listen, as well as how our PTs can critique the clinical interview structure as they enact it. Finally, we discuss our ideas of what an "ideal" clinical interview would look like.

> **Jessica:** Theodore, do you feel like clinical interviews help PTs dismantle deficit views of children?
>
> **Theodore:** Yes, in my mind. I think what's cutting through all of this is the way CGI was defined in your program versus the way I've encountered CGI. My experience with CGI is very much about uncovering children's thinking in ways not prescribed or structured. Clinical interviews really force PTs into the art of listening deeply to student mathematical thinking. And from what I hear, the approach espoused in your methods course situated CGI as very diagnostic. And that's not how I see it at all. It pains me, Jessica, that you were told to use a CGI clinical interview in a deficit-focused approach to diagnose what a student does not know as opposed to a focus on teacher learning through listening. Maybe I have my own rose-tinted glasses about CGI. But I'm really troubled by this idea that people use CGI in such diagnostic and non-child-centered ways.
>
> **Jessica:** Even when we used the CGI clinical interviews, the PTs seemed to really want to get at that sociocultural analysis of children's mathematical thinking. It scratches the surface of equity and I think these are the kind of practices that create access for more students and highlight students' mathematical brilliance. But I don't know that it digs nearly as deep as you would want it in terms of equity-based teaching.
>
> **Theodore:** I shared this vignette to showcase two points: (a) Some PTs still espoused a deficit view even after the clinical interview, and (b) Even though a PT might hold a deficit view of her student, the PT is starting to understand the importance of listening to children. So we might be getting somewhere; there is some impact we see.
>
> **Stephanie:** At the end of your vignette I thought this assignment really helped your PTs think differently about kids and families,

but you sounded disappointed about where they were in terms of their understanding of mathematics instruction.

Theodore: I grapple with whether or not these interviews reinforce students' deficit perspectives rather than forcing them to confront those perspectives. I ask my PTs to work with a child who is different from them on multiple levels (e.g., race, gender, home language), and my fear is that PTs will generalize what they learned from the interview for all Black children or all Latinx[1] children. It goes back to our point earlier that students might say amazing things in your course about equity and culturally responsive pedagogy, and then a year later they will turn around and say all these racist things about their students. I worry that our PTs are just writing what we want to hear as critical mathematics teacher educators.

Jessica: I wonder if it's less that PTs will make assumptions about mathematical abilities of certain groups of students, but rather that PTs will decide that certain groups of students will be able to connect to a mathematics problem based upon racialized or gendered stereotypes. Is it really authentic to students' voices if you are just inserting something your students are interested in into these problems? All we're doing is altering the language versus digging into real problems within a student's community. It's an interesting dilemma, particularly when looking at the CGI framework where we have these specific structures of story problems.

Theodore: I completely agree. Going back to Civil's (2007, 2009) work with community knowledge or Aguirre and colleagues' (Aguirre, Turner, et al., 2012; Aguirre, Zavala, & Katanyoutanant, 2012) work on helping elementary mathematics teachers connect to children's community funds of knowledge, how do our story problems even tangentially touch the spaces where our children live? Savannah was really proud that the child generated the problem himself, particularly how he brought his interests in Xbox games into the problem. But there's a part of me that still dismisses it as a simplistic story problem.

Jessica: When our PTs are sitting with children during an interview, how much of that CGI framework is running through their head to classify student responses instead of genuinely listening and following the child's lead? Reading through my interview transcripts, I can see that when a child gives my PT an answer that fits nicely into a CGI strategy category, the PT does not probe the child's thinking any further. I think this framework can be a really useful tool, but does it also hurt what I'm trying to do as a teacher educator?

Theodore: Yes, the first time my PTs enact the clinical interview, all they are worrying about is following the script, figuring out what they're supposed to do or say next. My PTs enact the clinical interview three times, and it's not until that third interview that they really start to understand what they're doing.

Jessica: If you're doing just one interview with a child you just met in a library, you don't really get deep into understanding a child's mathematical thinking. What PTs are practicing instead is how to take and classify student responses.

Theodore: Yes, diagnosing children does not get to what learning to be a mathematics educator is. I see the real value of the CGI clinical interview as helping our PTs learn to be flexible and listen deeply, so that even though you might have preconceived categories you're trying to fit children into, they might fit or they might not.

Stephanie: Being critical teacher educators and helping our PTs become critical teachers involves critiquing the tool itself. So how can we get PTs to critique the CGI clinical interview structure? Can they ask: What am I learning using this tool? What am I not learning and not attending to while using this tool?

Jessica: There will never be a perfect curriculum, so we should always be critiquing. For us as MTEs, we're never going to have a best clinical interview. So we should be critiquing that too. We sometimes have this idea that young mathematicians can't grapple with bigger issues and context in terms of social justice. I wonder how we can use the clinical interview to help our PTs connect to issues of equity and critical mathematics in their future mathematics teaching.

Stephanie: Yes. I want to make sure our PTs are critiquing and looking at the affordances and constraints of using a clinical interview with children and also modeling what it looks like for a PT to question everything we give them: How does it help children? What does it hide? And what ways might deficit views of children be reinforced or exposed using the clinical interview?

Theodore: I like that. To me, the clinical interview is not just a one-shot structure, trying to figure out how to interview and get to know a student in one way. What I want our PTs to realize is that it's really hard to know what's inside a child's head until you develop a relationship with the child, until you learn about who they are outside of the classroom environment, until you learn about their interests, until you talk to them as a fellow human being, as a person.

Jessica: For me, a perfect case study would not just have listening, but observing. I think about how different the clinical interview would look if instead of coming in with, "I'm going to ask you math question," we try, "I'm going to take a back seat and I'm going to watch you. I'm going to listen and observe how you interact with your friend or I'm going to listen and observe what's happening when you're cooking with your mom or when you are on the playground or the conversation you're having in the lunchroom." And then rather than asking students mathematics first, teachers ask themselves what mathematics is the child doing, and framing it that way.

FINAL THOUGHTS

All three of us have used the clinical interview in different ways, and none of us would say it is easy or even routine. Each course we teach, the nuances of individual PTs we work with, the students that our PTs choose to interview, and the settings that the PTs work within all change the nature of the interview and how we navigate its outcomes. Much like Stephanie and Jessica found, the work of the clinical interview can unravel in student teaching. Therefore, should we also think of using the clinical interview as an activity for PTs to enact while they are student teaching? In Jessica's experience, the clinical interview focused too heavily on diagnosing children rather than helping teachers learn to listen. She was able to use that structure to model creative insubordination for her PTs, empowering them to question the purpose of the clinical interview. And while Theodore's experience using the clinical interview was more positive, he questioned how this activity still resulted in PTs holding mechanistic and deficit views about mathematics learning.

We end with several tips for using the clinical interview in mathematics methods courses, stemming from our conversations. First, the best clinical interviews involve working with the same student over time in order to elicit family and community knowledge (Aguirre, Turner, et al., 2012; Aguirre, Zavala, & Katanyoutanant, 2012; Civil, 2007), which Theodore was able to do. Second, this work must be collaborative. PTs should talk to each other between interviews to discuss what they are noticing and learning about their students (DuFour, 2004; Van Es & Sherin, 2008), which Theodore was able to do because PTs had multiple interviews to reflect upon. Third, we as MTEs must continue talking about how we use the clinical interview to help PTs enact a sociopolitical teaching identity (Gutiérrez, 2013), much as we do in this conversation. Otherwise, we end up in situations in which a powerful activity for teacher learning becomes another assessment used to diagnose children rather than to listen to them. Fourth, even as we use the clinical interview in our mathematics methods courses, we should continue to question and critique this activity as Jessica had her PTs do.

We end with a note of hope. As much as we critiqued the ways we used the clinical interview, we noted that we did see PTs who, for first time, saw that mathematics teaching and learning can be critical, reflecting that knowing more about a child's lived experiences helped them better see the child's unique mathematical thinking. For this reason alone, we see the clinical interview as a potentially powerful sociopolitical activity for MTEs to enact in their mathematics methods teaching.

NOTE

1. We use the term *Latinx* so as not to dichotomize gender identity as either male or female.

REFERENCES

Aguirre, J. M., Turner, E. E., Bartell, T. G., Kalinec-Craig, C., Foote, M. Q., McDuffie, A. R., & Drake, C. (2012). Making connections in practice: How prospective elementary teachers connect to children's mathematical thinking and community funds of knowledge in mathematics instruction. *Journal of Teacher Education, 64*, 178–192. doi: 10.1177/0022487112466900

Aguirre, J. M., Zavala, M. del R., & Katanyoutanant, T. (2012). Developing robust forms of pre-service teachers' pedagogical content knowledge through culturally responsive mathematics teaching analysis. *Mathematics Teacher Education and Development, 14*(2), 113–136.

Carpenter, T. P., Fennema, E., Franke, M. L., Levi, L., & Empson, S. B. (1999). *Children's mathematics: Cognitively guided instruction.* Portsmouth, NH: Heinemann.

Civil, M. (2007). Building on community knowledge: An avenue to equity in mathematics education. In N. S. Nasir & P. Cobb (Eds.), *Improving access to mathematics: Diversity and equity in the classroom* (pp. 105–117). New York, NY: Teachers College Press.

Civil, M. (2009). A reflection on my work with Latino parents and mathematics. *Teaching for Excellence and Equity in Mathematics, 1*, 9–13.

DuFour, R. (2004). What is a "professional learning community"? *Educational Leadership, 61*(8), 6–11.

Empson, S. B., & Jacobs, V. (2008). Learning to listen to children's mathematics. In D. Tirosh & T. Wood (Eds.), *The handbook of mathematics teacher education: Tools and processes in mathematics teacher education* (pp. 257–281). Rotterdam, The Netherlands: Sense.

Felton-Koestler, M. D. (2015). Mathematics education as sociopolitical: Prospective teachers' views of the what, who, and how. *Journal of Mathematics Teacher Education, 20*(1), 1–26. doi:10.1007/s10857-015-9315-x

Foucault, M. (1997). On the genealogy of ethics: An overview of work in progress. In P. Rabinow (Ed.), *The essential works of Michel Foucault, 1954–1984* (Vol. I, pp. 253–280). New York, NY: New Press.

Freire, P. (1970). *Pedagogy of the oppressed* (30th anniversary ed.). New York, NY: Continuum.

Ginsburg, H. P. (1997). *Entering the child's mind: The clinical interview in psychological research and practice.* Cambridge, England: Cambridge University Press.

Graue, M. E., & Walsh, D. J. (1998). *Studying children in context: Theories, methods, and ethics.* Thousand Oaks, CA: SAGE.

Gutiérrez, R. (2013). The sociopolitical turn in mathematics education. *Journal for Research in Mathematics Education, 44,* 37–68.

Piaget, J., & Inhelder, B. (1969). *The psychology of the child* (2nd ed.). New York, NY: Basic Books.

Van Es, E. A., & Sherin, M. G. (2008). Mathematics teachers' "learning to notice" in the context of a video club. *Teaching and Teacher Education, 24*(2), 244–276.

Vygotsky, L. S. (1978). *Mind in society: The development of higher psychological processes.* Cambridge, MA: Harvard University Press.

CHAPTER 9

SITUATING LEARNING FOR SECONDARY MATHEMATICS PROSPECTIVE TEACHERS WITHIN THE CONTEXT OF REHEARSALS

Challenges and Resulting Adaptations

Fran Arbaugh
The Pennsylvania State University

Anne E. Adams
University of Idaho–Moscow

Dawn Teuscher
Brigham Young University

Laura R. Van Zoest
Western Michigan University

Robert Wieman
Rowan University

Building Support for Scholarly Practices in Mathematics Methods, pages 133–148
Copyright © 2018 by Information Age Publishing
All rights of reproduction in any form reserved.

Almost three decades ago, Ball (1988) argued that prospective teachers (PTs) need to "unlearn to teach" (p. 40) and that one of the inherent dilemmas of mathematics teacher education is to counter a vision of teaching and learning mathematics that PTs have developed through engaging in apprenticeships of observation (Lortie, 1975). As mathematics teacher educators (MTEs), we ask our PTs to learn to enact instruction (e.g., a launch, explore, summarize sequence) that is different from what they may have experienced, requiring a depth of mathematical understanding that they may not have yet developed. Our desire to provide educational experiences that advance PTs beyond their existing ways of thinking about teaching and learning mathematics influences the ways that we design and implement mathematics methods courses.

Through small-group discussions at the Scholarly Inquiry and Practices Conference (Sanchez, Kastberg, Tyminski, & Lischka, 2015), the five authors of this chapter discovered many commonalities in our work with secondary PTs; most pertinent to this chapter is that we all use rehearsals with secondary mathematics PTs. As conversations progressed, we realized that we were attracted to using rehearsals as an activity in our methods courses because we tended towards a situated perspective on learning (as described by Lave & Wenger, 1991). Our conversations also made clear that rehearsals provided a context in which many of the challenges of educating secondary mathematics PTs became explicit. We talked about how our PTs' predominantly procedural mathematical understandings sometimes compromised the PTs' pedagogical learning that we wanted to occur during rehearsals as well as how their beliefs about teaching and learning affected rehearsals. We also talked about organizational challenges that we face when implementing rehearsals. These talks began as discussions of challenges and evolved into stories of how, over time, we adapted rehearsals to combat these challenges.

This chapter contains our stories of adaptations made in response to our early uses of rehearsals—when we were attempting to enact rehearsals as they had been described by Lampert and her colleagues (Lampert, Beasley, Ghousseini, Kazemi & Franke, 2010; Lampert et al., 2013; Lampert & Graziani, 2009). We organize our adaptation stories around these three challenges: (a) PTs' content knowledge, (b) PTs' beliefs about teaching and learning, and (c) organizing rehearsals in our contexts. We begin, however, with a short discussion of how rehearsals and a situated perspective on learning naturally work together.

REHEARSALS AND A SITUATED PERSPECTIVE ON LEARNING

One of the most gratifying aspects of attending the Scholarly Inquiry and Practices Conference (Sanchez et al., 2015) was the opportunity for the five of us to talk as a group about our MTE practices. As we described in the chapter introduction, through our conversations we acknowledged that our tendencies towards a situated perspective of learning had drawn us to using rehearsals in the first place. It was not until the crafting of this chapter, however, that we were able to articulate explicitly how rehearsals and a situated perspective on learning naturally work together. Thus, the work that we have engaged in together in this project has been generative for our own understandings of our teacher education practices and the decisions we make as MTEs.

Like Lampert et al. (2013), we seek to engage secondary mathematics PTs in rehearsals to support their learning about and abilities to enact "intellectually ambitious instruction" (p. 226). Lampert and her colleagues described rehearsals by contrasting them with what they described as "run-throughs of novice-designed lessons that are sometimes conducted in methods courses" (p. 229). What distinguishes rehearsals from these other pedagogies of teacher education is the dual role of a teacher educator (TE; note that Lampert et al. used *TE*, whereas in this chapter we refer to mathematics teacher educators in particular and use *MTE*):

> The TE acts as both coach and simulated student, enabling both the rehearsing novice and the others in the group to investigate the actions a teacher might take in response to student performance. The TE has the opportunity to stop the action and coach the novice as he or she deliberately practices moves that are responsive to specific and multifaceted student actions. She can also lead a discussion among the group of novices in which different possible moves are weighed for their appropriateness and potential effectiveness. (Lampert et al., 2013, p. 229)

Rehearsals, as described by Lampert and colleagues, are embedded in *cycles of enactment and investigation* (CEIs) that comprise phases of observation, collective analysis, preparation, rehearsal, classroom enactment, and collective analysis. The glue that holds the phases together is a combination of targeted core teaching practice(s) and the same or similar *instructional activity* (IA; Lampert et al., 2013). IAs are the mathematical tasks or series of tasks that K–12 students undertake to learn particular mathematical content.[1] During the CEI observation phase, PTs engage in *observing a representation of practice* (Grossman et al., 2009) that can take the form of, for example, a video, narrative case, or TE-led activity. PTs then *collectively analyze* that representation of practice, using a core teaching practice as a lens. PTs then *prepare to instruct* (plan) the same or similar IA, *rehearse* that instruction in the mathematics methods course, and then *enact their plan* in a real classroom, which is recorded in some way. In the last phase of a CEI, PTs *collectively analyze* their classroom enactment. Although some of us do not engage PTs in full CEIs, we each embed rehearsals in a context of learning to plan for and enact ambitious mathematics instruction.

Lave and Wenger (1991) opened their seminal book on situated learning with this statement:

> Learning viewed as a situated activity has as its central defining characteristic a process that we call *legitimate peripheral participation*. By this we mean to draw attention to the point that learners inevitably participate in communities of practitioners and that the mastery of knowledge and skills requires newcomers to move towards full participation in the sociocultural practices of a community. (p. 29, emphasis in original)

Our use of rehearsals as a pedagogy of teacher education is consistent with Lave and Wenger's description of situated learning—we invite newcomers to the profession of teaching mathematics by engaging them in activity that supports their learning of the practices of the community of mathematics teachers. And so, our use of rehearsals—engaging PTs in *approximations* of the practices of the profession using particular *decompositions* of practice (see Grossman et al., 2009, for a full description of these terms)—allows us to "situate" their learning in the practices of teaching, thus moving them towards the practices of the profession.

OUR ADAPTATION STORIES

Rehearsals have allowed us to address some of the challenges inherent in mathematics teacher education, and we discuss specific benefits of using rehearsals with our secondary mathematics PTs in the last part of this chapter. The sections that follow each begin with a short description of an MTE challenge, followed by one or two stories of how we adapted rehearsals, or the activities in which PTs engage around rehearsals, in response to those challenges.

The Challenge of PTs' Content Knowledge

We surmise that the vast majority of MTEs in practice today would still agree with Ball, Lubienski, and Mewborn's (2001) contention that there exists an "unsolved problem of teachers' mathematical knowledge" (p. 433). As secondary MTEs, we certainly agree that this problem still exists, and in our individual contexts we use a number of strategies that push at the boundaries of our PTs' mathematical understandings (e.g., engaging PTs in doing mathematical tasks, analyzing student work). We see a need to rebuild their existing mathematical understandings in ways that will support their future students' mathematical understandings. Further, we believe it is critical that our PTs understand mathematics content deeply enough to effectively plan for and enact ambitious teaching practices, specifically so that they can (a) identify the mathematical goal(s) of the lesson and (b) anticipate, elicit, and make use of students' mathematical thinking in ways that support students' learning of the mathematical goal(s) of the lesson.

Laura and Dawn have explicitly faced this challenge in their rehearsal contexts. When they first began to implement rehearsals with PTs, it became clear that the PTs' mathematical content knowledge was interfering with their learning about *teaching mathematics*. As a result of these dissatisfactions with early rehearsals, Laura and Dawn now incorporate explicit attention to developing mathematical content knowledge within the rehearsal context and have made particular adaptations to rehearsals and associated activities in order to accomplish this goal. Despite their similar goals, they have adapted rehearsals in very different ways.

Laura's Story

My story involves the use of rehearsals to prepare PTs to support their students' development of CCSSM's (Common Core State Standards Initiative, 2010) Standard for Mathematical Practice #7: *Look for and make use of structure* (MP7). The IA I used in rehearsals is called *Contemplate then Calculate* (CtC; Kelemanik & Lucenta, 2015). The mathematics at the center of a CtC IA is a calculation problem designed to support students' recognition and use of mathematical structure (e.g., $81 - 72 + 63 - 54 + 45 - 36 + 27 - 18 + 9$). In a CtC, the teacher first displays the problem for only a few seconds and then asks students what they noticed about the problem that might help them calculate the answer in their head quickly and easily. After reporting their noticings, the teacher shows students the problem again and asks them to work in pairs to use their noticings to develop strategies for calculating the problem without pencil, paper, or calculator. Students then share their strategies and collectively analyze the mathematical reasons behind them. The CtC ends with students sharing individual reflections on what they learned about looking for and making use of structure. In an early implementation of this type of rehearsal, I modeled the CtC (as the teacher with my PTs as the students) as a way for the PTs to become familiar with the pedagogical components of the IA that they would rehearse in methods class and then use in a middle school classroom. What I discovered was that PTs had difficulty preparing for the rehearsal because they did not recognize the mathematical structure embedded in the procedural mathematics they knew so well.

The particular PT content knowledge challenge I encountered had two components: (a) the PTs had not experienced looking for and using structure as school students, thus everything about the CtC teaching sequence, including the purpose, was new to them; yet (b) because of their extensive experience doing arithmetic, the mathematics problem at the CtC center seemed familiar, even mundane. When I engaged my PTs in their initial CtC around 25 × 29, several of them expressed resistance when asked to solve the problem without using any tools. They knew how to get an answer by using paper and pencil with the standard algorithm or using a calculator, and they were frustrated when those tools were not allowed. The PTs had not yet developed the skill—nor understood the value—of looking for and recognizing structure; thus they were not yet able to focus on how to support their students to develop MP7. The PTs' feelings of frustration and lack of understanding of mathematical structure had to be dealt with before they could productively prepare for the rehearsal.

I responded to this challenge by expanding the preparation for rehearsal in two specific ways. First, rather than immediately expecting the PTs to engage with CtC as a teaching tool, I expanded the modeling component of the rehearsal preparation to use it as an opportunity for the PTs to learn important mathematics that might not have been part of their past experiences. I now use CtC as an instructional tool with the PTs as students multiple times *before* introducing it as the focus of rehearsals. Initially I had modeled the CtC once with the PTs as students and then given each small group a different computational task to use to rehearse CtC with their peers. In retrospect, it was unrealistic to think that the PTs could concentrate both on learning new content and the teaching of that content at the same time.

Second, because the student-thinking-centered nature of the IA is so different from most PTs' past pedagogical experiences, I am more intentional in asking them to expand their thinking about instruction. Thus, rather than simply modeling CtC and expecting the PTs to infer the intent and meaning behind the carefully constructed components of the IA, I deliberately engage the PTs in deconstructing the components of the IA so that they are better able to internalize both the mathematical and the pedagogical intents of a CtC. We have explicit conversations about teaching practices such as making student thinking public for the class to consider, documenting ideas as they are discussed, prompting students to listen and respond to each other's thinking, and summarizing the important mathematical ideas during class discussion that are built into CtC and ways in which these teaching practices support student learning. Not only do these changes support PTs' rehearsals, but they also contribute to other goals I have for the PTs, such as developing skill in the CCSSM mathematical practices and recognizing the benefits of intentional instruction.

Dawn's Story

Early in my use of rehearsals, I found that PTs struggled to plan for instruction that focused on the fundamental mathematics concept (FMC) for the lesson because they could not identify these concepts during planning. The FMC is the core building block of the learning goals that drive the selection and implementation of worthwhile mathematical tasks. FMCs are *not* procedures but are principles that underlie why a procedure works—conceptual content knowledge that many PTs may not possess. Without a clear articulation of the FMC, I have found that PTs typically develop and enact a lesson that is procedurally driven.

To address this challenge and have PTs rehearse lessons that elicit student thinking and promote productive struggle, I have modified the CEI (as presented by Lampert et al., 2013) by replacing the observation phase with an hour-long meeting with each group of PTs to discuss the FMC of the lesson they are planning. In this context of planning for instruction, I play the role of "coach." Prior to this meeting, PTs submit their lesson FMC through an online system that all PTs can view, and I provide feedback in the form of questions to assist PTs in conceptually thinking about the mathematics. The subsequent meeting allows for both clarifying the lesson FMC and exploring the PTs' understandings of the focal mathematics. My coaching prompts discussions about the FMC and makes explicit the conceptual underpinnings for the mathematical procedures in the lesson. For example, PTs who prepared a lesson on the period of trigonometric functions submitted the following FMC: "The period of a trigonometric function is the distance of a complete revolution divided by the time to complete the revolution. The distance of the revolution is 2pi radians. The period is expressed as the radians per second." This FMC is typical of what I receive—very definition-like and not illustrative of the conceptual underpinnings of the mathematics. My feedback was intended to push the PTs to think more deeply about the mathematics:

> I am a little confused by your first sentence—you state that the period is defined by the distance divided by time. If I have a trig function, isn't my input an angle measure and my output a distance (vertical or horizontal)? How can I change angle measure to time? Also, if I have different size circles, won't my distance be different? But you say the distance is 2pi—how is this so if different size circles have different distances?

After our meeting the PTs updated the FMC:

> In order for a trigonometric function to complete a cycle, the argument of the function must vary by 2pi radians, the total radians in any circle. The period is the interval of input needed to complete a full cycle of output values such that every output value is achieved in the interval. The period of the function varies by the number of cycles in 2pi radians, which is the coefficient of the input variable in the argument. This can also be thought of as the com-

posite of two functions: a function of angle measure with respect to time, and a trigonometric function of outputs with respect to angle measure.

With their lesson FMC updated, PTs then write a lesson plan that explicates details of the instructional sequence (e.g., launch, exploration, and discussion) of the lesson, the mathematical task, anticipated student thinking to that task, and responses to the anticipated student thinking. During the implementation stage, one PT from the group is selected to teach the lesson to their peers. The other PTs from that group are observers during the lesson. The teaching PT has 30 minutes to implement the lesson with classmates; the lesson is videotaped for use during the reflection stage. I have chosen not to coach during this phase; however, I reengage in coaching during the reflection stage.

Immediately after the lesson, I conduct a 15-minute debriefing of the lesson with the whole group. The teaching PT first responds orally to my questions (e.g., what was the FMC for your lesson, what went well in your lesson, what would you change if you were to teach this lesson again?), and then the debriefing is opened up to all PTs to ask questions. This debriefing is a way for all the PTs to engage in the reflection phase and learn from each other. I also use the debriefing time as a coaching time to assist the PTs in identifying strategies that could have been used to produce different outcomes. Over time, I have observed that my focus on improving FMCs has positively impacted the PTs' lessons, in both the planning and the rehearsal phases.

The Influence of Beliefs about Teaching and Learning

Secondary mathematics PTs are likely to arrive in a mathematics methods class with a vision of "good" teaching as providing clear explanations along with valuing memorization of procedures, rules, and facts (Weldeana & Abraham, 2014); they typically expect to teach mathematics in a similar fashion. PTs at the secondary level have largely felt successful in such learning environments and may be surprised to learn that large numbers of students have not experienced similar success. Yet the mathematics practices in the CCSSM call for students to be actively engaged in making sense of the mathematical ideas they learn. A correspondence between a teacher's beliefs about mathematics teaching and learning and the experiences the teacher offers students is well documented in the literature (Benken & Wilson, 1998; Charalambous, 2015; McClintock, O'Brien, & Jiang, 2005). Hence it is important in teacher education to address unproductive beliefs about teaching practices. Mathematics methods courses can help PTs build new visions of mathematics teaching and develop understanding and skill in new pedagogies—and rehearsals provide a mechanism for doing so.

Anne's Story

I use rehearsal as a tool for PTs to build a broader vision of mathematics teaching and to support them in developing skills in eliciting and responding to student thinking in ways that advance the mathematical discussion and goals of the lesson. Early in my use of rehearsals it became obvious that my PTs' strong drive to teach through telling was interfering with their abilities to enact teaching practices that build on students' mathematical ideas.

To support PTs in adopting a focus on student thinking, I asked all PTs to prepare to teach the same lesson to their field classroom students. To begin, I engaged them as students in the lesson, and we then unpacked the mathematics of the lesson and the pedagogy I used to teach it—identifying key concepts, where and how in the lesson those became evident, what I had asked PTs to do and think about, at what point in the lesson I had done so, and what the purposes of my actions were. I then provided PTs with a detailed and completed plan for the lesson. This plan indicated places in the lesson to pause and solicit student ideas and provided examples of questions to ask for this purpose, including potential follow-up questions.

I thought that working from a detailed plan designed to elicit student ideas would help the PTs focus on listening to and extending students' thinking during instruction. By following the lesson plan as written, PTs "just" needed to focus on listening to student ideas, deciding which ideas to pursue, and asking for clarification and extension of thinking. I surmised that if PTs followed such plans, they would elicit student thinking and have an opportunity to respond. During rehearsal, and despite providing these detailed plans, I was surprised to see how little my PTs elicited student thinking. Some of my PTs had such a strong drive to provide a good explanation (as teachers) that, after eliciting an initial student idea, the PTs completed the thought or provided a thorough explanation, rather than asking questions that would encourage the students to think about and explain mathematical concepts and relationships.

To counter this tendency for teacher explanation, I adapted the rehearsals to use coaching breaks that focused the PTs on building on student thinking during the lesson. One huge challenge for me as the coach was to step in and suggest an alternative approach without impacting the PTs' positive self-efficacy for teaching. My PTs were moderately nervous about these initial teaching experiences and had a strong desire to do well. They wanted positive feedback. I did not want to tell them they had taken a "wrong" approach, but I did want them to shift the responsibility for explaining from themselves to the students. To coach in a way that allowed the PTs to feel positive about their teaching while at the same time asking them to replay an interaction with a more student-focused approach has required care, caution, and knowledge of each individual PT. Through numerous iterations, I have developed a style of coaching that I think accomplishes my goals. I begin by first offering positive comments, then asking the teaching PT to think about what the episode would look like if the goal was to have students learn to explain their thinking. I then ask for a replay of the episode using that approach. For example, during a coaching break in the lesson, I might say:

> That was a beautiful explanation. I really liked the way you used the diagram and showed how to connect it with the equation. If, instead, a teacher wanted the students to work on giving such explanations, what might that teacher not want to say? What question(s) might that teacher ask to enable students to think about these connections and begin to articulate them?

PTs' initial discomfort at being interrupted while teaching dissipated. In each subsequent rehearsal, the PTs became better at building on student thinking by requesting elaboration, clarification, and reflection of student ideas. Each rehearsed lesson was better than the one before, although the teaching PTs were different each time. On end-of-term course evaluations, PTs pointed to the rehearsals as one of the most important parts of the course and as a context in which they were aware of enormous growth in skill in a short period of time.

Originally, rehearsals had seemed limited by PTs' beliefs about good teaching. Their image of teaching through telling prevented them from practicing and developing skill in eliciting and building on students' thinking. By changing the structure of rehearsals, I was able to give PTs opportunities to learn to enact effective teaching practices even if they continued to equate "good" teaching with providing good explanations.

Addressing Organizational Messiness

When Rob and Fran first started using rehearsals, each had to spend a great amount of time figuring out how to organize the logistics of rehearsals. They grappled with a number of questions, including How long should rehearsals be? How much of our methods class time can we afford to dedicate to rehearsals in one semester? What mathematical tasks or activities will form the foundation of the rehearsals? How can we organize rehearsals to make them most effective for the PTs? Early rehearsals provided the context for Rob and Fran to try out different organizational structures and learn from those implementations.

Rob's Story

I use rehearsals in my methods classes in conjunction with planning full-period lessons that PTs teach in practicum placements. These lessons can involve different content, but they all follow a launch, explore, discuss format and feature students working to solve a cognitively demanding task (Stein, Grover, & Henningsen, 1996). Originally, PTs planned these lessons, rehearsed them, received coaching and feedback, and then taught them in practicum. PTs used rehearsals to refine their plans in very specific ways. Rehearsals, in this light, led to improved lessons and an appreciation of detailed plans. Despite these improvements, when PTs taught at their placements, they consistently struggled with the discussion part of the lesson. Rehearsals did not seem to help PTs facilitate discussions focusing on mathematical connections between different student solutions.

As I had set them up, there were two problems with using rehearsals to support PTs in learning how to enact discussions. First, in order to facilitate a discussion of student ideas, PTs needed to elicit a variety of student solutions. In order to elicit multiple solutions, we focused our limited time for in-class rehearsal on the launch and explore, leaving little time to rehearse the discussion. Second, planning effective discussions also depended on successfully anticipating specific student solutions. When PTs taught the lesson in their practicum, the student solutions did not always align with the solutions PTs had predicted.

In response to these two problems, I split the practicum lessons into two parts and had PTs plan, rehearse, and enact them separately. The PTs planned and rehearsed the launch and explore phases first. Then they went to their practicum classrooms and implemented these parts of the lesson and collected student work, *without* leading a final discussion. In the next meeting of our methods course, PTs sorted their collected student work and then chose which solutions to share, in what order, and which aspects of the solution to focus on. Finally, they planned specific questions they would ask to focus student thinking on the connections among solutions. We then rehearsed the discussions. Because these rehearsals were based on actual student work from their practicum classes, PTs knew that the discussion they rehearsed would build on their students' thinking. Focusing on the discussion enabled the PTs to concentrate on questions and moves for that part of the lesson.

This adaptation had three concrete results. First, PTs learned about mathematics and students' thinking by analyzing student work in the context of planning for a discussion. PTs had to understand and verbalize mathematical connections between solutions. Second, instead of rehearsing discussions of solutions that they thought students *might* create, PTs were able to discuss solutions that they knew students *had actually created*. Consequently, these rehearsals resulted in feedback and suggestions that were clearly applicable to the discussion they would lead. Third, this adaptation resulted in classroom discussions where PTs' moves and questions, developed as a result of rehearsals, focused student thinking on the important mathematical connections among solutions. By situating the planning and rehearsing of the discussion within the context of actual student solutions, I was able to support PTs in learning how to facilitate rich, complex discussions more effectively.

Fran's Story

When I was first conceptualizing how to do rehearsals with my PTs, I planned to do three CEIs within one semester. I wanted the approximations of practice to get closer to real practice as the semester progressed, and I wanted the PTs to have multiple times to rehearse the pedagogical focus of the course—eliciting and making use of student thinking. So organizationally in that first semester, I had to figure out how to do three rounds of the CEIs with 18 PTs. I knew I would have to break with how rehearsal was defined in the literature, mostly around how much time to dedicate to each PT's individual rehearsal. I also knew that my PTs were not rehearsing in preparation for teaching secondary students in near time, as my methods course does not have a concurrent field component (an adaptation to CEIs as described by Lampert et al., 2013). In essence, they were rehearsing pedagogical moves that they could use in the future, which also influenced how I decided to organize the experience.

For one of the earliest rehearsals, the instructional team (myself and three doctoral students, Ben Freeburn, Duane Graysay, and Nursen Konuk) chose six different and true number-theory conjectures (e.g., the product of any two perfect squares is a perfect square; the sum of any two positive consecutive odd numbers is divisible by 4) as the mathematical tasks for the rehearsal. The mathematical goal for the tasks was to create a valid mathematical explanation (or proof) for the conjecture. For each task, Ben, Duane, and Nursen then created a set of three incomplete and different student responses, which resulted in 18 total student responses, matching the number of PTs in the class. PTs worked in groups of three to plan for instruction (organized by task). On the day of rehearsal, each PT worked with one "student" (Ben, Duane, or Nursen) around the student's incomplete response. The pedagogical goal of the rehearsal was to use assessing questions (Smith, Bill, & Hughes, 2008) to elicit how their assigned student was thinking about the problem and then ask advancing questions (Smith et al., 2008) and make appropriate use of telling (Lobato, Clarke, & Ellis, 2005) to move the student towards the mathematical goal of writing a valid justification for the task. Each rehearsal was limited to 3 minutes (an adaptation to the definition of rehearsal) so that we could complete all the rehearsals within a 75-minute class period.

On rehearsal day, we projected the "student" work onto a whiteboard so that the nonteaching PTs could see the student work during the rehearsal as well as anything the teaching PT and student added to that work during the episode. The teaching PT and the student stood on either side of the projected work at the front of the room, and the other 17 PTs sat facing the whiteboard. We set a video camera in the back corner of the room and focused the video to capture the whiteboard, the teaching PT, and the student. During the rehearsal, I played the role of coach—stopping the action, for example, for the teaching PT to rephrase a question; helping the teaching PT when (s)he seemed stuck on what to do next; or steering the teaching PT in directions I thought would be fruitful for the goal. All action stopped when the timer rang, indicating that 3 minutes had passed.

Organizationally, this 3-minute rehearsal solved my initial problems of having 18 PTs and wanting to do three CEIs in the semester. Pedagogically, this 3-minute rehearsal had benefits and also raised some dilemmas. On the one hand, limiting the time spent with the individual student to 3 minutes does reflect a likely classroom scenario of walking up to a student's desk, seeing what the student has done so far, asking about what the student is thinking, and then asking questions or telling the student something to help them move toward the mathematical goal. As MTEs, we felt good about how we had approximated practice. On the other hand, the PTs were often frustrated by being told to stop before the student had reached closure with the problem. This frustration, however, led to fruitful conversation about our visions of "successful teaching" of mathematics, a productive topic for a methods course. Our organization also differed from illustrations we had seen in the literature or at conferences of others' rehearsals; the 3-minute time limit and back-to-back rehearsals did not leave time for whole-group discussions to occur during the rehearsal class period. We have repeated this type of rehearsal in subsequent semesters and continue to find value in this kind of short rehearsal.

DISCUSSION AND CONCLUSIONS

Through the lens of our situated perspective, we have seen many benefits of using rehearsals with our secondary mathematics PTs. Rehearsals situate PTs' work within an approximation of practice that closely mirrors what they will do with students. We can also situate our support, class discussions, PT planning, reflection, and other activities in approximations of practice that are much closer to what they will do in the classroom than many other common pedagogical approaches (e.g., analyzing cases, watching video, or reading articles). Although we believe that situating PT learning in the practice of teaching is important, we see limitations of engaging PTs, particularly those in our methods courses, in (real) classroom-based teaching. When working with (real) secondary students, it is more challenging for MTEs to interrupt and give in-the-moment feedback, as doing so disrupts the flow of the lesson and shifts the focus from the student-learners to the PT-learners. The complexity of teaching makes it very difficult to create opportunities for work on specific pedagogical learning goals; the (real) lesson may simply go in a different direction. Finally, mistakes made by PTs while teaching may have negative consequences for (real) students in secondary classrooms. Rehearsals have allowed all of us to situate teacher education in practice while minimizing many of these risks; we have found rehearsals to be a "safe" environment for analyzing and enacting teaching practices and a productive way to prepare PTs for working with (real) students.

One unanticipated benefit of the situative nature of rehearsals has been how PTs have used them to begin to develop their teacher identity. In all our programs, PTs spend extended time and energy preparing for teaching without actually engaging in the practice of teaching. They are often anxious about teaching and have uncertain and naïve understandings of who they will be as teachers. Rehearsals allow them to practice holding and exercising authority, making decisions, and reflecting on those decisions. Rehearsals give them a chance to see themselves as teachers. A second unexpected benefit of our use of rehearsals has been the opportunity for us, as MTEs, to use them for formative assessment. Unlike other contexts for assessment, rehearsals afford in-the-moment assessment of our PTs situated in authentic acts of teaching. When we observe and reflect on what PTs do during rehearsals, we are able to see what they know, understand, and can do, demonstrating how they are processing course content. It allows us to assess our MTE practices—to see how effectively we have taught them to teach. We believe that rehearsals are especially powerful formative assessment vehicles for two reasons. First, because they are performance based, rehearsals allow us to assess what PTs *do*. Second, because we can engage in coaching, we are able to act immediately on the information we get from these formative assessments.

Finally, the coaching we do in the context of rehearsals and the discussions that occur with the PTs during debriefing sessions provide opportunities to bring other aspects of our methods courses—such as work on tasks, discourse, and mathematics itself—to bear on the PTs' teaching in a situation where they are open to learning how this information can actually help them become better mathematics teachers. Thus, the time spent on rehearsals enhances the rest of the methods course and allows us to bridge theory and practice by creating a situation where they are seen as mutually supportive—practice gives rise to issues that can be resolved by applying theory and theory provides a lens for analyzing and making sense of practice. Through the use of rehearsals, PTs recognize that the methods course is not simply a program requirement but an opportunity to acquire knowledge and develop skills that will support them as teachers.

NOTE

Note that in this chapter, we distinguish the phrase *instructional activity* from the one-word descriptor *activity*, with "activity" meaning the work in which we engage our PTs. So, rehearsal is an activity and "can involve novices in publicly and deliberately practicing how to teach rigorous content to particular students using particular instructional activities (IAs)" (Lampert et al., 2013, p. 227).

REFERENCES

Ball, D. L. (1988). Unlearning to teach mathematics. *For the Learning of Mathematics, 8*(1), 40–48.

Ball, D. L., Lubienski, S., & Mewborn, D. (2001). Research on teaching mathematics: The unsolved problem of teachers' mathematical knowledge. In V. Richardson (Ed.), *Handbook of research on teaching* (4th ed., pp. 433–456). New York, NY: Macmillan.

Benken, B. M., & Wilson. M. (1998). The impact of a secondary preservice teacher's beliefs about mathematics on her teaching practice. In S. B. Berenson, K. R. Dawkins, M. Blanton, W. N. Coulombe, J. Kolb, K. Norwood, & L. Stiff (Eds.), *Proceedings of the Twentieth Annual Meeting of the North American Chapter of the International Group for the Psychology of Mathematics Education* (pp. 595–600). Columbus, OH: ERIC Clearinghouse for Science, Mathematics, and Environmental Education. Retrieved from http://eric.ed.gov/?id=ED428941

Charalambous, C. Y. (2015). Working at the intersection of teacher knowledge, teacher beliefs, and teaching practice: A multiple-case study. *Journal of Mathematics Teacher Education, 18*(5), 427–445. doi:10.1007/s10857-015-9318-7

Common Core State Standards Initiative. (2010). *Common Core State Standards for Mathematics. Common Core State Standards (College-and Career-Readiness Standards in English Language Arts and Math).* National Governors Association Center for Best Practices and the Council of Chief State School Officers. Retrieved from http://www.corestandards.org

Grossman, P., Compton, C., Igra, D., Ronfeldt, M., Shahan, E., & Williamson, P. W. (2009). Teaching practice: A cross-professional perspective. *Teachers College Record, 111*(9), 2055–2100.

Kelemanik, G., & Lucenta, A. (2015). Contemplate then Calculate. *Teacher Education by Design* [Activity]. Retrieved from http://tedd.org/?tedd_activity=contemplate-calculate-submitted-bpes-boston-teacher-residency-program

Lampert, M., Beasley, H., Ghousseini, H., Kazemi, E., & Franke, M. (2010). Using designed instructional activities to enable novices to manage ambitious mathematics teaching. In M. K. Stein & L. Kucan (Eds.), *Instructional explanations in the disciplines* (pp. 129–141). New York, NY: Springer Science + Business Media.

Lampert, M., Franke, M. L., Kazemi, E., Ghousseini, H., Turrou, A. C., Beasley, H.,... Crowe, K. (2013). Keeping it complex using rehearsals to support novice teacher learning of ambitious teaching. *Journal of Teacher Education, 64*(3), 226–243.

Lampert, M., & Graziani, F. (2009). Instructional activities as a tool for teachers' and teacher educators' learning. *The Elementary School Journal, 109*(5), 491–509.

Lave, J., & Wenger, E. (1991). *Situated learning: Legitimate peripheral participation.* New York, NY: Cambridge University Press.

Lobato, J., Clarke, D., & Ellis, A.B. (2005). Initiating and eliciting in teaching: A reformulation of telling. *Journal for Research in Mathematics Education, 36,* 101–136.

Lortie, D. (1975). *Schoolteacher: A sociological study.* London, England: University of Chicago Press.

McClintock, E., O'Brien, G., & Jiang, Z. (2005). Assessing teaching practices of secondary mathematics student teachers: An exploratory cross case analysis of voluntary field experiences. *Teacher Education Quarterly, 32*(3), 139–151.

Sanchez, W., Kastberg, S., Tyminski, A., & Lischka, A. (2015). *Scholarly inquiry and practices (SIP) conference for mathematics education methods.* Atlanta, GA: National Science Foundation.

Smith, M. S., Bill, V., & Hughes, E. K. (2008). Thinking through a lesson: Successfully implementing high-level tasks. *Mathematics Teaching in the Middle School, 14*(3), 132–138.

Stein, M. K., Grover, B. W., & Henningsen, M. (1996). Building student capacity for mathematical thinking and reasoning: An analysis of mathematical tasks used in reform classrooms. *American Educational Research Journal, 33*(2), 455–488.

Weldeana, H. N., & Abraham, S. T. (2014). The effect of an historical perspective on prospective teachers' beliefs in learning mathematics. *Journal of Mathematics Teacher Education, 17*(4), 303–330. doi:10.1007/s10857-013-9266-z

CHAPTER 10

REHEARSING FOR THE POLITICS OF TEACHING MATHEMATICS[1]

Rochelle Gutiérrez, Juan Manuel Gerardo, Gabriela E. Vargas
University of Illinois at Urbana–Champaign

Sonya E. Irving
Kent State University

As mathematics methods course instructors seek to more tightly connect course-work with the kinds of activities that mirror the complexity of teaching, some have chosen to use "rehearsals" where prospective teachers (PTs) practice the real work of teaching. Rehearsals often require the candidate to actively teach peers or students and receive feedback in real time or in a short debriefing session. These rehearsals are seen to more adequately prepare PTs for the kinds of "in the moment" decisions they might have to make and the "performance" aspect of teaching that is not captured by simply developing lesson plans, con-structing a philosophy statement, or creating a classroom management strategy.

However, teaching requires much more than preparing for the act of de-veloping students' understanding of mathematical concepts. In an era of

Building Support for Scholarly Practices in Mathematics Methods, pages 149–164
Copyright © 2018 by Information Age Publishing
149

high-stakes education, beginning teachers need to develop the knowledge and skills to negotiate the politics of the teaching profession (e.g., working with colleagues who may have deficit views of students; responding to a school administrator's emphasis on raising students' test scores to the exclusion of other indicators of success; a school culture that frowns upon students working in groups; or students who want teachers to be the "arbiters of mathematical 'correctness'"; Stein, Engle, Smith, & Hughes, 2008, p. 315). This chapter offers one way for mathematics teacher educators (MTEs) to prepare mathematics PTs for the political nature of teaching. We begin with a review of the field (e.g., how rehearsals are carried out, for what purpose, and with what results) and argue for greater attention to the political nature of teaching. From there, we share our work as three MTEs of color supporting primarily White, secondary mathematics PTs to develop their knowledge and skills for teaching through something we call In My Shoes. We end with implications for other researchers and mathematics methods instructors who want to adopt or adapt similar rehearsals in other settings.

REHEARSALS IN MATHEMATICS METHODS COURSES

Mathematics teacher educators seek to prepare PTs for the complexity of teaching mathematics through a variety of approaches (Horn, 2010; Inoue, 2009; Lampert et al., 2013; Tyminski, Zambak, Drake, & Land, 2014; Virmani, 2014). Rehearsals provide one way for PTs to teach mathematics lessons, receive feedback in real time, and then act upon that feedback. Methods courses provide a low-stakes environment for them to learn. In this manner, PTs are developing skills in a group setting where, together, they support the development of effective practices of mathematics teaching.

Researchers have reported that mathematics PTs are prepared for rehearsals by first observing examples of mathematics teaching, such as videos, classroom placements, and lessons taught by PTs (Lampert et al., 2013) or by rehearsing a mathematical concept that was selected by instructors (Inoue, 2009). As part of a mathematics methods course, many PTs develop a mathematics lesson and rehearse by teaching this lesson to their peers as if they are students, opening up their decision-making process to others in the room and learning to enact particular teacher moves. Another implementation has mathematics PTs develop summaries of mathematical problems they identify as important, including good explanations of the mathematical concepts (Inoue, 2009). These lessons tend to involve a specific mathematics concept, such as multidigit addition (Tyminski et al., 2014) or proportional reasoning (Inoue, 2009), or a pedagogical approach, such as promoting mathematics discussions (Lampert et al., 2013; Tyminski et al., 2014). For some, rehearsals offer opportunities for elementary PTs to teach a lesson promoting mathematics discussions by focusing on analyzing student work and planned questions (Inoue, 2009; Tyminski et al., 2014). For others, rehearsals provide the space for PTs to give 15-minute presentations of mathematical concepts (Inoue, 2009). As a result of these short presentations, PTs seem to help students make connections between different strategies and extend students' thinking through the use of visual representations and clear presentations of mathematical concepts.

With the exception of Virmani (2004), who has worked with secondary mathematics PTs, much of the current research on rehearsals focuses on the preparation of elementary mathematics PTs. From this research, we learn that rehearsals provide opportunities for PTs to "learn to navigate the social and intellectually complex demands of teaching" (Lampert et al., 2013, p. 238) and "be given feedback on how to improve their explanations in relation to specific content" (Inoue, 2009, p. 57). During rehearsals in mathematics methods courses, feedback can be provided in multiple ways. Some methods instructors encourage peers to give real-time feedback involving an interruption of the rehearsal so that the PT can implement an almost instantaneous informed decision to their rehearsal (Lampert et al., 2013). Others do not allow interaudience explanations during the rehearsal from peers or MTEs and, instead, provide written feedback afterwards that highlights the clarity of mathematical concepts presented (Inoue, 2009). Lampert and colleagues (2013) analyzed the exchanges between the instructor and PTs of 90 rehearsal videos. They documented that most exchanges were initiated by the instructor to suggest the next teaching move and that the most common discussion involved how to elicit student thinking. The expectation was to help the PTs develop skills to respond to the demands of teaching mathematical content.

Although these researchers encourage ambitious teaching to "all students" (Lampert et al., 2013) and encourage clear communication of mathematical concepts to students (Inoue, 2009), the politics of teaching (e.g., responding to a student comment about not being able to do mathematics because she is Black) seems to be overlooked in rehearsals. When the technical aspects of teaching are privileged, PTs can leave with the unintended message that mathematics teaching is a universal endeavor. Yet, educational researchers highlight the importance of student identity, arguing that effective teachers do not expect students to "park their identity at the door" (Gutiérrez & Irving, 2012, p. 10). Instead, effective teachers integrate the multiple knowledge bases of students, such as culture, language, and lived experiences (Turner et al., 2012). Moreover, we have identified few opportunities in the literature on rehearsals for PTs to interrogate what mathematics is, which mathematical knowledge is privileged, who determines this, and who benefits from this knowledge? Although students may benefit from scoring well on standardized exams and learning mathematics recommended by the Common Core State Standards in Mathematics (National Governors Association Center for Best Practices, 2010), limiting rehearsals to mathematics pedagogy and content will not fully prepare mathematics PTs to navigate the current political climate of teaching or advocate for marginalized students (Gutiérrez, 2013). We see rehearsals as useful activities in mathematics methods courses and seek to expand their use to include the kinds of political scenarios that teachers will face inside and outside of the classroom.

IN MY SHOES

In My Shoes is a kind of rehearsal that grew out of a more basic version Rochelle had used with her secondary mathematics PTs since 2006. In that version, groups of four students would share scenarios they had confronted during student teaching and would help each other brainstorm ways of dealing with the problems if they had been *in each other's shoes*. The focus was not explicitly on the politics of teaching. Rather, secondary mathematics PTs were encouraged to discuss scenarios where they were stopped in their tracks while teaching because they were unsure of how to respond. They tended to report mathematical content they had memorized and were unable to explain (e.g., a student asking, "Why is a negative times a negative a positive?") or classroom management issues (e.g., students who refused to do homework). Of course, deciding whether to teach students to memorize mathematics procedures or understand things conceptually is a form of politics. However, such mathematical content and explanations were not recognized as political by the secondary mathematics PTs. Even so, some political scenarios also arose (e.g., a student claiming, "I'm Black. I don't do math."). We discussed these scenarios, in small groups and as a whole class, providing suggestions of possible teaching moves that could be productive. Although secondary mathematics PTs were encouraged to consider the scenarios from as many angles as they could, these discussions were not rehearsed. Instead, they tended to be collective brainstorming and advice-giving sessions where members of the class helped their fellow PTs consider a range of options for the future.

Expanding the use of role-playing for political situations, Rochelle received a grant, and Sonya, Juan Manuel, and Gabriela joined the research team. The focus of the project was to (a) prepare teacher candidates to be effective mathematics teachers who could advocate for historically marginalized/colonized students (e.g., students who are Latinx,[2] Black,[3] emergent bilinguals,[4] historically looted[5]) and (b) understand what kinds of knowledge bases, skills, and stances are required to become effective teachers and advocates for students. (See, for example, Gutiérrez, Chapter 2 in this volume, for a broader framing.)

We work at a large Midwestern university where both undergraduate mathematics majors working towards a minor in education and master's students who already hold degrees in mathematics or a math-based discipline are seeking their credentials for grades 6–12 mathematics teaching. Some secondary mathematics PTs interviewed for and were selected into the R1 Scholars Program that was the focus of the grant. They committed to teach in high-needs schools upon graduation. During their 2 years with us, the scholars were involved in more than just In My Shoes. They participated in the regular teacher education program, including field observations, lesson planning, and portfolio development for state teaching credentials. As part of the grant, they also participated in a cluster of equity-based activities that included a 3-hour biweekly seminar focused on such things as rigorous and creative mathematics; social justice teaching; strategies for supporting youth who are Black and Latinx, students who historically have been looted, emerging bilinguals, and students who identify as LGBTQ; and strategies for negotiating teaching in an era of high-stakes testing.

Scholars were also engaged in a variety of critical professional development activities, including conferences, movie viewings, community events, and summer boot camps that all provided deeper understandings of students in a diverse society and how to make mathematics teaching more conceptual and meaningful with them. Scholars visited our partnership teacher Philip, who works in a large city school system, to observe his teaching, and he joined us by way of Google Hangout every other week to participate in the discussions of readings and activities. Philip apprenticed them into the field of teaching through role-playing and mathematics lessons that connected their university mathematics coursework with high school concepts. He also modeled creative insubordination in practice with examples of how he handled the politics around grading, school paperwork, and the design of a precalculus course. Being a scholar also meant developing activities for and volunteering in a weekly, after-school bilingual mathematics club for middle school students who are Black and Latin@/x, and attending regular one-on-one mentoring sessions with members of the research team. This larger context of equity-based activities complemented our use of In My Shoes. When responding to a scenario, scholars could reflect on the knowledge they had built (e.g., through readings, seminars, professional development sessions, summer boot camp, mathematics club) about the ways in which individual teachers could productively respond to complex situations. Among other things, mentoring sessions were a place to express concerns and reflect on how adults interacted with students in their observation settings (e.g., deficit-based perspectives of students, implicit treaties where students were not required to work).

The scholars represented a variety of ethnic and racial backgrounds and included 11 Whites, 5 Asians, 1 Latinx, and 2 with multiracial backgrounds;

8 identified as men, 11 as women; and within the group, two of the scholars identified as LGBTQ and 17 identified as straight. All were unmarried. Participants' language backgrounds varied, with 6 having been raised in bilingual households (Spanish, Vietnamese, Korean, and Hindi), and one participant had achieved functional language proficiency through coursework towards both a teaching credential and minor in Spanish. Almost all of the participants were undergraduate mathematics majors; two were master's degree students who already possessed a mathematics-related bachelor's degree before entering the teacher education program. Although the regular teacher education program offered classes separately to cohorts of students, we chose to collapse them into one group such that scholars participated together in the same R1 activities as either first-year students new to the Scholars Program or as second-year students who were continuing in the R1 Scholars Program and were familiar with the activities, something we discuss later.

As part of the R1 scholars program, Rochelle adapted In My Shoes to help scholars focus more specifically on the political situations that can arise in teaching (Gutiérrez, 2012). These scenarios ranged from incidents that occur with students in classrooms, to conversations with colleagues, to interchanges at parent-teacher meetings, to interactions with administrators, to comments made in faculty meetings. For example, an individual might report on a racist comment made by a colleague about the home lives of students who are emerging bilinguals. We do not consider In My Shoes an activity; it is more of a cluster of learning opportunities, a language, and a way of connecting where the role-playing is just one part. In this sense, it is a larger social justice approach that is grounded in the particular context in which we work. The focus of this chapter is on the role-play portion.

In this version, In My Shoes begins with a practicing teacher presenting a brief scenario (usually less than 1 minute) of something troubling that has happened in her or his teaching and then inviting other scholars to respond with what they might do if they were "In My Shoes." Prompted usually by more veteran members, scholars spend 5 to 10 minutes asking clarifying questions (e.g., How well do you know this colleague? Was this the first such comment that they had made? Were you alone with this person, or were others present and could they hear?). This phase allows the scholars to gain more context about the situation so that they can act in more informed ways. Then, the scholars begin to offer their responses (what they would do if they were in the presenter's shoes). In the case of the racist comment, some suggest what they might say to that colleague; others argue that they would bring the issue to the attention of another colleague who is an accomplice in antiracist teaching; some advise raising the general issue at a faculty meeting; others say they would do nothing; and so on. After several different responses are on the table and elaborated upon (15–30 minutes), one is chosen and is rehearsed. The presenter then serves as the antagonist who, in this case, made the racist comment; the person whose response is being rehearsed now acts as the teacher (protagonist) in the rehearsal. The situation is role played for approximately 5 minutes and then debriefed. The protagonist offers his or her perspective on how they felt during the rehearsal as well as how well their strategy worked. Other scholars provide feedback on how they thought it went, pointing out specific things they liked and commenting on critical points in the rehearsal (e.g., if the person had done something different at a particular point, might the scenario have played out differently?). Finally, the presenter shares what she or he did in that scenario when it actually occurred. Many times, the real-life situation still had not been resolved, perhaps because a follow-up meeting between a teacher and student had been planned but had not yet occurred or because the presenter was not satisfied with what they had done in the moment, and was considering additional action. Each In My Shoes rehearsal and debriefing lasted approximately 1 hour of the 3-hour seminar.

Not only did In My Shoes evolve over the time Rochelle used it with all secondary mathematics PTs, it evolved during the time we used it within the R1 Scholars Program. Overall, during their 2 years in the program, four cohorts of scholars were exposed to increasingly more rehearsals. For example, the first year had four formal scenarios, all of which were led by our partner teacher, Philip. He selected and brought to our seminar scenarios from his practice, things that had happened to him earlier that day or week. While rehearsing the scenario, Philip would play devil's advocate and try to make the situation difficult so that the scholars had to think deeply about their strategy. Our goal was to highlight that dealing with the politics of mathematics teaching is akin to playing chess, where teachers need to think beyond that first move (Gutiérrez, 2016). Philip probed for scholars to ask him questions, which led to them slowly realizing that they needed to ask questions in subsequent rehearsals and model this practice for future scholars. Over time (during Cohorts 3 and 4), instead of relying on Philip, some of the scholars began to offer scenarios they faced in their field placements as well as during student teaching, or as practicing teachers once they had graduated. Again, the scholars and teachers, not the research team, selected the scenarios, though we shared the role of being the devil's advocate during role-plays, something we discuss later in this chapter. In selecting scenarios, the aim was not to choose the scenarios that would be most difficult, most important, or even most likely for our scholars to encounter in their first years of teaching. Rather, we valued the process of scholars grappling with what they might say or do in any given scenario and to become more familiar with the tensions and intense emotions that arise when one stands up for one's students or oneself. For the most part, scenarios were seen as viable spaces to help understand an aspect of practice by collaborating with others. Some scenarios were described to us or to other scholars but were never formally role-played; others were spontaneously incorporated into seminars because of their perceived value to the community ("If this bothered me, it probably will bother others and we should discuss it").

SCHOLARS' REACTIONS TO IN MY SHOES

As scholars participated through In My Shoes, we began to notice patterns in the things that they voiced or showed through actions. When we first started implementing In My Shoes with Cohorts 1 and 2, Philip would push the scholars to ask more questions in order to gather information about the scenario before giving suggestions of what action to take. Over time, after the scenario was presented, scholars would first sit and think about what information they needed before giving advice or flatly stating what should be done. This was an important experience for new scholars who joined the group. As Cohorts 3 and 4 joined the program, they were not given a list of expectations or directions of what should be done during In My Shoes. Rather, previous cohorts modeled asking questions about the context of the situation (e.g., Who were you with? How well do you know the person? Is this the first time a comment like that has been made?), which allowed the junior scholars to learn the norms of In My Shoes, to understand the importance of attending to context, and to recognize that these kinds of scenarios could arise in their future teaching. More importantly, scholars learned there was no one "right" way to handle the scenario and that they also needed to consider their own identities in carrying out this work (e.g., Is confronting a colleague in a lunchroom something I see myself doing? Does the response reflect my values, beliefs, lived experiences, and personality?). At the same time, we encouraged scholars to step out of their comfort zones and take a stand on something they might not normally do (e.g., challenging a colleague), while staying true to themselves. Having various suggestions of how to respond to the scenario both before the role-play and afterwards in debriefing allowed the scholars to learn from each other a variety of stances and approaches, including using particular language that they could take into their future teaching.

More than just a role-play or rehearsal, In My Shoes became a language for the kinds of political scenarios that scholars would see in their teaching or observations. With this language, scholars became more observant of political scenarios that would play out in their student teaching and observations. After witnessing such a scenario, scholars would express via email, text, or vocally in mentoring sessions that they "have an In My Shoes" they would like to present to the group. Scholars would either wait to present the scenario to the rest of the group or spontaneously role-play the scenario in a mentoring session if it was a timely issue that they would like to resolve during the next couple of days. In those cases, the scholar would take on the role of the protagonist while the mentor became the antagonist of the scenario, pushing back on them.

In My Shoes became a part of the scholars' professional development, both as participants in the program and after they became professional teachers. During their first years of teaching, alumni brought In My Shoes scenarios to the group, wanting to rehearse the different scenarios and hear how other scholars would handle the various situations. Some alumni mentioned that In My Shoes helped them develop a stance on their teaching and become advocates for both their own students and others. Scholars also commented on the fact this was the only part of their teacher education program in which they were exposed to the political nature of teaching or being prepared to successfully negotiate it (Gutiérrez, 2013, 2015). Furthermore, when asked for advice about which components of the R1 scholars program should be integrated into the regular teacher education program after the grant ended, alumni indicated that In My Shoes was an important way to prepare for teaching because it helped them develop a stance and practice acting on it. Overall, In My Shoes became an opportunity to practice handling and navigating a political situation with a group of colleagues who were not simply supportive in the ways some schools of education promote "safe spaces," but who were critical friends willing to take on the role of devil's advocate in order to challenge and push each other to think critically about the situation and consider possible ways of approaching the scenario if they were to face it again.

WHO WE ARE MATTERS

We designed and carried out the activities in this grant with an eye towards family, recognizing that scholars would benefit most by building knowledge/*conocimiento* together (Vargas et al., 2017). That is, we positioned the second-year scholars as "older brothers and sisters" who were partly in charge of caring for their "younger siblings" (first-year scholars). Over the 5 years of the grant, new members entered and then graduated, some later participating as alumni through In My Shoes. This intergenerational family helped create a space to try out new things, knowing that someone else could help guide you, and where older siblings could model for younger siblings or give advice about how to handle things on the basis of their own experiences. This family culture also took the weight off of the research team from being the experts or authorities during activities and allowed us to engage in moments of nos/otrxs relationships (i.e., in solidarity) with our scholars (Gerardo, Vargas, & Gutiérrez, 2013). Even so, with the exception of alumni, the research team members often were the ones creating the more hostile environment (playing devil's advocate) during rehearsals. We did so because in our experiences of standing up for the rights of others, those with whom we were speaking typically did not agree easily with our arguments or evidence. As such, it was important for scholars to practice speaking to people who were challenging them.

Reflecting upon our experiences with In My Shoes, we wondered about the contribution of the facilitator to the dynamics and outcomes of the activity. How did our lived experiences and ethical stances inform why we created In My Shoes and how we would participate in it? How did our lived experiences affect how the scholars responded? How did promoting a culture of family contribute to the participation of scholars? How might In My Shoes differ if our research team were not from a historically marginalized or colonized group?

We comment here on the lived experiences of the research team in order to ground our claims. Raised in a Chicanx activist family, Rochelle has drawn on her passion for social justice and her interest in opening up a dialogue about what counts as "mathematics" while working with secondary mathematics PTs over the past 21 years. A Black woman who was raised in the South, Sonya's experiences teaching marginalized students in the U.S. and abroad troubled her ideas about the traditional goals she held for her students. She now works with PTs, supporting them to consider broader notions of success for their own students. Juan Manuel is from a working-class family and was undocumented. His experiences as a student, teacher, and youth organizer influence his interest in understanding mathematics teacher–student relationships as a means for investigating mathematical concepts. Gabriela, a first-generation college student and mathematics major, developed a critical perspective from her ethnic studies courses that allowed her to develop an advocacy stance for historically marginalized or colonized students. As a research team, from similar yet distinct lived experiences, we planned, participated, and debriefed In My Shoes in order to privilege the complexity, challenges, and perspectives of those who experience daily microaggressions inside and outside of schools.

In contrast to the "coach" or "simulated student" roles played by other MTEs during rehearsals (Lampert et al., 2013), our role was "facilitator." Sometimes, we simply observed scholars enacting a scenario and debriefed with them by asking those involved in the role-play to explain how they felt and what strategies they used to respond to the scenario. Our goal was to position the scholars as experts who could respond to each other, rather than simply to us. During these scholar-to-scholar conversations, the research team typed notes to document their exchanges. Other times, similar to the "student" roles enacted by other MTEs during rehearsals, we participated during In My Shoes as antagonists. For example, at times, we acted as faculty and staff members in a department meeting who made comments we had heard before that reflected stereotypes about communities. Because we share a common experience as historically colonized peoples and have a wealth of knowledge about the common ways people justify deficit views or are not open to hearing the perspectives of others, in general, we did not consult with each other before role-playing. We simply knew one of us could spontaneously come up with an appropriate comment on the spot. This served to challenge the protagonists in the rehearsal to have to think beyond their first move. Moreover, because Rochelle had worked with Philip for over a decade, we did not screen his selections or help him decide what would be most beneficial for our scholars. Overall, we would characterize our work as rigorous improvisation to highlight its organic and spontaneous nature while attending to our firm social justice principles. That is, we were not simply making things up; we were improvising from prior experiences.

We were able to challenge scholars' stances during In My Shoes and to make the scholars uncomfortable because we had developed a community like "familia." That is, working through In My Shoes was a collective effort. When an In My Shoes role-play became difficult, scholars sometimes drew upon each other in a kind of "life line" manner, seeking help from peers who had similar ideas. If stuck, they could look to a peer who would trade seats with them and continue the role-play. We did not do this on a regular basis because we wanted the protagonist to fully experience the feelings and emotions of trying to think of something convincing to say in the moment. Struggling through an initial stage of being uncomfortable was worthwhile. As a research team, we felt it was best to simulate a degree of discomfort for scholars to prepare them for the unpredictable and challenging encounters they were likely to face as mathematics teachers with colleagues who may not see it as their role to advocate for students.

Even though In My Shoes seemed to be productive for the scholars, we grappled with a series of questions. Were the scholars always prepared to deconstruct the main political issue that the presenter had intended? For example, during one In My Shoes that involved the racialized and negative comment made by a presenter at a professional workshop that showed a video of a student who was Black responding to another student who was Black, the research team expected the scholars to discuss how they would respond to the presenter who had made the racist comment. Instead, the scholars focused on the how they would respond to the student as their teacher. We also wondered whether the scholars were responding to us as facilitators. Although we privileged their sense making over any correct answers or strategies, they were well aware of our stances on social justice issues and our Chicanx and Black identities and therefore perhaps tried to tell us what we wanted to hear. In reflecting upon the somewhat organic and constantly evolving nature of our work, we also wondered how MTEs who might not share our lived experiences, teaching backgrounds, and stances regarding social justice would prepare for taking on the role of the devil's advocate. Would they need to screen the kinds of scenarios that were launched? If working with a partner teacher, could they rely upon that person to convey the messages they valued? Would they benefit from reflecting with others on the variety of perspectives that opponents might hold before they carried out the role play? Might they need to decide ahead of time what each member of the research team could say, given the scenario? If they did not have a wealth of discriminatory experiences upon which to draw, to what degree could they play the role of a devil's advocate and push back on PTs spontaneously while role playing? We do not mean to imply that White teacher educators are incapable of discussing the politics, ethics, or morals surrounding the teaching of mathematics. However, we surmise that In My Shoes would differ in purpose and sense making for mathematics PTs who had different facilitators.

RECOMMENDATIONS FOR MATHEMATICS EDUCATORS

Over the past 6 years, we have learned a number of things about the value, complexity, and challenges in preparing teachers for the political nature of mathematics teaching. Therefore, we offer some guidelines for mathematics educators who seek to carry out In My Shoes with their mathematics PTs.

Before carrying out In My Shoes, PTs need opportunities to understand why they might be role-playing situations that involve the politics of teaching mathematics. In other words, there should be support for PTs to have explored articles, educational blogs, or other media that highlight the fact that mathematics teaching is, in fact, political, that it requires negotiating one's values and practices with others, including administrators, colleagues, parents, and students. Without this background knowledge, PTs might feel the facilitator is pushing an agenda that they are unlikely to face in their future teaching. In that sense, In My Shoes becomes nothing more than a theatrical performance. In My Shoes also requires that someone in the room be willing to play the devil's advocate in an authentic manner. In other words, the devil's advocate perspective should be believable, not contrived or a parody. During role play, everyone should stay in character, as either the protagonist or antagonist, and not break the fourth wall (e.g., asking the facilitator, "Is it okay if I assume X right now?"). Staying in character allows the protagonist to be deeply immersed in the tensions and emotions that arise in negotiating one's practice with others (Gutiérrez, 2009).

Early on, PTs need opportunities to understand how context (e.g., curriculum, students, a school's mission statement or school improvement plan, a teacher's identity) affects teaching. In this way, PTs can be encouraged to recognize that any response to a political scenario requires first understanding more about the context so that the response is appropriate and the protagonist is not simply dismissed as naïve. PTs also need opportunities to deconstruct the kinds of negative messages that society and schools send to students about what they are capable of, as well as having had opportunities to position students as experts. This will better ensure that during the initial discussion of the scenario, participants can identify the salient political issue at hand rather than being distracted by other aspects of the scenario.

When carrying out In My Shoes, MTEs need to be conscious of their role and should aim to be a facilitator rather than a coach. The goal should be allowing PTs to grapple with complex issues, not find *the* correct approach or strategy for addressing the situation. Some situations require further thought and alliances with others before acting. So, PTs should feel they could plan for how they will respond with others before actually enacting a role-play.

Overall, In My Shoes should not be thought of as a stand-alone activity, but rather part of a larger social justice approach that supports PTs to deconstruct the deficit-based narratives and institutional structures that maintain an achievement gap and fail to foster opportunities for students to have rigorous, creative, and meaningful mathematical experiences. Other activities and measures of assessment in a mathematics methods course should support the idea that teaching is a political endeavor that requires professional knowledge and practice.

Since the grant ended, Rochelle has collaborated with mathematics educators across the country to carry out In My Shoes in their mathematics methods courses or in workgroups with teachers. One streamlined adaptation that has worked well if no partner teacher or alumnus can offer their own In My Shoes scenarios from their everyday practice is to provide a variety of written scenarios and place students in groups to discuss how they might address the given situation if they were in the protagonist's shoes. Then, after some time developing a plan and identifying one member of the group to be the protagonist for the role-play, the PTs switch papers with another group and learn of a different situation. In this new situation, they develop their strategy for becoming the devil's advocate and develop their own details of the context to make it a more hostile environment. During this phase, the facilitator visits different groups and helps them develop authentic devil's advocate perspectives. Finally, the facilitator asks for a volunteer who is willing to role play their scenario, and that individual becomes the protagonist. The antagonist draws from the group that developed the devil's advocate position from the related scenario. The rest of the role play continues as has been described earlier.

Collaborating with other mathematics educators allows us to continue to develop a database of political scenarios that mathematics teachers face (Gutiérrez & Gregson, 2013). For example, because our work has focused on secondary mathematics teachers, engaging with MTEs who work with elementary PTs allows us to see which kinds of political scenarios overlap and which seem to differ by grade level, institutional context (e.g., type of university or program, extent to which social justice issues are highlighted in mathematics methods), or identities of the facilitators.

We continue to learn immensely from the teachers with whom we work. They tell us that in an environment of high-stakes testing, new teacher evaluations, and Common Core State Standards for Mathematics (National Governors Association Center for Best Practices, 2010), *all* teachers need opportunities to prepare for not only the kinds of mathematical lessons they will carry out but also the political aspects of the profession that often get ignored in mathematics methods courses. We remain convinced that opportunities like In My Shoes can go a long way towards helping PTs prepare for the full complexity of teaching mathematics. In this way, we can more adequately position teachers not as technicians, but as professionals who can reclaim the profession.

NOTES

1. This research was funded by the National Science Foundation, Grant # 0934901. Thank you to the teachers who so graciously shared their teaching struggles and accomplishments with us.
2. We use the term Latinx to indicate solidarity with those who identify as lesbian, gay, bisexual, transgender, questioning, or queer. Latinx represents a de-centering of the patriarchal nature of the Spanish language whereby groups of males and females are normally referred to with the "o" (male) ending. Our choice to use these terms reflects a rejection of the gender binary, our understanding of gender fluidity, the ability for "x" to stand for any gender performance, and our respect for how people choose to name themselves.
3. We use the term Black, as opposed to African American, to highlight the fact that many Black students living in the United States have ancestry in the Caribbean, South America, and Asia.
4. We use "emergent bilingual/multilingual" instead of "English learner" or "English language learner" to decenter the idea that English should be the standard by which we measure students. The term highlights that such students already have facility in one or more languages.
5. We use the term "historically looted" to emphasize the fact that certain students and their families are not just "low income"; they have not been able to accrue wealth or have had their wealth stolen as a result of governmental policies such as those developed by the Federal Housing Administration.

REFERENCES

Gerardo, J. M., & Gutiérrez, R. (2013). Negotiating Nos/otr@s relationships in an after-school mathematics club. In M. Martinez & A. Castro Superfine (Eds.), *Proceedings of the 35th annual meeting of the North American Chapter of the International Group for the Psychology of Mathematics Education* (p. 919). Chicago, IL: University of Illinois at Chicago.

Gerardo, J. M., Vargas, G. E., & Gutiérrez, R. (2017). *Constructing conocimiento in community through In My Shoes.* Manuscript in preparation.

Gutiérrez, R. (2009). Embracing the inherent tensions in teaching mathematics from an equity stance. *Democracy and Education, 18*(3), 9–16.

Gutiérrez, R. (2012, October). *Developing political knowledge for teaching: One way of making classrooms more equitable for all students.* The inaugural webinar given in the Association of Mathematics Teacher Educators' webinar series.

Gutiérrez, R. (2013). Why (urban) mathematics teachers need political knowledge. *Journal of Urban Mathematics Education, 6*(2), 7–19.

Gutiérrez, R. (2015). Nesting in Nepantla: The importance of maintaining tensions in our work. In N. M. Joseph, C. Haynes, & F. Cobb (Eds.), *Interrogating Whiteness and relinquishing power: White faculty's commitment to racial consciousness in STEM classrooms* (pp. 253–282). New York, NY: Peter Lang.

Gutiérrez, R. (2016). Strategies for creative insubordination in mathematics teaching. *Teaching for Excellence and Equity in Mathematics, 7*(1), 52–60.

Gutiérrez, R., & Gregson, S. (2013, April). *Mathematics teachers and creative insubordination: Taking a stand in high-poverty schools.* Paper presented at the annual meeting of the American Educational Research Association, San Francisco, CA.

Gutiérrez, R., & Irving, S. E. (2012). *Latino/a and Black students and mathematics.* Washington, DC: Jobs for the Future. Retrieved from http://www.studentsatthecenter .org/sites/scl.dl-dev.com/files/Students%20and%20Mathematics.pdf

Horn, I. S. (2010). Teaching replays, teaching rehearsals, and re-visions of practice: Learning from colleagues in a mathematics teacher community. *Teachers College Record, 112*(1), 225–258.

Inoue, N. (2009). Rehearsing to teach: Content-specific deconstruction of instructional explanations in pre-service teacher training. *Journal of Education for Teaching, 35*(1), 47–60.

Lampert, M., Franke, M. L., Kazemi, E., Ghousseini, H., Turrou, A. C., Beasley, H., Cunard, A., & Crowe, K. (2013). Keeping it complex: Using rehearsals to support novice teacher learning of ambitious teaching. *Journal of Teacher Education, 64*(3), 226–243.

National Governors Association Center for Best Practices. (2010). *Common Core State Standards for Mathematics.* Washington DC: Author.

Stein, M. K., Engle, R. A., Smith, M. S., & Hughes, E. (2008). Orchestrating productive mathematical discussions: Five practices for helping teachers move beyond show and tell. *Mathematical Thinking and Learning, 10,* 313–340. doi:10.1080/10986060802229675

Turner, E. E., Drake, C., McDuffie, A. R., Aguirre, J., Bartell, T. G., & Foote, M. Q. (2012). Promoting equity in mathematics teacher preparation: A framework for advancing teacher learning of children's multiple mathematics knowledge bases. *Journal of Mathematics Teacher Education, 15*(1), 67–82.

Tyminski, A. M., Zambak, V. S., Drake, C., & Land, T. J. (2014). Using representations, decomposition, and approximations of practices to support prospective elementary mathematics teachers' practice of organizing discussions. *Journal of Mathematics Teacher Educator, 17,* 463–487.

Vargas, G. E., Gutiérrez, R., Gerardo, J. M., & Irving, S. E. (2017). *Social justice, risk taking, and identity: Mathematics teachers preparing to take a stand.* Manuscript in preparation.

Virmani, R. (2014). Rehearsal and enactment for teaching in urban school settings: A qualitative study investigating the connections between a math methods course and fieldwork. *Doctoral Dissertations.* Paper 114. http://repository.usfca.edu/diss/114

CHAPTER 11

ACTIVITIES AND A COGNITIVE PEDAGOGY FOR FOSTERING PROSPECTIVE TEACHERS' CONCEPT-DEVELOPMENT PRACTICES IN MATHEMATICS METHODS COURSES

Barbara Kinach
Arizona State University

Stephen Bismarck
University of South Carolina–Upstate

Wesam Salem
University of Memphis

Mathematics teachers are most often called upon to develop three types of knowledge in students: concepts, problem-solving strategies, and justifying arguments (Perkins & Simmons, 1988). Supporting students' development of each of these knowledge types requires a different logic of instruction or

Building Support for Scholarly Practices in Mathematics Methods, pages 165–180
Copyright © 2018 by Information Age Publishing
165

set of teacher moves. Thus, one of the goals of mathematics teacher educators (MTEs) is to develop activities for mathematics methods courses that support prospective teachers' (PTs) practice for developing desired knowledge types in students.

This chapter presents a cognitive perspective on concepts, concept teaching, and how to prepare PTs to teach concepts. On the one hand, the chapter addresses the pedagogical content knowledge (PCK) PTs need for concept teaching: knowledge about what a concept is generally, how students learn concepts, and, for specific concepts, the most effective ways of representing and scaffolding learning activities to support the development of students' naïve understanding of concepts into robust mathematical notions (Shulman, 1987). On the other hand, MTEs need activities and an instructional approach for developing PTs' PCK related to concept teaching along with a real-time assessment of PTs' concept understanding and concept-teaching beliefs.

With these needs in mind, the chapter develops a cognitive pedagogy and three institutional case studies that describe the concept-development activities that MTEs currently use in mathematics methods courses (hereafter referred to as "methods") to transition PTs from their acknowledged preference for direct-instruction teaching methods toward teaching practices that foster the rich type of mathematical understanding called for by the National Council of Teachers of Mathematics (NCTM, 2000) and the Common Core State Standards for Mathematics (CCSSM; National Governors Association, 2010).

Below, we define *concept* as a specific type of mathematical knowledge and situate our work on concept-development practices for both K–12 students and PTs within the literature on mathematics teacher preparation task design and teacher knowledge-transformation processes. A cognitive pedagogy and three institutional case studies describing activities to foster PTs' concept-development teaching practices follow. The chapter concludes with recommendations for a theory of PCK development related to concept teaching.

UNDERSTANDING CONCEPTS RELATIONALLY VERSUS INSTRUMENTALLY

We begin by defining the term *concept* to distinguish it as a specific type of mathematical knowledge requiring a logic-of-teaching different from that needed by other knowledge types (e.g., information, skills, problem-solving practices, argumentation). Broadly, a concept is a generalization derived from particular instances. Typically, single words such as *osmosis, freedom, variable, derivative, area, inequality,* or *integer* are used to convey a concept.

Concepts are the content of thought; we use them to categorize and structure our reality.

Knowledge of concepts is not to be confused with *conceptual understanding*, a construct that the NCTM (2014) and researchers such as Simon, Tzur, Heinz, and Kinzel (2000) have employed to describe the quality of understanding called for by the CCSSM (National Governors Association, 2010). In contrast to procedural understanding, where isolated bits of information dominate, conceptual understanding is a web of knowledge where relationships link facts and propositions into a logical network of ideas (Hiebert & Lefevre, 1986). The contrast between conceptual and procedural knowledge aligns with Skemp's (2006) notions of relational and instrumental understanding in mathematics, and as such, it is important for MTEs to note that one can have relational or instrumental understanding of concepts and other forms of mathematical knowledge. In this regard, PTs' typical preference for direct-instruction teaching practices is problematic in that such practices foster an instrumental and not the desired relational understanding of school mathematics concepts. Because pedagogical preferences are linked to epistemological beliefs (Gill, Ashton, & Algina, 2004), we hypothesize that challenging PTs' preferences for direct-instruction teaching methods through fostering relational concept-teaching practices that go beyond rule-based teaching will assist PTs in developing the desired relational understanding of concepts in their future students as well as for themselves.

SITUATING OUR WORK

Cognitive Perspective and Tasks Research

The cognitive perspective on learning has a long history rooted in the debate over views of knowledge, understanding, and teaching methods in mathematics. Broadly, we use cognitive learning theory in the sense of using thinking to learn. Simon (2013) has defined the cognitive perspective for mathematics teacher education as focusing on the cognitive activity of individual teachers including their conceptions of mathematics, teaching, and learning. The implementation of the cognitive perspective within mathematics teacher preparation has typically involved whole-group teaching experiments whose aim is to transition teachers from instrumental to relational views of mathematics, learning, and teaching (Heinz, Kinzel, Simon, & Tzur, 2000; Simon et al., 2000). To launch these experiments, the researcher or MTE models specific activities or teaching practices that emphasize making mathematical connections through providing opportunities to identify mathematical relationships first hand (Simon et al., 2000). Heinz and colleagues presented three characteristics that are central to the design of activities that provide opportunities for knowledge growth and transformation during whole-group teaching experiments. First, "mathematics is created through human activity" (p. 86). Second, "what we see, understand, and learn are constrained and afforded by what we currently know by...our assimilatory schemes" (p. 86). An individual's assimilatory schemes are their current views of conceptual structures that have been informed by prior experiences and their environment. Third, "mathematical learning is a process of transformation, [called accommodation], of our mental structures" (p. 86). This is the process of adjusting our mental structures to take into consideration different mathematical ideas or approaches.

Integrating the cognitive perspective and the characteristics of knowledge-transformation activities presented above, we have developed whole-group teaching experiments centering on activities that foster PTs' relational concept-development teaching practices in methods for the purpose of transitioning them from their preferred instrumental teaching practices. In so doing, we expand the extensive literature on the use of concept-development tasks in mathematics teacher education for the purpose of deepening the mathematical understanding of PTs (Watson & Mason, 2007) to include the use of concept-development tasks by MTEs to support PTs' concept-teaching ability.

A Cognitive Pedagogy for Fostering Relational Concept Teaching

The cases in this chapter employ a cognitive pedagogy that is grounded in the whole-group teaching experiment approach (Heinz et al., 2000; Simon et al., 2000) of the cognitive perspective. The three-part pedagogy guiding the design of our whole-group teaching experiments consists of devising and planning, situating PTs as learners of teaching methods and concepts, and finally situating PTs as teachers of concepts and reflective practitioners.

Devising and Planning

This phase draws upon the MTE's knowledge of school mathematics concepts (e.g., area, inequalities, integers) and how to teach them for relational understanding. According to the intended learning goal for the PTs, teaching concepts relationally, the activity is constructed to provide PTs opportunities to perceive mathematical relationships while experiencing relational concept-teaching methods. To plan effective activities to foster relational concept-teaching, it is important that MTEs understand their PTs' assimilatory schemes. For example, an MTE developing an activity for teaching the concept of area of a circle relationally must be aware that the majority of PTs' understanding of the concept and how to teach it will be instrumental, specific to identifying and applying a formula.

Situating PTs as Learners of Teaching Methods and Concepts

This phase includes the MTE's enactment of an activity with PTs who are situated first as learners of mathematics teaching and then as learners of relational mathematical concepts to evoke an understanding of (a) how to teach mathematical concepts relationally and a relational understanding of the targeted mathematical concept and (b) the logical challenges students encounter when first learning the concept. This connects to human activity, the first characteristic of the whole-group teaching experiment.

Situating PTs as Teachers of Concepts and Reflective Practitioners

During this phase, MTEs support PTs' reflection on the pedagogy experienced during the relational concept-learning activity. Discussion centers on the nature of the teaching method and the evolution of mathematical insight with emphasis on how PTs' prior knowledge and beliefs have shaped their learning. PTs then design a relational concept-development activity on an assigned or self-chosen topic. This connects to the second and third characteristics of the whole-group teaching experiment. The reflection helps PTs recognize the constraints of their existing assimilatory schemes and provides them opportunity to adjust their mental structures from an instrumental to a relational understanding of how to teach concepts.

THREE CASE EXAMPLES OF RELATIONAL CONCEPT DEVELOPMENT

Case 1: Arizona State University (ASU), Special Education Graduate Program

Context

Mathematics Methods and Assessment is the sole mathematics/mathematics education course in the special education graduate program at ASU. Students enrolled in the course are second-career seekers whose procedural background in mathematics typically predicts a preference for direct-modeling teaching methods and a view of mathematical knowledge as simple, certain, and fixed (Gill et al., 2004). Thus, it is common for PTs to believe that students learn mathematics best from teachers' demonstrations and explanations, and therefore that teachers should model how to solve problems before students are allowed to solve them (Simon et al., 2000). The "show and tell" conception of mathematics learning and teaching that PTs bring to the methods course is at odds with the relational understanding and nonprocedural approach to mathematics learning and teaching called for by the CCSSM. Strong preparation for mathematics teaching demands that MTEs design methods learning activities that engage and ultimately transform PTs' traditional conceptions of mathematics, mathematics teaching, and mathematics learning, as the necessary changes in knowledge and beliefs likely will not occur in the program's nonmathematics coursework.

Devising and Planning

The activity described here centers on the concept of area. As Barbara's students most often hold instrumental views of mathematics teaching and mathematics, her goals for the activity are to provide PTs with an experience of learning and teaching concepts inductively, to contrast inductive learning and teaching with direct instruction, and to shift PTs' instrumental understanding of the area concept to a relational one.

One of the first things she seeks to do in this activity is to help PTs uncover and confront their beliefs on how the concept of area should best be taught, as well as their current conception of area. Over years of working with this particular population, and novice mathematics teachers in general, Barbara understands that PTs' preferences for direct instruction are usually grounded in their procedural conception of area as a formula, namely "area is length times width." If asked to describe or create a lesson plan for teaching area, PTs will typically outline a direct instruction lesson in which the teacher (a) introduces the new concept word (area) and its definition, (b) provides examples and nonexamples of area, (c) asks students to select instances of area from a collection of examples and nonexamples, and (d) provides a worksheet of practice problems. In opposition to this deductive approach to concept teaching, the preferred approach for teaching concepts relationally should be an inductive one. The relational concept-development activity below illustrates such an inductive approach.

The Task-Based Activity

> **T:** *Suppose this card represents the kitchen floor in your home* (see Figure 11.1). *You want to upgrade your kitchen by retiling it. How many color tiles does it take to cover the 4-by-6 index card floor?*

After students cover the card with color tiles, the teacher introduces area as the name of the concept just illustrated and has students write: "It takes ____ color tiles to cover the card, therefore we say: 'The AREA of the card measures ____ color tiles.'"

> **T:** *Suppose the floor-tile changes to post-its; how many post-its will it take to cover the floor?*

Figure 11.1   AREA floor covering activity (color tiles and post-its).

After students cover the card with post-its, the teacher says/students write: "It takes _____ post-its to cover the card, therefore we say: 'The *area* of the card measures _____ post-its.'" Continue with similar floor-covering scenarios as needed concluding with a counterexample.

   T: *Now the tile shape is a circle. How many circles will it take to cover the floor?* (See Figure 11.2.)

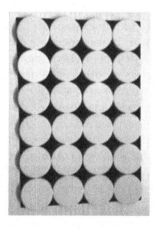

Figure 11.2   AREA floor covering activity (circles).

After students cover the card with circles, the teacher says/students write: "It takes _____ circles to cover the card, therefore we say: 'The *area* of the card measures (approximately) _____ circles.'"

> **T:** *OK, now let's review what we've done. Given these problem-based examples and nonexamples of area, define* area *in your own words.*

Situating PTs as Learners of Concepts

Assuming the role of teacher, the MTE enacts the area concept-discovery activity above so PTs can experience the logic of learning abstract mathematical concepts both relationally and inductively. For many, participation in this task is the first time they have experienced learning a new concept through generalizing patterns (as opposed to being given the concept definition by the teacher—or a book). For each of the floor-covering tasks, PTs physically cover their index cards with color tiles, post-its, and other floor "tiles" that tessellate the surface. After each task, the MTE introduces the *word* (i.e., the verbal representation) for the concept (*area*) just experienced as a concrete representation. After multiple positive examples of area, the MTE introduces a nonexample. PTs identify commonalities and differences across the area examples and nonexamples to generalize a definition of *area*. After PTs share their definitions, the MTE uses whole-class discussion to negotiate consensus on a mathematically accurate definition for *AREA*.

Situating PTs as Teachers of Concepts and Reflective Practitioners

Next, PTs assume the learning design aspect of a teacher's role by creating a concept-development activity for a different concept, specifically *perimeter*. Beyond understanding the logic of the *area* activity structure, PTs must understand its pedagogical rationale. Specifically, the fact that the examples are not isolated but arranged in a series of single-type problems is significant as the sequence of like examples makes it possible for students to perceive commonalities and differences for the purpose of generalizing a definition of the targeted concept (Krutetskii, 1976). Further, inclusion of the counterexample assists in focusing student perception. Because certain properties of the counterexample will not exist in the positive concept examples (e.g., the circles covering the card do not cover it completely, leaving parts of the area unaccounted for by the area measurement of approximately 24 circles), this omission provides opportunity for students and PTs to identify a property of area as a continuous flat space without holes. Additionally, the pairing and sequencing of concrete and verbal representations in this concept-discovery activity models how to implement cognitive theories of mathematics learning (Bruner, 1961) that maintain that students learn abstract mathematical ideas not from the abstractions themselves but by generalizing the intended abstraction from multiple concrete representations of it. As such, the activity provides opportunity for PTs to learn the preferred order of mathematics and vocabulary instruction for all students, but especially second-language learners, to be introduction of concept experiences prior to vocabulary. Finally, one reflection characteristic of all PTs reveals growing insight into the benefits of inductive teaching over direct instruction:

> This course gave me a whole new understanding on how students truly learn new math concepts instead of just the steps. Learning about the differences between direct instruction and inductive teaching has impacted my ideas about teaching math in numerous ways. I now focus much more on enduring understanding of content rather than procedural steps.

Case 2: University of South Carolina–Upstate: Undergraduate Middle/Secondary Mathematics Program

Context

Historically, the secondary mathematics methods course at the University of South Carolina–Upstate focused on developing teacher candidates' relational understanding through completing, planning, and modeling different mathematical tasks from various mathematical domains (numbers and operations, algebra, geometry, and probability and statistics) and with the use of various materials (manipulatives and technology). Prior to taking the course, PTs had not had exposure to the development of relational understanding through the teaching of mathematics, nor had they considered their own relational understanding. In earlier course iterations, the PTs did not see the importance of developing one's relational understanding of mathematics topics; nor did they see the impact relational understanding could have on their teaching of mathematics. PTs' focus was on instrumental understanding, and they expressed that they "already know how to do math" and that "we're not learning how to teach." After reexamining the program of study for secondary mathematics teacher candidates, the focus of the methods course shifted. The program of study now presents a clear division between content knowledge and pedagogical knowledge, which indicates that the methods should develop teacher candidates' knowledge of the union of both pedagogical content knowledge and specialized content knowledge (Hill, Rowan, & Ball, 2005). The instructional approach taken in the course has shifted, focusing on supporting PTs' concept-development teaching. Different models for instruction (Estes, Mintz, & Gunter, 2010) are used as the grounding for the course, and teacher candidates learn and analyze different instructional methods for developing students' instrumental understanding (direct instruction with guided practice) and relational understanding (concept development/attainment, task-based, and inquiry-based). This approach also supports the historical focus of the course: development of the PTs' relational understanding of mathematics.

Devising and Planning

This activity centers on the concept of inequalities. Through prior experience with PTs, Stephen has determined that PTs' knowledge of inequalities constitutes an instrumental understanding and is often limited to the identification of inequality symbols, the algebraic manipulation of inequalities involving variables, and procedures for graphing inequalities involving one or two variables. The goals of this activity are for Stephen's PTs to analyze the implementation of a task-based activity as a way to build students' relational understanding of inequalities, to develop PTs' logical reasoning within a realistic problem context to shape their perception of the concept of inequality, and for PTs to develop their relational understanding of inequalities through multiple representations.

The Task-Based Activity

You and a group of your friends go to the Pizza Hut Express after school. Small pizzas are $5 and cheese sticks are $3. All together you have $22. What can you buy?

Situating PTs as Learners of Concepts

The problem is implemented using the *5 Practices for Orchestrating Productive Mathematics Discussions* (Stein & Smith, 2011). The MTE monitors, selects, and sequences PTs' solutions. Through the monitoring practice, the MTE helps the PTs focus on the question of "what can you buy?" rather than "what combination of pizza and breadsticks will total $22?" This distinction is subtle but critical because the question "what can you buy?" involves the concept of inequalities, rather than the concept of equations. This subtle distinction gives rise to relational understanding for inequalities through the different representations students produce for their solutions (guess and check, tables, literal inequalities, and graphs). The activity continues as the MTE supports the development of connections between different representations as they relate to the concept of inequality.

Finally, PTs analyze the problem and its implementation and then compare this method for instruction with direct instruction. The MTE emphasizes to the PTs that the purpose of beginning a class with this instructional model is for students to develop their relational understanding of a topic through prior knowledge and multiple representations. The PTs are then situated as learners for a second task that involves a realistic context to shape their relational understanding for systems of linear equations. At the conclusion of analyzing both tasks, PTs are asked to identify characteristics for the concept-development teaching method and provide a rationale for teaching in this manner.

Situating PTs as Teachers of Concepts and Reflective Practitioners

This is the first time that many of the PTs experience and reflect on the use of instructional models other than direct instruction with guided practice. Examining an instructional model wherein the central focus was the concept of inequalities rather than the manipulation of procedures associated with solving inequalities provides PTs with a view of instructional methods that support relational understanding of mathematics. PTs are asked to analyze how the inequality task and concept-development teaching can help students develop an understanding of the concept. Here is one typical PT's written comment:

> Inequalities encompass a multitude of answers while equations provide one answer. This method will help students develop problem-solving skills. This task is an example of real-world applications and connects to better understand graphing and boundaries (domain/range).

A byproduct of the concept-development teaching and its analysis is the development of the PTs' own relational understanding of inequalities. This activity allows PTs to view the concept of inequalities as a range of solutions rather than just the symbolic representation associated with inequalities (greater than or less than). Through the development of the concept, PTs can make clear connections between multiple representations for inequalities (tables, literal inequalities, and graphs) and the rationale for examining these representations.

At the completion of the lesson and discussion, PTs are given time to apply concept-development teaching to a topic from their unit plan (course assignment) but are allowed to choose any topic found in the middle/secondary school mathematics curriculum. The PTs present a short outline and rationale for their concept-development lesson and work on developing the lesson with MTE feedback. Approximately 5 weeks later, the PTs model their concept-development lessons and teaching in the methods class.

Case 3: The University of Memphis, Graduate Elementary Education Program

Context

The program of study for the K–6 master's program at the University of Memphis includes one methods course that centers on developing PTs' relational understanding of mathematical concepts in the K–6 school curriculum. During this course, PTs are placed in the school system for 10 hours of clinical experience to apply their knowledge of teaching gained in methods. Watson and Mason (2007) have argued that most teacher education programs engage PTs in mathematical thinking to bring their attention to effective tasks and strategies they could use in their classrooms. These activities cultivate teachers' understanding of students' thinking and the challenges students may encounter when learning new mathematical concepts or skills. Therefore, the course combines a focus on developing PTs' relational understanding of mathematical concepts as well as their pedagogical knowledge of teaching mathematics concepts relationally using modified activities from the K–6 school curriculum.

Devising and Planning

This activity focuses on understanding integer subtraction and illustrates how fostering PTs' concept-development practices are supported in the graduate-level methods course. Wesam and the course instructor, Dr. Angeline Powell, discussed entering PTs' instrumental views of teaching and conceptions of mathematical concepts. Though instrumental understanding of mathematics is deemed to be inadequate according to the CCSSM, PTs continue to rely on and feel comfortable with mathematical rules as being appropriate instructional explanations for teaching students mathematical concepts (Kinach, 2002). Consequently, planning and devising targeted activities that require PTs to reason and make sense of mathematical concepts using manipulatives and real-world representations facilitates the development of teaching practices that intend to develop K–12 students' relational understanding of concepts (Reeder & Bateiha, 2010). These opportunities also contribute to PTs' knowledge of students' thinking and reveal the challenges students may encounter as they learn new concepts (Carpenter, Fennema, Peterson, Chiang, & Loef, 1989) as well as PTs' own relational understanding of mathematics concepts.

The goals of this activity are to (a) provide PTs with task-based practices that enable them to plan for teaching the integer concepts relationally in their classroom and (b) assess and contribute to the PTs' current mathematical knowledge of integers by affording them the opportunity to learn the concept relationally using representations. The activity builds from the work of Mitchell, Charalambous, and Hill (2014), who explored knowledge demands required for teachers to use representations successfully when teaching integers to "challenge students' incomplete understanding" (p. 39) and create a base for building future understanding. In addition, the activity draws from the notion that number line and cancellation (i.e., use of colored chips) are the most common models for teaching addition and subtraction with negative numbers relationally (Bofferding & Richardson, 2013; Mitchell et al., 2014).

The Task-Based Activity

Part 1: *Use the concept of Add Opposites to write five representations for zero using two-color chips.* **Part 2:** *While playing football, Ivan's team gained 6 yards on the first run but lost 8 yards on the second run. (a) What was the team's net gain? (b) What was the team's net loss? Does your answer to (a) differ from your answer to (b)? Explain why.*

Situating PTs as Learners of Concepts

The MTE began by asking PTs to solve simple number problems with integers: $3 + 4$, $4 + 3$, $4 - 3$, $-3 + -4$, $3 - 4$, $4 - (-3)$, $-3 - (-4)$. Out of 15 PTs, a few answered $4 - (-3)$, and $-3 - (-4)$ incorrectly. When PTs explained their answers they shared rules such as "minus minus equals plus." Similarly, when PTs were asked how they would explain addition and subtraction of integers to students, many referred to these rules as appropriate instructional methods. Some referred to the number line as a tool to teach students addition and subtraction of integers by moving right on the number line when adding and moving left on the number line when subtracting. PTs added that the movement on the number line is reversed when the number added or subtracted is negative. These responses contributed to the MTEs' knowledge of PTs' assimilatory schemes and revealed PTs' instrumental understanding of addition and subtraction of integers.

The MTE enacted the two-part activity outlined above thereby situating the PTs as learners of the concepts related to integer addition and subtraction. After the MTE initiated a discussion about using the two-color chips to represent zero, many PTs wondered how to do this because zero "represents no quantity." Afterwards, in groups and with Wesam's probing questions, they were able to represent zero with equal numbers of yellow and red chips which neutralize each other (Stephan & Akyuz, 2012), a "pivotal mathematical idea" to understanding integers relationally (Mitchell et al., 2014, p. 44).

Next, PTs were asked to solve the problem in Part 2, which is intended to represent the number sentence $6 + (-8)$ and use the two-color chips and the number line to represent the answer. This part aims at promoting PTs' knowledge of the affordances and the limitations of each of the representations and how each representation can scaffold students' learning and understanding of the mathematical concepts relationally. PTs, working in groups of three, struggled with the meaning of the negative sign. They did not differentiate between "gain" and "loss" in terms of integer representation. One group stated that the team's net gain was zero, rather than -2. They also shared that the loss was -2, rather than 2. They reasoned, "We can't say he gained negative 2." All groups were successful in using two-color chips and the number line to represent the solution; however, the precise use of mathematical language was absent when they were prompted to explain their representations. For example, some PTs referred to the negative value as "take away," some as "subtract," and others as "minus." It is important that PTs help students make sense of the multiple roles of the "−" symbol to fully conceptualize the distinction between the operation of subtraction, a negative number, and the notion of opposite within the context of integer addition and subtraction (Stephan & Akyuz, 2012). All PTs reported they found the number line more appropriate to represent this problem than the two color chips. One PT stated that the number line was an "easier" and "more familiar" approach to use when teaching students integer addition and subtraction. One group stated that the directionality of the movement on the number line is almost identical to the movement of the football team but could not use "gain" and "loss" accurately.

Situating PTs as Teachers of Concepts and Reflective Practitioners

At the conclusion of the above two-part activity, the MTE led a discussion of representational concept-development practices that PTs could use in their classrooms to teach integers relationally. Further, based on their own experience doing the activity, PTs discussed the challenges students encounter while learning about integers, possible misconceptions, and inaccurate use of mathematical language such as "negative" instead of "minus." PTs also modeled using two-color chips and a number line to teach addition and subtraction of integers to students relationally. They discussed how proper use of mathematical notation and language supports students' gradual detachment from representations and aids to develop relational and abstract understanding (Mitchell et al., 2014). One group shared, "We have to consider that our practices should aim at developing students' abstract thinking at later grades."

DISCUSSION AND CONCLUSION

At first glance, the activities in these three cases may be perceived as distinctly different experiences for PTs in a methods course. A comparative analysis of the activities, however, identifies a common learning goal and pedagogy undergirding all three cases. Providing PTs with an opportunity to experience, reflect upon, and apply a relational concept-development activity is the common thread emphasized in each case. As PTs go through the experience, they are also able to develop their own relational understanding of school mathematics concepts.

Although these cases illustrate a common cognitive pedagogy, the logical structure of the concept-teaching practices that MTEs model for PTs' future use with K–12 students differs dramatically. Specifically, Barbara's activity required PTs to identify commonalities in a sequence of visual area problems from which students generalized a definition of area. Stephen situated the inequality concept within a realistic context that could be misinterpreted as representing the equality concept but from which the inequality concept could be inferred. Wesam contrasted the concepts of integer addition and integer subtraction through debate over the multiple meanings of the "–" symbol in the school curriculum. Each of the cases concluded with PTs reflecting and applying the accommodations to their assimilatory schemes for the teaching of mathematics concepts. Providing PTs with these types of concept-development teaching experiences is a first step in getting PTs to see their value and ultimately implement similar concept-development lessons in their classrooms.

In closing, these cases illustrate how common goals and pedagogy are enacted by MTEs in a variety of ways to model the teaching of relational concept-development activities. Collectively, the cases lay the foundation for developing a "logic of instruction" for teaching concepts. The approach used by these three MTEs presents a clear opportunity to shift the teaching of mathematics from fostering instrumental to relational understanding.

REFERENCES

Bofferding, L., & Richardson, S. E. (2013). Investigating integer addition and subtraction: A task analysis. In M. V. Marinez & A. C. Castro Superfine (Eds.), *Proceedings of the 35th annual meeting of the North American Chapter of the International Group for the Psychology of Mathematics Education* (pp. 111–118). Chicago, IL: University of Illinois at Chicago.

Bruner, J. S. (1961). *The process of education*. Cambridge, MA: Harvard University Press.

Carpenter, T. P., Fennema, E., Peterson, P. L., Chiang, C., & Loef, M. (1989). Using knowledge of children's mathematics thinking in classroom teaching: An experimental study. *Educational Research Journal, 26*(4), 499–531.

Estes, T. H., Mintz, S. L., & Gunter, M. A. (2010), *Instruction: A models approach* (6th ed.). Boston, MA: Pearson Education.

Gill, M. G., Ashton, P., & Algina, J. (2004). Authoritative schools: A test of a model to resolve the school effectiveness debate. *Contemporary Educational Psychology, 29*(4), 389–409.

Heinz, K., Kinzel, M., Simon, M., & Tzur, R. (2000). Moving students through steps of mathematical knowing. An account of the practice of an elementary mathematics teacher in transition. *Journal of Mathematical Behavior, 19*(1), 83–107.

Hiebert, J., & Lefevre, P. (1986). Conceptual and procedural knowledge in mathematics: An introductory analysis. In J. Hiebert (Ed.), *Conceptual and procedural knowledge: The case of mathematics* (pp. 1–27). Hillsdale, NJ: Erlbaum.

Hill, H. C., Rowan, B., & Ball, D. L. (2005). Effects of teachers' mathematical knowledge for teaching on student achievement. *American Educational Research Journal, 42*(2), 371–406.

Kinach, B. M. (2002). Understanding and learning-to-explain by representing mathematics: Epistemological dilemmas facing teacher educators in the secondary mathematics "methods" course. *Journal of Mathematics Teacher Education, 5*(2), 153–186.

Krutetskii, V. A. (1976). *The psychology of mathematical abilities in school children.* Chicago, IL: University of Chicago Press.

Mitchell, R., Charalambous, C. Y., & Hill, H. C. (2014). Examining the task and knowledge demands needed to teach with representations. *Journal of Mathematics Teacher Education, 17*(1), 37–60.

National Council of Teachers of Mathematics. (2000). *Principles and standards for school mathematics.* Reston, VA: Author.

National Council of Teachers of Mathematics. (2014). *Principles to actions: Ensuring mathematical success for all.* Reston, VA: Author.

National Governors Association Center for Best Practices, Council of Chief State School Officers. (2010). *Common Core State Standards (Mathematics).* Washington, DC: Author.

Perkins, D. N., & Simmons, R. (1988). Patterns of misunderstanding: An integrative model for science, math, and programming. *Review of Educational Research, 58*(3), 303–326.

Reeder, S., & Bateiha, S. (2010). Examining prospective teachers' conceptual understanding of integers. In P. Brosnan, D. B. Erchick, & L. Flevare (Eds.), *Proceedings of the 32nd annual meeting of the North American Chapter of the International Group for the Psychology of Mathematics Education.* Columbus, OH: The Ohio State University.

Shulman, L. (1987). Knowledge and teaching: Foundations of the new reform. *Harvard Educational Review, 57*(1), 1–23.

Simon, M., Tzur, R., Heinz, K., & Kinzel, M. (2000). Characterizing a perspective underlying the practice of mathematics teachers in transition. *Journal for Research in Mathematics Education, 3*(5), 579–601.

Simon, M. A. (2013). Promoting fundamental change in mathematics teaching: A theoretical, methodological, and empirical approach to the problem. *ZDM Mathematics Education, 45*(4), 573–582.

Skemp, R. (2006). Relational understanding and instrumental understanding. *Mathematics Teaching in the Middle School, 12,* 88–95.

Stein, M. K., & Smith, M. (2011). *5 practices for orchestrating productive mathematics discussions.* Reston, VA: National Council of Teachers of Mathematics.

Stephan, M., & Akyuz, D. (2012). A proposed instructional theory for integer addition and subtraction. *Journal for Research in Mathematics Education, 43*(4), 428–464.

Watson, A., & Mason, J. (2007). Taken-as-shared: A review of common assumptions about mathematical tasks in teacher education. *Journal of Mathematics Teacher Education, 10*(4), 205–215.

SECTION IV

ACTIVITY DEVELOPMENT

CHAPTER 12

AN ILLUSTRATION OF SCHOLARLY INQUIRY FROM THE COGNITIVE PERSPECTIVE

The Development of an Integer Activity for Prospective Elementary or Middle School Teachers

Nicole M. Wessman-Enzinger
George Fox University

Wesam Salem
University of Memphis

We contend that Tobias et al. (2014) have provided a framework that mathematics teacher educators (MTEs) can utilize to construct activities that support prospective teachers' mathematical knowledge for teaching integer subtraction in the context of temperature (Thanheiser et al., 2016). We follow Thanheiser et al. (2016) in using the work of Tobias et al. (2014) to adapt a mathematics task designed to support children's learning of

Building Support for Scholarly Practices in Mathematics Methods, pages 183–197
Copyright © 2018 by Information Age Publishing
All rights of reproduction in any form reserved.

integers, for use with prospective elementary and middle school teachers (PTs). To illustrate our process, we begin by describing our cognitive perspective as it relates to supporting mathematical knowledge for teaching (Ball, Thames, & Phelps, 2008). Because theoretical perspectives have explanatory power (e.g., Thompson, 2013), theory should be used to help guide MTEs' work. Aligned with the cognitive perspective (Simon, 2008), we use Tobias et al. (2014) as our theoretical framework for structuring our activity development.

BACKGROUND OF LITERATURE

The Cognitive Perspective and PTs' Conceptual Understanding for Teaching

Ball et al. (2008) have illustrated that *mathematical knowledge for teaching* (MKT) involves more than being able to solve mathematical problems. MKT includes the development of *specialized content knowledge* (SCK), such as making sense of the nuances of children's mathematics, which is often different from how adults solve mathematical tasks. Learning goals that MTEs create and activities they design influence the MKT that PTs construct, which supports their future teaching (Stein & Lane, 1996). Simon (2008) explained that in order for "instruction to be effective, the knowledge to be learned must be clearly identified" (p. 19)—pointing to the importance of clearly defined learning goals. Simon also pointed to defining learning goals as an essential step in supporting PTs' pedagogical knowledge, such as learning to facilitate discourse that elicits children's thinking. Often, learning goals for pedagogical knowledge (e.g., facilitating mathematical discourse) or specialized content knowledge (e.g., making sense of children's mathematics) are neglected. Thus, MTEs are responsible for clearly articulating established learning goals during activity development within methods courses.

After learning goals are set, MTEs may facilitate raising PTs' awareness of the depth of elementary mathematical content through their use of children's thinking (e.g., Thanheiser et al., 2016). Incorporating children's thinking into activities facilitates opportunities for PTs to "understand the process through which learners develop new concepts" (Simon, 2008, p. 26). We desired to build up a conception-based perspective (Simon, 2008) with the PTs. This entails understanding the extant conceptions that PTs hold and facilitating more advanced conceptions. That is, building a conception-based perspective with PTs happens by first understanding the extant conceptions that PTs hold and then extending those initial conceptions.

Tenets of the Tobias et al. (2014) framework for task design include exploring the mathematics in children's tasks and establishing clear learning goals—in alignment with the cognitive perspective. The framework supports modifying children's mathematics tasks to increase cognitive demand so PTs experience learning mathematics in ways similar to those children face when constructing mathematical concepts (Thanheiser et al., 2016). Increasing the cognitive demand of children's tasks promotes PTs' thinking about conceptions and builds PTs' MKT.

FRAMEWORK FOR MODIFYING TEMPERATURE TASK

Reasoning about integers in general is challenging (cf. Piaget, 1948), with additional challenges in the context of temperature (cf. Piaget, 1952). Yet, in our culture we often utilize integers in the context of temperature (Altiparmak & Özdoğan, 2010). For these reasons, we used the Tobias et al. (2014) framework and a children's task about temperature to create an activity for PTs aimed at developing their MKT for teaching integer operations in the context of temperature. Further, we selected Tobias et al. (2014) to frame our work because it aligns with the cognitive perspective (Simon, 2008) and can be used to structure scholarly inquiry (Lee & Mewborn, 2009). Thanheiser et al. (2016) illustrated the potential of the framework for structuring scholarly inquiry by using a children's fraction comparison task as the basis for building activities for PTs. We assert the Tobias et al. (2014) framework can also be used to support MTEs' design of mathematics tasks for PTs in other domains, such as integer operations.

Building mathematical activities for PTs outlined by Tobias et al. (2014) includes:

1. "selecting a children's mathematical task" (p. 183)
2. "analyzing the children's task and identifying learning goals" (p. 184)
3. "considering PTs' experience and knowledge in modifying the task" (p. 185)
4. "extending the task to develop mathematical knowledge for teaching" (p. 185)
5. "implementing, reflecting on, and further modifying the task" (p. 186)

These tenets of the framework guide MTEs in task development by beginning with tasks from school mathematics curricula and modifying children's tasks in a way that builds an activity for PTs. We illustrate that Tobias et al. can be used to structure scholarly inquiry and task design by describing our development of an activity to support the teaching and learning of integer subtraction in the context of temperature.

Activity Development and Selecting a Children's Mathematical Task

To support PTs' MKT in the domain of integer subtraction, we looked for a children's task using integer subtraction in the context of temperature. Integer instruction often focuses on rules and procedures rather than conceptualizations and uses contrived contexts (e.g., Ball, 1993), but we sought tasks with contexts that support making meaning with integers. In addition, we knew that existing research on integer learning suggests explicit attention to meanings of the minus sign (symmetry, unary, binary) and integers as directed magnitudes support development of integer concepts (Bofferding, 2014). The use of temperature as a context supports attending to meanings of the minus sign and the use of integers as directed magnitudes (Schwarz, Kohn, & Resnick, 1993). Temperatures are measured on a conventional scale (symmetry), relative to each other (unary), and the changes or differences in temperature facilitate subtraction (binary). Yet, few PTs use temperature to model integers and integer operations (e.g., Wessman-Enzinger & Tobias, 2015).

To motivate the use of temperature as a model of integers and integer operations, we turned to the *Connected Mathematics Project* (CMP) curriculum. CMP is reform-based, follows Common Core State Standards (CCSS) recommendations, and focuses on conceptual understanding as well as procedural fluency. Within *Accentuate the Negative* (Lappan, Fey, Fitzgerald, Friel, & Phillips, 2006), we found a task that utilized temperature as a context for solving and modeling integer subtraction. The multipart task included "sketching number lines and writing number sentences" (p. 11) given two temperatures with an unknown change:

What is the change in temperature when the thermometer reading moves from the first temperature to the second temperature? Write an equation for each part.

1. 20° F to –10° F   2. –20° F to –10° F   3. –20° F to 10° F
4. –10° F to –20° F   5. 20° F to 10° F   6. 10° F to 20° F   (p. 11)

Analyzing the Children's Task and Identifying Learning Goals

Following the Tobias et al. (2014) framework, we each solved the children's mathematics task from CMP and together established learning goals for PTs. We met to discuss underlying mathematical concepts of the task and identified difficulties children or PTs might have when working on the task.

We noticed that all the problems in the CMP task involved finding a difference between two temperatures. Wessman-Enzinger and Tobias (2015) identified this problem type as *State-State-Translation* (SST). SST problems provide two temperatures (20° F, –10° F), indicate a direction for the temperature change (from the first temperature to the second temperature), and then ask for the change between the given temperatures. An example of an SST problem is:

> The temperature in Boston is 2 degrees. The temperature in Chicago is –5 degrees. How much warmer is it in Boston compared to Chicago? (2 – –5 = □).

PTs struggle with SST problem types (Wessman-Enzinger & Tobias, 2015). Yet such problems help PTs move beyond a take-away conception of subtraction to develop a distance conception, which can be challenging with integers (Bofferding & Wessman-Enzinger, 2017; Selter, Prediger, Nührenbörger, & Hußmann, 2012). Although both conceptions of subtraction are important, PTs need more experience with distance because they do not often draw upon this conception in the context of temperature (Wessman-Enzinger & Tobias, 2015).

Solving the children's task and identifying conceptual hurdles PTs have with integer subtraction (Bofferding & Richardson, 2013), particularly with SST problem types (Wessman-Enzinger & Tobias, 2015) and temperature as a context (Altiparmak & Özdoğan, 2010; Schwarz et al., 1993), influenced two of the three learning goals we developed for the activity:

1. Explore subtracting a negative number conceptually in the context of temperature.
2. Utilize SST problem types in the context of temperature.

Discussion in the cognitive subgroup of the Scholarly Inquiry and Practices Conference (Sanchez, Kastberg, Tyminski, & Lischka, 2015) focused on supporting PTs to attend to and make sense of children's thinking by providing them with opportunities to build concepts associated with such attention. These discussions influenced our third learning goal:

3. Unpack the mathematical accomplishments and challenges of Grade 5 children's responses to a SST problem type in the context of temperature.

Considering PTs' Experience and Knowledge in Modifying and Extending the Task

To adapt a children's task for use with PTs, the Tobias et al. (2014) framework suggests making the task cognitively demanding by building from PTs'

Pose a story for each of the following number sentences:

$$-17 + 12 = \square \qquad -12 - -6 = \square \qquad -2 - 3 = \square$$

Figure 12.1 Activity opener.

experiences and knowledge. We reasoned that because PTs find the SST problem type challenging and as such do not utilize it often (Wessman-Enzinger & Tobias, 2015), we could increase the cognitive demand of the task by drawing PTs' attention to such problem types.

To make the CMP task more demanding, we added an "opener" in which PTs are given number sentences and asked to pose stories for them (see Figure 12.1). From previous experiences (Wessman-Enzinger & Mooney, 2014), we thought that this segment of the activity increased the cognitive demand, making it more appropriate for PTs as stipulated by Tobias et al. (2014). Furthermore, we thought that having the PTs pose problems and revisit them later in the activity might provide an opportunity for conceptual change.

To decide which number sentences to include in the opener we considered the types of problems PTs typically pose for various number sentences. Wessman-Enzinger and Tobias (2015) found that PTs typically write State-Translation-State problems (STS) such as "It was –17 degrees in the morning. In the afternoon the temperature rose 12 degrees. What is the temperature now?" PTs less often write SST problems such as "It is –12 degrees in the morning and –6 degrees in the afternoon. How much did the temperature increase?" Even when number sentences like "–12 – –6 = $\square$" do not support the STS problem type, PTs will pose STS problems (i.e., starting with a temperature and then increasing or decreasing in temperature) with unrealistic scenarios such as "It is –12 degrees in the morning and dropped –6 degrees in the morning." The opener provides opportunities for the PTs' misconceptions about posing integer stories to emerge and be explored during a whole class discussion. The inclusion of story posing facilitates PTs' use of integers within a temperature context (Learning Goal 1). The inclusion of the number sentence "–12 – –6 = $\square$" supports the use of SST problems types (Learning Goal 2), because STS problem types are not appropriate for –12 – –6 = $\square$ in the context of temperature.

We further modified the children's task for use with PTs by reducing the number of problems and focusing on specific number types in temperature comparison problems. PTs are asked to compare pairs of temperatures, given a skeleton sentence (see Figure 12.2).

The numbers in A, B, and C (see Figure 12.2) create different situations when modeled on a number line. A and C involve crossing zero, whereas B does not. We included this problem type in light of research that found crossing zero to be more challenging than not crossing zero on the number

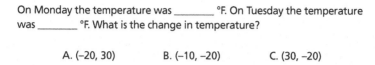

On Monday the temperature was _____ °F. On Tuesday the temperature was _____ °F. What is the change in temperature?

A. (–20, 30)   B. (–10, –20)   C. (30, –20)

Figure 12.2 The temperature sentence modification to the children's task.

line for subtraction problems (e.g., Bishop, Lamb, Philipp, Whitacre, & Schappelle, 2014). The question "What is the change in temperature?" avoids suggesting a direction for the temperature change. We anticipated PTs would produce the number sentence –20 – 30 rather than 30 – –20 for Part A (–20, 30). Our PTs were unfamiliar with creating number sentences for contexts such as those in Figure 12.2. This modification provided PTs with an opportunity to use a temperature scale to think about integer subtraction as distance, supporting movement beyond an algorithmic conception of integer subtraction. Including the temperature sentence addressed the use of SST problem types (Learning Goal 2).

To target Learning Goal 3 (attending to children's thinking), we included a video clip of Grade 5 children solving an SST problem. In addition, we added a whole class discussion of what the PTs noticed in the video to build connections between attention to children's thinking and PTs' knowledge of mathematics (Jacobs, Lamb, & Philipp, 2010).

In the temperature video clip, three children solve and discuss an SST problem similar to those PTs have already solved (see Figure 12.3). Using the same problem type provides PTs with the opportunity to extend the domain of their mathematical thinking to make sense of children's thinking.

We included the temperature video to support PTs' attention to children's thinking as they solved an SST problem type (Learning Goal 3). In the video, Kim, Alice, and Jace are working in a small group with a teacher. The teacher gives the problem to the children and asks them to solve it independently first. As the video clip begins, the children have solved the problem and are sharing number sentences they wrote for the problem. Jace thought the answer was 4 and wrote the number sentence –5 + 9 = 4. Alice thought the answer was 14 and wrote two number sentences, 5 + 9 = 14 and 5 – –9 = 14. Kim thought the answer was 4 or 14 and wrote the number sentence 5 + 9 = 14. The video focuses on Kim and Alice's debate about the answer 14 and the number sentence 5 + 9 = 14.

The warmest recorded temperature of the North Pole is about 5° Celcius. The warmest recorded temperature of the South Pole is about –9° Celcius. Which place has the warmest recorded temperature? And, how much warmer is it?

Figure 12.3 Temperature problem the children in the video solved.

We hoped the PTs would notice that the children determined the solution but struggled to write a subtraction number sentence. Children do not utilize integers and write number sentences as adults do (Whitacre et al., 2015). In particular, in the video, children write addition number sentences. Thus, exploring children's thinking with the PTs and discussing ways to transition to the use of integer subtraction is important. To build this opportunity, we included prompts such as "What questions might you pose to help a child to transition from writing an addition number sentence to writing a subtraction number sentence?"

Implementing, Reflecting on, and Further Modifying the Task

The final step in the Tobias et al. (2014) framework is implementing the activity. Based on MTEs' reflections on PTs' experiences with the activity, further modifications may be made. Our refinements, in alignment to learning goals, are described next.

Refinements Supporting the Context of Temperature and the SST Problem Type

PTs struggled with the activity opener, posing stories for the number sentences, because they focused on the lack of a physical embodiment of the negative integers. PTs reflected, "No matter what story that they posed they would not get a negative answer." For example, given "$-17 + 12 = \square$," a PT wrote, "I had a gas tank that was 17 gallons short of gas and I filled it up 12 gallons so it was 5 gallons short." The PTs argued a 17-gallon tank existed and could be 17 gallons short. None of the PTs posed temperature stories, which was a challenge later in the task when they were asked to reflect on the stories they posed in relationship to the integer temperature task from the video. We asked PTs the following question in Part III: In what ways are the stories you posed in Part I similar or different from the task in the temperature video clip? Because they did not pose temperature stories, they noticed the temperature context as a difference whereas we had aimed for a deeper discussion—such as identifying SST problem types. Although discussing physical embodiment with PTs is important, it was not the focus of the activity and was not aligned with our learning goals. We decided to refine the opener in Part I to "Pose a *temperature* story for the following number sentences"—explicitly drawing reference to the context of temperature (Learning Goal 1).

After piloting the activity and noticing that the PTs did not make connections to the SST problem type, the following day we asked the PTs to pose stories again and to use temperature as a context. After exploring children's thinking, PTs noticed that STS stories could be posed for both "–17 + 12 = □" and "–2 – 3 = □." Also, they were able to pose SST stories (e.g., comparing two different temperatures) for both "–12 – –6 = □" and "–2 – 3 = □." Having PTs pose temperature stories up front, rather than any story, draws more attention to the SST problem type (Learning Goal 2).

PTs worked on writing number sentences for the SST problem type (see Figure 12.2) in the skeleton sentence longer than anticipated. We planned 10 minutes, but it took about 20 minutes. For example, PTs wrote a variety of number sentences for the problem: "On Monday the temperature was –20°F. On Tuesday the temperature was 30°F. What is the change in temperature?" Some wrote a number sentence consistent with what was expected in the CMP curriculum (i.e., "30 – –20 = □"), while others did not (e.g., "–20 + □ = 30", "–20 – 30 = □", "–20 + 50 = □"). Like the children in the video clip, some PTs wrote addition number sentences (i.e., "–20 + 30 = □"), and others wrote missing-addend addition problems (i.e., "–20 + □ = 30"). Of these number sentences, some fit well ("–20 + □ = 30"), whereas others did not fit as well because they did not include all the numbers ("–20 + 50 = □") or did not utilize the operations appropriately ("–20 – 30 = □"). Some of the PTs wrote a subtraction number sentence that did not reflect the magnitude of the change from the context correctly (i.e., "–20 – 30 = □"), but very few wrote "30 – –20". The subtraction sentence "–20 – 30" does not support viewing the temperature change as an increase like "30 – –20" does. And while "–20 + □ = 30" works, some PTs did not think that "30 – –20" would also work. To better support using temperature with integers (Learning Goal 1) and the SST problem type with temperature (Learning Goal 2), we also refined the activity to include more time for this process. Very few of our PTs used a number line or thermometer during any point in the activity and likewise did not suggest that children would use them. This behavior further supports the need to explore the temperature context upfront.

Refinements That Supported Attending to Children's Thinking

One MTE colleague provided feedback on our activity after implementation. She found relating Learning Goal 3 (attending to children' thinking when solving SST problem types) to Learning Goal 2 (solving SST problem types) challenging. To address this challenge, we shared the temperature problem from the video (see Figure 12.3) with the PTs prior to showing the video.

We also shared some context for the video with the PTs and asked them to solve the problem given to the children:

Kim, Alice, and Jace are in Grade 5. They are working in a small group with a teacher. The teacher gave the problem to the children and asked them to solve it independently first. The group is discussing their solutions. This video clip is an excerpt from a larger conversation (20 minutes) about this problem. At the beginning of the video, the children have solved the problem and are sharing number sentences they wrote for the temperature problem. Think about how you solved the problem and think about the things that children are drawing, writing, and saying in this clip.

Our colleague was also concerned about the quality of the video. We added a transcript for the video to improve PTs' understanding of the video content (see Figure 12.4).

To help PTs make connections between their thinking and that of the children, we added the children's written work to Part II and III of the activity (see an example of the work in Figure 12.5). Kim's work was significant because she wrote an addition number sentence rather than a subtraction number sentence to find the correct answer. Again, Kim was unaware that subtracting a negative integer is equivalent to adding; rather, she conceptualized the problem as a sum of distances. It is this conceptualization the PTs had difficulty with during this activity.

When we piloted the new version of the activity in our classes, PTs still struggled to connect their number sentence writing within the temperature

Kim:	Um . . . I just added. Without a negative, I added five plus nine. Because . . . (makes a face). I don't know. I don't know how to explain this.
Teacher:	You don't know how to explain it?
Kim:	Technical difficulties.
Alice:	Technical . . .
Teacher:	But you think it's 14?
Kim:	I am positive it is 14.
Teacher:	You are positive it's 14, but are struggling to explain it . . . So where do you think it's warmer at?
Alice/Kim:	North Pole.

Figure 12.4 Excerpt from video transcript provided to PTs.

Figure 12.5 Kim's work on an SST problem.

context (Learning Goals 1 and 2) to what the children did in the video (Learning Goal 3). The discussion question, meant to prompt PTs' attention to the children's additive approaches ("What questions might you pose to help a child transition from writing an addition number sentence to writing a subtraction number sentence?"), appeared too late in the activity. Also, very few PTs suggested using a number line or thermometer, which we thought would facilitate connections between the children's additive approaches and integer subtraction. To motivate this discussion we adjusted the question ("After watching the video clip, what would you do next as a teacher? What questions might you pose to help discussion and learning?") and moved the prompt to follow the exploration of the children's work. Despite their lack of use of thermometers, as PTs reflected on the children's thinking in Part III, they suggested asking children to create a visual representation, again without drawing any themselves. Because they did not draw thermometers, we suspected that a number line or thermometer scale would have facilitated attending to children's thinking about the SST problem type (Learning Goal 3). Also, we thought the use of a number line or scale could support the discussion about why Kim's strategy (adding the temperatures) does not work when both temperatures are negative or both temperatures are positive.

Trying to connect Learning Goal 2 (utilizing the SST problem type) to Learning Goal 3 (children's thinking about the SST problem type), we posed the following discussion question, "How does the number sentence that Kim wrote relate to what you did in #2? In what ways were you surprised?" The PTs all stated that Kim knew the rule that "two negatives make a positive." Because the PTs were procedurally focused, they did not discuss that Kim may have been able to think about adding the distances even though she was not able to write a subtraction number sentence. Making sense of distances, even with addition, is an important component of developing an understanding of the SST problem type (Learning Goal 2). Consequently, we moved the questions "After watching the video clip, what would you do next as a teacher? What questions might you pose to help discussion and learning?" and "How does the number sentence that Kim wrote relate to what you did in #2? In what ways were you surprised?" to the end of Part II. We thought this might give PTs an opportunity to draw a number line or thermometer representation earlier in the activity. The second question in Part II asks PTs to solve the problem like a child—providing the opportunity to draw a number line. We thought this would give more opportunity to work with a number line and thinking about the distances. Finally, to ensure that PTs did not assume the children in the video knew or were using the procedure for finding differences (two negatives make a positive), we reminded them several times that the children produced their written work prior to instruction.

We also revised our directions: "Watch the temperature video clip (transcript included)" to "Examine the written work from the children and watch the temperature video clip (transcript included)." The written work demonstrates that Alice and Kim obtained the correct solution (see, e.g., Figure 12.5). Kim wrote an addition number sentence, and Alice wrote both an incorrect subtraction and a correct addition number sentence. The video clip is challenging without knowing the children were engaged in this type of problem for the first time and examining the children's written work. These were important points to emphasize to support PTs' attention to children's thinking (Learning Goal 3).

CONCLUSIONS

Using a framework for activity development (Tobias et al., 2014) and literature about children's and PTs' thinking about integers (e.g., Bofferding, 2014; Wessman-Enzinger & Tobias, 2015), we developed a cognitively demanding activity for PTs through a series of steps, one of which included adapting a children's task. PTs engaging in the activity had opportunities to build conceptions in the direction of our learning goals: Explore addition and subtraction of integers in the context of temperature, utilize the SST problem type, and attend to children's thinking as they solved a similar problem. The activity we created (see Appendix A) and the work of Thanheiser et al. (2016) demonstrate that MTEs can use the Tobias et al. (2014) framework to develop activities for PTs through scholarly inquiry.

The Scholarly Inquiry and Practices Conference (Sanchez et al., 2015) and this chapter provided the opportunity to reflect on the importance of learning goals and utilizing children's thinking to leverage the development of MKT in alignment with the cognitive perspective (Simon, 2008). The discussions of the cognitive subgroup at the Scholarly Inquiry and Practices Conference challenged us to create activities with intentional, explicitly aligned learning goals that supported PTs attending to children's thinking. Investigating the literature, reflecting on learning goals, and utilizing a research-based framework (Tobias et al., 2014) provided the opportunity to illustrate the creation of an activity for PTs by modification of a children's task.

APPENDIX A
Activity Developed Using the Tobias et al. (2014) Framework

PART I (Opener)

Pose a temperature story for each of the following number sentences:

$-17 + 12 = \square$
$-12 - -6 = \square$
$-2 - 3 = \square$

PART II (Explore & Discussion)

On Monday the temperature was _____°F. On Tuesday the temperature was _____°F. What is the change in temperature?

   A. (–20, 30)   B. (–10, –20)   C. (30, –20)

1. For each of these sets of temperatures (A–C), write a number sentence that illustrates the story.
2. Suppose this is an introductory problem for integer operations that children (late elementary or early middle school) are given. These children have not had a formal lesson on integer operations yet. Solve each of these problems like you think a child may initially solve it.
3. Examine the written work from the children and watch the temperature video clip (transcript included).
   a. What was Kim's strategy? And, what mathematical topic were Alice and Kim debating about?
   b. Does Kim's strategy in the video work for each of the temperature sets given? Why or why not?
   c. After watching the video clip, what would you do next as a teacher? What questions might you pose to help discussion and learning?
   d. How does the number sentence that Kim wrote relate to what you did in #2? In what ways were you surprised?

PART III (Extension)

1. How would you change your previous stories in Part I?
2. In what ways are the stories you posed in Part I similar or different from the task in the temperature video clip?
3. Implement the temperature task from Part II at your school (clinical placement, child you tutor, etc.). Try to give this task to a child who has not learned about integer operations in school yet. Report on the responses from children.

REFERENCES

Altiparmak, K., & Özdoğan, E. (2010). A study on the teaching of the concept of negative numbers. *International Journal of Mathematical Education in Science and Technology, 41*(1), 31–47.

Ball, D. L. (1993). With an eye on the mathematical horizon: Dilemmas of teaching elementary school mathematics. *Elementary School Journal, 93*(4), 373–397. doi:10.1086/461730

Ball, D. L., Thames, M. H., & Phelps, G. (2008). Content knowledge for teaching: What makes it special? *Journal of Teacher Education, 59*(5), 389–407.

Bishop, J. P., Lamb, L. L., Philipp, R. A., Whitacre, I., & Schappelle, B. P. (2014). Using order to reason about negative integers: The case of Violet. *Educational Studies in Mathematics, 86*, 39–59. doi:10.1007/s10649-013-9519-x

Bofferding, L. (2014). Negative integer understanding: Characterizing first graders' mental models. *Journal for Research in Mathematics Education, 45*(2), 194–245. doi:10.5951/jresematheduc.45.2.0194

Bofferding, L., & Richardson, S. E. (2013). Investigating integer addition and subtraction: A task analysis. In M. Martinez & A. Superfine (Eds.), *Proceedings of the 35th annual meeting of the North American Chapter of the International Group for the Psychology of Mathematics Education* (pp. 111–118). Chicago, IL: University of Illinois at Chicago.

Bofferding, L., & Wessman-Enzinger, N. M. (2017). Subtraction involving negative numbers: Connecting to whole number reasoning. *The Mathematics Enthusiast, 14*, 241–262.

Jacobs, V. R., Lamb, L. L., & Philipp, R. A. (2010). Professional noticing of children's mathematical thinking. *Journal for Research in Mathematics Education, 41*, 169–202.

Lappan, G., Fey, J. T., Fitzgerald, W. M., Friel, S. N., & Phillips, E. D. (2006). *Connected mathematics 2: Accentuate the negative.* Boston, MA: Prentice Hall.

Lee, H. S., & Mewborn, D. S. (2009). Mathematics teacher educators engaging in scholarly practices and inquiry. In D. S. Mewborn & H. S. Lee (Eds.), *Scholarly practices and inquiry in the preparation of mathematics teachers* (AMTE Monograph 6, pp. 1–6). San Diego, CA: Association of Mathematics Teacher Educators.

Piaget, J. (1948). *To understand is to invent: The future of education.* New York, NY: Viking Press.

Piaget, J. (1952). *The child's conception of number.* London, England: Routledge & Kegan Paul

Sanchez, W., Kastberg, S., Tyminski, A., & Lischka, A. (2015). *Scholarly inquiry and practices (SIP) conference for mathematics education methods.* Atlanta, GA: National Science Foundation.

Schwarz, B. B., Kohn, A. S., & Resnick, L. B. (1993). Positives about negatives: A case study of an intermediate model for signed numbers. *Journal of the Learning Sciences, 3*(1), 37–92. doi:10.1207/s15327809jls0301_2

Selter, C., Prediger, S., Nührenbörger, M., & Hußmann, S. (2012). Taking away and determining the difference—A longitudinal perspective on two models of subtraction and the inverse relation to addition. *Educational Studies in Mathematics, 79*, 389–408.

Simon, M. A. (2008). The challenge of mathematics teacher education in an era of mathematics education reform. In B. Jaworski & T. Wood (Eds.), *The international handbook of mathematics teacher education* (Vol. 4, pp. 17–29). Rotterdam, The Netherlands: Sense.

Stein, M. K., & Lane, S. (1996). Instructional tasks and the development of student capacity to think and reason: An analysis of the relationship between teaching and learning in a reform mathematics project. *Educational Research and Evaluation, 2,* 50–80.

Thanheiser, E., Olanoff, D., Hillen, A., Feldman, Z., Tobias, J., & Welder, J. M. (2016). Reflective analysis as a tool for task redesign: The case of prospective teachers solving and posing fraction comparison problems. *Journal of Mathematics Teacher Education, 19,* 123–148.

Thompson, P. T. (2013). In the absence of meaning. In K. Leatham (Ed.), *Vital directions for mathematics education research* (pp. 57–93). New York, NY: Springer.

Tobias, J. M., Olanoff, D., Hillen, A., Welder, R., Feldman, Z., & Thanheiser, E. (2014). Research-based modifications of elementary school tasks for use in teacher preparation. In K. Karp & A. R. McDuffie (Eds.), *Annual perspectives in mathematics education 2014: Using research to improve instruction* (pp. 181–192). Reston, VA: National Council of Teachers of Mathematics.

Wessman-Enzinger, N. M., & Mooney, E. S. (2014). Informing practice: Making sense of integers through story-telling. *Mathematics Teaching in the Middle School, 20*(4), 202–205.

Wessman-Enzinger, N. M., & Tobias, J. (2015). Preservice teachers' temperature stories for integer addition and subtraction. In K. Beswick, T. Muir, & J. Wells (Eds.), *Proceedings of the 39th annual meeting of the International Group for the Psychology of Mathematics Education* (Vol. 4, pp. 289–296). Hobart, Australia: Psychology of Mathematics Education.

Whitacre, I., Bishop, J. P., Lamb, L. L. C., Philipp, R. A., Bagley, S., & Schappelle, B. P. (2015). 'Negative of my money, positive of her money': Secondary students' ways of relating equations to a debt context. *International Journal of Mathematical Education in Science and Technology, 46*(2), 234–249. doi:10.1080/002073 9X.2014.956822

CHAPTER 13

ENHANCING ACTIVITIES IN MATHEMATICS METHODS COURSES TO ACHIEVE SOCIOPOLITICAL GOALS

Brian R. Lawler
Kennesaw State University

Raymond LaRochelle
San Diego State University

Angela Thompson
Governors State University

Since the 1970s, the mathematics education community has increased attention toward equity issues, but seemingly to little avail. As we experience advances in achievement across the board, shameful gaps persist (for a critical review, see Gutiérrez, 2008). The community of mathematics educators and researchers is beginning to understand that these gaps have deep bases in the social and economic inequalities of society (Gutiérrez, 2008; Martin,

Building Support for Scholarly Practices in Mathematics Methods, pages 199–214
Copyright © 2018 by Information Age Publishing
All rights of reproduction in any form reserved.

2003) and that the institution of mathematics education serves as an instrument to maintain the power and privilege of the dominant, White society (Bishop, 1990; Martin, 2015).

SOCIOPOLITICAL GOALS FOR MATHEMATICS METHODS COURSE DESIGN

The sociopolitical perspective examines how power and privilege impact the way mathematics and mathematics education operate (Gutiérrez, 2013a). It is a critical perspective; it foregrounds the notion that knowledge not only is a product of social interactions but also is shaped by power as it operates both in the classroom and in the larger society. As researchers and educators, we are compelled to recognize that power is everywhere (Foucault, 1991), to engage in dialogue that seeks to notice and disrupt, and to admit our own implication in power dynamics. This perspective encourages questions such as *Why this mathematics? Who benefits from this mathematics education?* and *Who is the child this mathematics education produces?* Sociopolitical work often probes the discourses that shape identity formation. It may also focus on interactions in the mathematics classroom that may promote healthy individuals, *conscientização* (Freire, 1970; Gutiérrez, 2013a), and a more just society.

At the 2015 Scholarly Inquiry and Practices Conference (Sanchez, Kastberg, Tyminski, & Lischka, 2015), a meeting of university-based mathematics teacher educators (MTEs), the sociopolitical challenges of mathematics education were discussed in the context of teacher preparation. An initial list of sociopolitical goals to be addressed in the preparation of mathematics prospective teachers (PTs) were identified:

1. Develop strategies for disrupting current mathematics education norms, and agency for pushing back
2. Become aware of and draw on knowledge of context in which PTs work, including families and communities
3. Develop a critical orientation to mathematics
4. Critique discourses of education (e.g., schools are failing, achievement gap is really about achievement)
5. Critically analyze and develop one's own mathematics teacher identity

These goals point toward supporting future mathematics teachers to develop a critical orientation toward (a) the foundational elements of mathematics instruction, (b) mathematics itself, (c) the socially held norms about what it means to be a mathematician and who is good at mathematics, and

(d) the teacher's own sense of who they are with regards to not only mathematics but also their positionality in a racist society (Bonilla-Silva, 2003). The goals emphasize drawing on children's and communities' funds of knowledge (Gutstein, 2006; Moll, Amanti, Neff, & Gonzalez, 1992) and the need to develop future mathematics teachers' identity, agency, and activism.

In light of these sociopolitical goals, mathematics methods teachers are now charged with additional instructional responsibilities. However, we feel these responsibilities are manageable and can be assumed at least in part through modifications or enhancements to preexisting activities in the mathematics methods course. In this chapter, we provide three examples of how to enhance an existing activity to include sociopolitical goals. Certainly, additional time in class will be needed for the suggested discourse and critique. However, it is our hope that these examples will spark MTEs to embrace the responsibility to prepare future mathematics teachers to instruct in ways that may reconstruct the mathematics education system via a revolution of values (Martin, 2015).

We provide three examples of activity enhancements, each based in our work in secondary mathematics methods courses. The first illustrates how engaging PTs in a generalization task taught through a student-centered approach can provoke interrogation of PTs' identity as a mathematics educator. The second example demonstrates how an activity intended to teach pedagogical techniques through content-specific instruction can be enhanced to develop a habit to question the unquestioned, such as the colorblindness of mathematics (Tate, 1994), by interrogating whose or which mathematics we have chosen to teach. And the third is an example in which PTs revise an earlier unit plan after the study of sociopolitical issues. Each example begins with a description of the foundational activity. Next, the enhancement is described, along with an analysis of how the change has potential to achieve a primary sociopolitical goal. Examples from PT work are included to illustrate PTs' experiences with each activity.

COMPARING NEW MATHEMATICAL EXPERIENCES WITH OLD: REMAKING ONE'S IDENTITY AS A TEACHER

Identity refers to one's sense of self and defines the complex relationship between how we perceive ourselves and what actions we take in situations (Gutiérrez, 2013a; Philipp, 2007). Teachers tend to alter new ideas to fit with their current identity as a teacher (sometimes in unproductive ways) or reject new ideas that do not align (e.g., Battey & Franke, 2008). Consequently, teachers' identities influence how they instruct and what they learn throughout their career. Hence, MTEs must support the development and critical analysis of the emerging mathematics teacher identity (Goal 5) of their PTs.

In this example, LaRochelle addresses PTs' beliefs, which manifest themselves through one's identity (Philipp, 2007), about empowering students. Ernest (2002) distinguished three ways of empowering students: mathematically, socially, and epistemologically. Mathematical empowerment relates to obtaining the necessary skills and knowledge to solve mathematical problems. Many believe this form of empowerment is the only form mathematics teachers should support in their students (Gutiérrez, 2002). Our sociopolitical goals lead us to argue, however, for the equal importance of social empowerment, which relates to feeling capable of using mathematics to improve local and global communities, as well as epistemological empowerment, which relates to growth in confidence through "a personal sense of power over the creation and validation of knowledge" (Ernest, 2002, p. 2). In this example, LaRochelle creates an opportunity for PTs to explicitly reflect on and critique beliefs about empowering students, thus supporting PTs to develop their mathematics teacher identities.

The Activity

Many PTs enter credential programs having experienced mathematics instruction comprised of lectures and procedural practice (Hiebert et al., 2005), which significantly influences PTs' pedagogical beliefs (Handal, 2003). To challenge such beliefs, MTEs often engage PTs in mathematics lessons taught using alternative, research-based pedagogies. LaRochelle creates such an opportunity by engaging PTs in "The Border Problem" (Boaler & Humphreys, 2005).

LaRochelle's instruction focuses on structure and generalization of concepts rather than procedures. To begin, he asks PTs to find the number of tiles in the border of a 10×10 grid (Figure 13.1), using any method to count the number of tiles.

LaRochelle then facilitates a discussion around PTs' methods of counting, emphasizing connections between the visual, verbal, and symbolic representations within each strategy, highlighting the underlying mathematical structures. In particular, he introduces the idea of representing calculations without performing the calculations. For example, a PT might say, "I added the top and bottom rows which were 10 units each, and then since I already added the corners, the left and right sides only had 8 units each, so I added 16 to 20 and got 36." This explanation would be represented with $10 + 10 + 8 + 8$, rather than simply writing 36. This representation of the sum highlights the structure $n + n + (n - 2) + (n - 2)$. Afterward, LaRochelle leads a discussion reflecting on his pedagogical moves and what

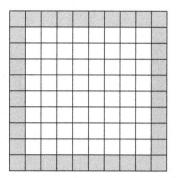

Task: Find the number of tiles that make up the border of this 10 × 10 grid.

Figure 13.1   The border problem.

value each may have. For homework, PTs interview a student using the Border Problem and write a reflection about the experience.

The Enhancement

LaRochelle noted that PTs' reflections about their interview have included disappointment that their student was only able to think of one method for counting the border squares. PTs created several methods for counting the number of squares in class, likely contributing to a view that students would generate multiple counting strategies on their own. Also, many PTs expressed remorse for not "getting to" the symbolic rule with their students. LaRochelle recognized an opportunity to critique PTs' pedagogical values and beliefs, foundational elements of identity. To do so, he would press them to question *Why are multiple solution routes valuable? Why is the symbolic rule important? What else may be important?* In particular, he would want PTs to consider ways past experiences shaped their present values and how mathematics teachers might utilize mathematics instruction to empower future students.

LaRochelle's first step in a planned future activity enhancement is to provoke PTs to identify their current pedagogical beliefs and values. Before engaging PTs in the Border Problem, LaRochelle would ask them to plan a lesson with generalization as a goal (see Figure 13.2). PTs draw upon their knowledge, beliefs, and values as they create such lessons; this first activity will expose present mathematics teacher identities. These lesson plans are set aside for a future discussion.

After discussing the pedagogical moves of the Border Problem, LaRochelle would have PTs reconsider how students may experience the

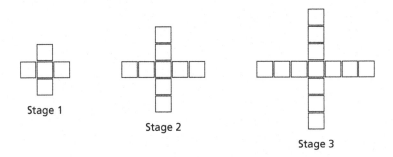

Imagine a lesson about *generalization* using the pattern above. You have two goals:

1. Help students see the underlying structure of this pattern.
2. Describe the pattern with algebraic symbols.

Task: In roughly 4–5 brief steps, create a general outline for how such a lesson might proceed.

Figure 13.2 Reflection question prior to border problem.

implementation of the first draft of their generalization lesson plan. Considering hypothetical students' experiences creates a space for PTs to reflect on two interrelated ideas. First, PTs consider their experiences as students in relation to their values and beliefs as teachers and consequently their nascent mathematics teacher identity. For example, PTs may claim their lesson teaches the necessary material in a shorter period of time, which opens doors for critiquing what counts as the "necessary" material and what larger systemic structures have forced teachers to move quickly through material (e.g., the testing regime). Second, because the Border Problem activity provides many opportunities for epistemological empowerment, PTs may reflect on student empowerment.

Some possible questions to promote PT reflection might include *What differences do you see between your lessons and the lesson we just completed? How do you think students will feel going through either lesson? How might students' sense of themselves as mathematicians be different?* and *What might students gain from either lesson?* By reflecting on differences in students' affect, students' identity, and other potential impacts of each lesson, PTs may come to recognize that mathematics lessons can empower students in ways beyond simply mastering mathematical skills. Through discussion, PTs may also begin to see how traditional mathematics pedagogy may actually hinder students' feelings of confidence and senses of themselves as mathematicians. In this way the discussion may support PTs to critique and develop their pedagogical beliefs and values and thus shape their mathematics teacher identity.

LaRochelle hopes PTs reflect on experiences they had in schools and how such experiences shaped their current identities. He encourages PTs to recognize different options for teaching mathematics, provoking them to analyze and critique their mathematics learning experiences. As a result, PTs may begin to develop mathematics teacher identities that do not perpetuate inequitable outcomes in mathematics education.

CRITIQUING THE EUROCENTRISM OF MATHEMATICS: WHOSE CONVENTIONS?

Lawler has regularly used a mathematical task from the Interactive Mathematics Program (IMP; Fendel, Resek, Alper, & Fraser, 2007) in his mathematics methods courses to demonstrate pedagogy that respects children's conceptions and processes of constructing logico-mathematical knowledge (Kamii, 1999; von Glasersfeld, 1995). He begins the activity by asking PTs *What is an angle?* Without exception, the (post-baccalaureate) students name static definitions for angle (Clements & Burns, 2000) including descriptions like "two rays sharing an endpoint" or "the corner of a polygon." He follows this discussion with an activity designed to achieve two goals: (a) to teach PTs that children need to develop a dynamic conception of angle and draw upon this to create representations for and measurements of angles; and (b) to reorient their mathematics instruction to center on the child and his or her ways of knowing (Kamii, 1999; Moll et al., 1992; Tate, 1994). Being charged with the sociopolitical goals that emerged at the Scholarly Inquiry and Practices Conference (Sanchez et al., 2015), in particular Goal 3: to develop a critical orientation to mathematics, Lawler enhanced this activity to reflect on the mathematical conventions involved in *angle*. He extended that conversation to question more broadly *Who decides on conventions? Who decides the mathematics to be taught, and why that mathematics? How might mathematics be different if an alternative or multiple mathematics were the standard?*

The Activity

Lawler uses this angle activity with his PTs by engaging them in each of the IMP sequence of tasks, spending less time doing the activity itself while increasing focus on the pedagogical moves. A first step in the unenhanced activity, designed to (re)introduce adolescents to angle, is for all students (or PTs) to stand facing front and then turn in complete and partial circles. Together, they name the angles by the amount of turn; for example a "half turn" is the equivalent of turning to face the opposite direction. In the second step, the instructor (Lawler when done with PTs) demonstrates a convention for representing (i.e., drawing) a specific angle that shows where the individual stands (vertex), the direction initially faced (ray), the direction faced after the turn (a second ray), and a mark to indicate the orientation of the turn. As a result, the static representation that emerges has an embodied and dynamic meaning.

To further develop the notion of angle measure, students recognize how everyday verbal descriptions such as a "half turn" can be indicated in the representation. While making connections between the dynamic actions, the symbolic representation, and verbal descriptions, the notion of fractional turns emerges and is readily understood. Finally, the standard unit of degree for angle measurement is compared to everyday language for amount of turn.

The IMP task "Pattern Block Investigations" advances students' notions of angle measure and makes meaning for the use of a protractor. First, Lawler instructs students, whether adolescents or PTs, to "explore different ways to arrange several pattern blocks so they come together at a single point." Next, students are prompted to imagine standing at this point and looking down one edge of one pattern block. If they turned to look down an adjacent edge of the same block, they would turn the amount of the angle defined by the two edges of that pattern block. With that orientation, students are instructed to draw upon the convention that a complete turn is named 360° to determine the degree measurement for each angle of the six different pattern blocks.

Students collaborate to determine the angles within each block. For example, by recognizing that six green (equilateral) triangles must be tiled together to make one complete turn and that a complete turn is 360°, each of the six congruent angles that come together at the point must be 1/6 of 360°, or 60°.

After computing all angle measurements, the final step is to learn how the protractor can be used to yield these angle measures. Lawler suggests that students select one angle of one block and replicate it on paper, then extend the edges as rays so they can be seen with the protractor. PTs' experiences with this activity, rethinking angle as an amount of turn and developing methods to determine the angles on pattern blocks without the use of the protractor, opens their eyes to pedagogy that allows mathematics to be (re)invented.

The Enhancement

Lawler extended the activity with an aim to provoke PTs to develop a critical orientation toward mathematics. He followed with the question *Who got to decide that we use and teach the convention that a complete turn should be 360°?* (Bishop, 1990). PTs responded with recollections of the Babylonians and that angles can also be measured in radians. PTs also noted that each measurement system was likely best in different contexts. The following topics did not spontaneously arise: *Who invented radians? Why was 360° used? Why did 360° become a standard?* and *Who got to decide 360° is what we teach in schools?* That is to say, in this context it seems Lawler's PTs did not have a *critical* orientation to mathematics.

After that initial question, Lawler presented information to provoke curiosity about possible origins of the division of the circle into 360 parts, including theories about early Babylonian, Indian, and Greek use. Greek (European) practices of angle measurement seem to have the strongest influence on modern-day mathematics, yet the ideas first emerged in ancient Babylonian (Arab) and Indian societies. This information emphasizes that mathematics is a history of human activity. Lawler also described a variety of additional angle measure systems, including the Furman, Binary degree, Angular Mil, and Gradian, recognizing that degree is a particular system selected to be taught in schools.

This first part of his angle measure activity enhancement strove to remind PTs the mathematics they teach is a particular mathematics, chosen from many (Powell & Frankenstein, 1997). Lawler built on this to consider how a particular mathematics becomes what is taught in schools (Joseph, 1987), and who benefits (Martin, 2003, 2015). *Which mathematics has been selected to make up the content and practice standards of our curriculum? To whose benefit?* To conclude, PTs discussed options they have as teachers to work in ways that do not reinforce the status quo. Many PTs indicated intentions to weave the history and humanity of mathematics into their instruction.

After these discussions, Lawler asked PTs to journal in response to two questions:

1. What are your immediate thoughts? and
2. Gutstein (2006) suggested that teaching mathematics well is complex, building on students' and communities' knowledge while simultaneously supporting the development of their mathematical competencies and critical awareness. How do you see his advice having some role in your career?

Responses included connections to prior aspects of the activity, but few made connections to the notion that normalized practices of mathematics education perpetuate the dominant culture and its inequities.

In response to the first question, roughly half the students recorded ideas about why degree is used as part of the school curriculum. "I figure we use 360° as one complete turn instead of the other systems in school because it has been so widely used throughout history." And similarly, with an economic rationale, "Maybe we learn about 360° as a way of reasoning because it is what's used most in the future job market." One student preferred not to think about such things, because "the answer is open to debate" and "I could never understand why someone chooses a standard convention and how we get everyone to follow it." These comments reflect little if any move from uncritical assumptions about mathematics education, preferring to not interrogate social or political dimensions of mathematics education. Other students were more critical. One student was "curious now to why/who chooses what's important to learn in school." Similarly, another "wonder[ed] what counts as classical knowledge," and a third recognized that "students receive only a small subset of possible knowledge and ways of knowing."

Students' responses to the second question were less varied. Almost each response connected Gutstein's (2006) message to drawing upon students' lives and ways of knowing. One student wrote, "He emphasized the importance of using students' lived realities and prior knowledge as assets in their learning. It tempers the use of classical knowledge as the only way of learning." Another student responded feeling challenged: "That concept of building on the community knowledge implies an acceptance of the community knowledge as truth. To then try and promote students who regard that same community critically could seem like a contradiction." This PT, and maybe others in the class, reflected some notion of wrestling with seeming contradictions and tensions in the identity they were developing as mathematics teachers (Gutiérrez, 2015).

The responses overall from the PTs did not fully express Goal 3, to develop a critical orientation to mathematics. Although PTs did recognize that *degree* was a particular choice for convention, a critical interrogation of the Eurocentric history and pathway for this convention to be the one taught was unconsidered. Further, teaching specific mathematical content for the benefit of serving the market or economy was identified but similarly undisturbed.

The Eurocentric mathematics of U.S. schooling is in many instances a mathematics borrowed from other cultures (Joseph, 1987), erasing the history of the human development of mathematics. This positioning is likely to contribute to a view for many people that mathematics is removed from their everyday lives and divorced from their cultures and histories. Lawler hopes that PTs begin to recognize that the mathematics we teach is not only the mathematics of the dominant society, but it also reflects ways of thinking, behaving, and reasoning that are both deeply rooted in and perpetuate that particular culture (Bishop, 1990; Lawler, 2012).

REVISING THE CURRICULUM TO FIT STUDENTS INSTEAD OF REVISING STUDENTS TO FIT THE CURRICULUM

Sometimes providing PTs an authentic experience becomes a chicken-and-egg problem. MTEs wonder how PTs can understand pedagogical theory without first having practical teaching experience, while at the same time realizing it is difficult to teach without a foundation of methods upon which to base curricular designs (Ball & Forzani, 2009). One way Thompson brings the two together is to revise unit plans previously written by PTs. The revision is based on new information about (in)equity in schooling and the large population of students who continue to be victimized by the systemic nature of those inequities (Benbow & Stanley, 1996). This activity enhancement primarily serves Goal 2, for PTs to become more aware of and draw on knowledge of the context from their students' perspectives, although the enhancement overlaps into Goals 4 and 5 as well.

The Activity

At Thompson's institution, secondary mathematics PTs take two mathematics methods courses, plus a third methods course that focuses on content area reading, social emotional learning, and ELL instruction. One of the major assignments in the first course is to design a unit plan that maps to local standards. This first experience writing curriculum can be challenging, as it is difficult to write without a specific set of students in mind. To mitigate this challenge, students are not overwhelmed with every aspect of excellent lesson plans. Instead, they are scored for their ability to map activities to standards; to incorporate pedagogical practices including cooperative learning, technology use, and differentiated instruction; and to clearly articulate planned instruction using a provided lesson-plan template.

In the third methods course, PTs learn about English learners, underserved populations of students, and children who do not seem to meet the expectations of mainstream public schooling. These experiences often result in a self-perception of failure (Ellis, 2008). PTs learn best practices for reaching these learners and examine authentic experiences of students who struggle along with the kinds of support teachers can provide. In addition to readings and discussion, PTs practice empathy by engaging in several activities that place them in the role of a subaltern, such as a student in a setting where they do not understand the dominant discourses. For example, Thompson teaches one lesson entirely in Mandarin. She explains in advance that because she will model using multiple redundant inputs and ask for multiple outputs from them, they will learn *something*, and that as the teacher, she will be able to measure their learning.

The Enhancement

Halfway through the third course, PTs revise three lessons from their previous unit plan, explicitly for English learners, social emotional learning, and Universal Design (Hitchcock, Meyer, Rose, & Jackson, 2002). Looking back at one's own ideas from the past, reflecting, and then making changes helps PTs develop a critical understanding of the nonhomogeneous nature of the children in a classroom, the push for homogeneity through schooling and standardized testing, and PTs' identity as a mediator between the two (National Council of Supervisors of Mathematics & TODOS, 2016). Students receive feedback on these revised lessons with a particular focus on

- Factors that influence teaching and learning, including development (physical, social, emotional, cognitive, linguistic), prior experience and knowledge, talents, economic circumstances, and community diversity
- How linguistic and cultural diversity shapes learning and communication
- Strategies for differentiating materials, pace, levels of complexity, and language
- Theories of learning and human development as they relate to the cultural, linguistic, cognitive, physical, and social and emotional student differences
- Serving all students and their families with equity and honor and advocating on their behalf, ensuring the learning and well-being of each child in the classroom

PTs' products and reactions from this extended assignment have been mixed. One PT claimed his lessons were already written for diverse learners and thus needed no improvement. Another disliked his original plans so much he asked if he could throw out the original lesson entirely and start over. Thompson declined his request because she intends that PTs reflect on curriculum and make decisions on *how to alter* it to improve the learning outcomes for all students by drawing on broader understandings of the student and family contexts of their classroom.

PTs' experiences, as represented in their revised plans, suggest that they moved toward the sociopolitical goal of this enhancement. Two examples of changes PTs made illustrated that they (a) worked to avoid humiliation of students, (b) acknowledged different ways for students to show their learning, and (c) provided students with an outlet for alternative models of thinking. For example, several PTs included revisions that eliminated "cold calling" and instead provided students with advance notice that they would be called on for a specific prompt. These revisions were made to give students who struggle (with reading, language, or anxiety) time to prepare a response and participate in discussions instead of facing humiliation and hesitation.

A number of PTs changed their assessments to performance assessments, allowing students to choose to explain a mathematical concept through writing, a speech, or a poster made with a partner. One PT made extensive revisions, entitling her new lesson "*Ms. Smith, can you explain it different?*" The new lesson urged students to come up with alternative ways to explain the classification of geometric shapes being studied. She made the lesson into a game where the winning team came up with the most creative alternative explanation, as voted on by the class. In both examples, PTs recognized the need for students to use communication styles in class that promote dignity and linguistic and cultural diversity. Therefore, the revision activity maps well to learning Goal 2, because allowing students to choose how and when they participate in class activities recognizes a variety of differences: cultural, linguistic, and social-emotional.

Revising self-authored curriculum with an aim to acknowledge each learner's experiences and perspectives and for each learner to recognize him- or herself in the lesson is an activity that allows PTs not only to develop such a lesson, but also to value children's and communities' funds of knowledge. This activity for PTs develops a sort of pedagogical empathy for students who may be different from themselves, by being able to recognize themselves in the potential classroom experiences of their future students. Overtly altering classroom activities to encourage different forms of discourse, provide a learning environment of equity and honor, and promote cultural and linguistic diversity are examples of drawing on the context in which the PT works. These strategies can help PTs to think about their role as an educator and their commitment to educate every student who arrives in their classroom.

PREPARING MATHEMATICS TEACHERS IS A SOCIOPOLITICAL ACTIVITY

Just as teaching mathematics is a cognitive, social, *and* political activity (Gutiérrez, 2013a), so is the teaching of future mathematics teachers. In addition to content knowledge for teaching mathematics, there is a need to understand the social context for teaching (both inside and outside the classroom) and how to maneuver in that space, a political knowledge. A mathematics teacher must develop an understanding of the ways in which the teaching of mathematics and mathematics itself has the potential to perpetuate current societal inequities (Martin, 2015).

In this chapter, we described how as MTEs we enhanced activities of our methods courses beyond cognitive and social goals in order to develop political *conocimiento* (Gutiérrez, 2013b) in our PTs, and we reported results of our initial efforts. It is our hope in sharing our ideas for activity enhancement that fellow MTEs will be emboldened to begin work toward sociopolitical goals in their methods courses, whether by taking up our specific examples or gaining insight on how to modify their own activities. Our PTs, future teachers of mathematics, are those who shape the school-based mathematical experiences of our emerging citizens. As MTEs, we must prepare PTs "to creatively resist a definition of the profession that unnecessarily limits the relationship between mathematics and historically underserved and/or marginalized youth" (Gutiérrez, 2013b, p. 14). Working toward sociopolitical goals in our mathematics methods courses is an important step toward Gutiérrez's appeal.

REFERENCES

Ball, D. L., & Forzani, F. M. (2009). The work of teaching and the challenge for teacher education. *Journal of Teacher Education, 60*(5), 497–511.

Battey, D., & Franke, M. L. (2008). Transforming identities: Understanding teachers across professional development and classroom practice. *Teacher Education Quarterly, 35*(3), 127–149.

Benbow, C. P., & Stanley, J. C. (1996). Inequity in equity: How "equity" can lead to inequity for high-potential students. *Psychology, Public Policy, and Law, 2*(2), 249–292.

Bishop, A. (1990). Western mathematics: The secret weapon of cultural imperialism. *Race & Class, 32*(2), 51–65.

Boaler, J., & Humphreys, C. (2005). *Connecting mathematical ideas: Middle school video cases to support teaching and learning.* Portsmouth, NH: Heinemann Educational Books.

Bonilla-Silva, E. (2003). *Racism without racists: Color-blind racism and the persistence of racial inequality in the United States.* Lanham, MD: Rowman and Littlefield.

Clements, D. H., & Burns, B. A. (2000). Students' development of strategies for turn and angle measure. *Educational Studies in Mathematics, 41*(1), 31–45.

Ellis, M. (2008). Leaving no child behind yet allowing none too far ahead: Ensuring (in)equity in mathematics education through the science of measurement and instruction. *The Teachers College Record, 110*(6), 1330–1356.

Ernest, P. (2002). Empowerment in mathematics education. *Philosophy of Mathematics Education Journal, 15*(1), 1–16.

Fendel, D., Resek, D., Alper, L., & Fraser, S. (2007). *Interactive mathematics program: Year 1.* Berkeley, CA: Key Curriculum Press.

Foucault, M. (1991). *Discipline and punish: The birth of a prison.* London, England: Penguin.

Freire, P. (1970). *Pedagogy of the oppressed.* New York, NY: Herder and Herder.

Gutiérrez, R. (2002). Enabling the practice of mathematics teachers in context: Towards a new equity research agenda. *Mathematical Thinking and Learning, 4*(2 & 3), 145–187.

Gutiérrez, R. (2008). A "gap gazing" fetish in mathematics education? Problematizing research on the achievement gap. *Journal for Research in Mathematics Education, 39*(4), 357–364.

Gutiérrez, R. (2013a). The sociopolitical turn in mathematics education. *Journal for Research in Mathematics Education, 44*(1), 37–68. [first published online 2010]

Gutiérrez, R. (2013b). Why (urban) mathematics teachers need political knowledge. *Journal of Urban Mathematics Education, 6*(2), 7–19.

Gutiérrez, R. (2015). Nesting in nepantla. In N. M. Joseph, C. Haynes, & F. Cobb (Eds.), *Interrogating whiteness and relinquishing power* (pp. 253–281). New York, NY: Peter Lang.

Gutstein, E. (2006). *Reading and writing the world with mathematics: Toward a pedagogy for social justice.* New York, NY: Routledge.

Handal, B. (2003). Teachers' mathematical beliefs: A review. *The Mathematics Educator, 13*(2), 47–57.

Hiebert, J., Stigler, J. W., Jacobs, J. K., Givvin, K. B., Garnier, H., Smith, M., ... Gallimore, R. (2005). Mathematics teaching in the United States today (and tomorrow): Results from the TIMSS 1999 video study. *Educational Evaluation and Policy Analysis, 27*(2), 111–132.

Hitchcock, C., Meyer, A., Rose, D., & Jackson, R. (2002). Providing new access to the general curriculum: Universal design for learning. *Teaching Exceptional Children, 35*(2), 8–17.

Joseph, G. G. (1987). Foundations of Eurocentrism in mathematics. *Race and Class, 28*(3), 13–28.

Kamii, C. (1999). *Young children reinvent arithmetic: Implications of Piaget's theory.* New York, NY: Teachers College Press.

Lawler, B. R. (2012). The fabrication of knowledge in mathematics education: A postmodern ethic toward social justice. In A. Cotton (Ed.), *Towards an education for social justice: Ethics applied to education* (pp. 163–190). Oxford, England: Peter Lang.

Martin, D. B. (2003). Hidden assumptions and unaddressed questions in *Mathematics for All* rhetoric. *The Mathematics Educator, 13*(2), 17–21.

Martin, D. B. (2015). The Collective Black and *Principles to Actions. Journal of Urban Mathematics Education, 8*(1), 17–23.

Moll, L. C., Amanti, C., Neff, D., & Gonzalez, N. (1992). Funds of knowledge for teaching: Using a qualitative approach to connect homes and classrooms. *Theory Into Practice, 31*(2), 132–141.

National Council of Supervisors of Mathematics & TODOS: Mathematics for All. (2016). *Mathematics education through the lens of social justice: Acknowledgement, actions, and accountability.* Retrieved from https://toma.memberclicks.net/assets/docs2016/2016Enews/3.pospaper16_wtodos_8pp.pdf

Philipp, R. A. (2007). Mathematics teachers' beliefs and affect. In F. K. Lester, Jr. (Ed.), *Second handbook of research on mathematics teaching and learning* (pp. 257–315). Charlotte, NC: Information Age.

Powell, A., & Frankenstein, M. (1997). *Ethnomathematics: Challenging the Eurocentrism in mathematics education.* Albany: State University of New York Press.

Sanchez, W., Kastberg, S., Tyminski, A., & Lischka, A. (2015). *Scholarly inquiry and practices (SIP) conference for mathematics education methods.* Atlanta, GA: National Science Foundation.

Tate, W. F. (1994). Race, retrenchment, and the reform of school mathematics. *The Phi Delta Kappan, 75*(6), 477–484.

von Glasersfeld, E. (1995). *Radical constructivism: A way of knowing and learning.* London, England: Routledge.

CHAPTER 14

SHIFTING FOCUS

Exploring the Evolution of the Learner Analysis

Jennifer Ward
University of South Florida

On my journey to address the sociopolitical perspective within a mathematics methods course (hereafter referred to as methods) for early childhood majors, I have revised activities across iterations of the course. My revisions resulted from shifts in my understanding of sociopolitical perspectives and how I can scaffold prospective mathematics teachers' (PTs) attention to focus on lived experiences of children inside and outside of school. I also wanted to explore how I facilitate PTs' attention to informal assessment methods, elements of access and equity in mathematics, and how they can construct learning experiences that leverage out-of-school experiences in their teaching.

SOCIOPOLITICAL PERSPECTIVE

Garii and Appova (2013) have explained that PTs do not often recognize mathematics as part of social and political processes; however, Gutiérrez

Building Support for Scholarly Practices in Mathematics Methods, pages 215–229
Copyright © 2018 by Information Age Publishing
215

(2002) has argued that much of school mathematics reflects the dominant culture and status quo of society. In working to develop connections between children's real-world experiences and mathematics content, teachers can support children in developing both an appreciation for and successful experiences with mathematics. Seeing how mathematics connects to larger political, economic, and societal structures can later support children in taking action against injustices (Frankenstein, 1983).

Gutiérrez (2013) has explained that not all children from the same culture have the same experiences. This has been echoed by Rogoff (2003), who situated children's experiences as being related to their learning within a community context. Events from children's lives outside of school need to be embedded within learning opportunities. However, Vomvoridi-Ivanović (2012) found that PTs seldom draw from children's lives in mathematics, especially because most curriculum materials do not reflect the diverse nature of classrooms. Moreover, PTs in primary-aged classrooms struggle to see connections between mathematics content they are teaching and mathematics outside the classroom space (Garii & Appova, 2013).

As I began working to incorporate this perspective into my own context, my goal was to have PTs focus on information with which I was familiar, mainly assessment scores and observations of children's thinking within the classroom. In particular, I asked PTs to look at beginning and end-of-year assessments, document interactions with children, and collect work samples. Upon reflection into the pivotal role the mathematics environment plays in learning, I asked PTs to reflect and document elements of the learning environment such as mathematical norms and the storage of manipulatives. The focus of these iterations rested largely on the pedagogical elements of the classroom, rather than supporting PTs in knowing their learners.

I wanted PTs to spend more time getting to know the actual children in their field placements. Looking back, this was something that I strived to do in my own classroom, so why not support PTs engagement in the same type of practice? At this point my goal became to have PTs look beyond simple labels such as test scores, demographic information, and categories that implied deficit thinking (such as "below level"). Upon reflection, I realized that I wanted the PTs in my course to deeply explore informal data they gathered while working with children as well as the experiences their children were having inside and outside the school setting with mathematics. These included not only what the child did while at home, but also within the community. Furthermore, I wanted PTs to begin to question how all children within their classroom were reflected in classroom tasks related to mathematics.

PROGRAM CONTEXT

The early childhood program at the University of South Florida supports PTs to work with children from birth to Grade 3. PTs who complete the program can apply for early childhood certification, which covers age 3 through Grade 3. PTs complete prerequisite courses in the College of Education focused on working with learners, using technology, background on content (including mathematics), and instructional design. Successful completion of these courses is required for admittance to any education program. These courses contain assignments to be completed in course-associated field experiences. Through these assignments, PTs apply pedagogies learned at the university in the field. PTs' field experience in the first semester is in a prekindergarten setting. Related courses focus around cognitive and creative experiences of young learners, child psychology, and emergent literacy. Their second semester field experience is in kindergarten; related courses highlight working with diverse populations, assessment, literacy instruction, and programs for young learners.

During the third semester of the program, PTs have a third field experience in a kindergarten, first-, or second-grade classroom. Methods courses related to mathematics, science, social studies, assessment, working with English to Speakers of Other Languages (ESOL) children, and management accompany these field experiences. At first, PTs are in the field 1 day per week but increase to all days each week with university-based course meetings suspended, a period referred to as full immersion.

Mathematics Methods Course Context

The methods course in the third semester is the only one PTs take in our program. Consequently, the course focuses on both content and pedagogy for early learners. These topics are discussed briefly in early field experience seminar meetings and through onsite coaching, as PTs begin to take over responsibilities for mathematics teaching in their placements. Explicit instruction focused on mathematics teaching is reserved for the methods course. Through course assignments, PTs explore state mathematics learning standards and examine learners in field experience through the learner analysis activity. Information from these sources is used to brainstorm lessons that address learning standards and draw from student backgrounds and interests. Finally, PTs construct, implement, and reflect on a lesson plan.

LEARNER ANALYSIS ACTIVITY

The learner analysis activity is part of all three methods courses (mathematics, science, and social studies). Although assignment descriptions overlap, unique requirements exist for each course. Overall, the purpose is to provide PTs with opportunities to unpack experiences in their field placement classrooms and prepare to plan for instruction. I came to believe there were components of the course that could better support PTs in providing equitable access to mathematics content. Examples of this included using children's outside experiences and merging them with mathematics content. PTs are asked to examine information such as children's files, systematic classroom observations, conversations with prior year teachers and their current collaborating teacher, and assessment information, both formal and informal, from current and previous years. The learner analysis activity also asks PTs to explore children's backgrounds, interests, and experiences through conversations and written narratives produced by the children. My work to revise the exploration of children's experiences and interests has undergone the most revision. My goal in revising the activity has been to support PTs to attend to and leverage children's voices through informal assessments, out-of-school experiences, and cultural backgrounds in mathematics teaching and learning. I further support the PTs to use these experiences and backgrounds to provide entry points for children into mathematics learning.

To craft my first iteration of the learner analysis activity, I drew from descriptions of the same activity in science and social studies courses and a description crafted by other faculty from the combined mathematics and science course. PTs are given the assignment description at the beginning of their field experience, approximately 2 weeks before formal classes begin at the university and 1 week before children arrive at school sites. Getting the assignment early means PTs have time to discuss ideas with cooperating teachers (CTs) prior to the children's first day of school. A description of the activity from the syllabus is provided (see Appendix A), further delineated by a bulleted list of ideas for the assignment (see Appendix B). Text in regular font represents the initial assignment. Text that has been struck through indicates items removed. Bolded text reflects changes made following my first use of the assignment. Italicized text represents further revisions to meet the goal of steering PTs towards the experiences children have outside of the school setting, as well as issues of access and equity in mathematics during their field placement.

Starting Off—Iteration 1

In the first iteration of the course, the goal I set was primarily for PTs to explore the prior learning and backgrounds of the children in mathematics.

The emphasis was largely on the use of student "data" that had been obtained from state and national testing, with limited attention to the cultural background or experiences of learners. I wanted PTs to review these data to plan for later instruction so they knew what content learners had been exposed to and mastered in earlier years. Furthermore, I wanted PTs to develop methods that support all learners in making connections to mathematics content. Knowing that PTs would rely on curriculum maps and textbook resources (Herbel-Eisenmann, 2007), I wanted to push them to search for other resources or to construct their own classroom lessons drawing from the experiences of children.

Assignment Outcomes

Although I believed PTs needed to attend to and leverage children's out-of-school experiences and cultural backgrounds, they relied on more formalized sources of evidence such as standardized assessments from the prior year and chapter tests from beginning-of-the-year units of study. PTs from earlier cohorts shared examples of learner analyses with my PTs. These examples did not draw insight about children's experiences from informal sources, such as teacher observations of children's interactions outside of school or information provided from meeting and conversing with families. As a result, most PTs relied heavily on children's prior assessment scores and brief observations or interactions with children in the classroom. However, one PT's work stood out as an example of getting to know the experiences outside of school and the interests of the children. She asked the children to complete an interest inventory and developed a narrative describing how the results could inform instruction. She said,

> I gathered data from having conversations with students in class and during lunch, as well as from work samples and two class journals that students can contribute to during Daily 5 Time: "Things We Like to Do" and "Our Favorites." I organized the data into a table using numbers in place of student names. Having an interest inventory for each of my students informs my instructional planning in that I can plan to engage children based on things that appeal to them individually and in group settings. For example, I can look at my table and see that a large portion of children have pizza as their favorite food, which tells me that I could incorporate pizza into my instruction about fractions or about recording students' favorite pizza toppings into a graph. Additionally, noting those individual differences gives me a way to engage students when I work with them one-on-one. For example, if I was working one-on-one with Student 12, I could choose to use LEGO characters in a story problem or bring them in as counter manipulatives because of his interest in LEGOs. (PT Assignment, September 2013)

PTs describing children's backgrounds largely used "favorites," reading levels, who was deemed "below level" in mathematics on the prior school year's report card, and those who receive support services from exceptional student education (ESE) personnel or ESOL teachers. Rarely was prior learning in earlier grades discussed.

Children's work samples were often student-completed textbook pages or entries from children's mathematics notebooks. PTs did not include child narratives. PTs reported limited discussion of classroom teaching practices or tools being used to support mathematics instruction. When discussion was provided, they suggested the cooperating teachers' (CTs) use of textbook pages as the only mode of instruction, with these pages being modeled on a document camera or torn out so that children could complete them independently or in groups. One PT included a description of the typical instruction provided in her field placement. She further included associated images of children's work but did not analyze them for proficiency with mathematics concepts.

PTs described using information about classroom teaching and the children to create lessons they would teach during the full-immersion period. The focus was on modifying lessons from the textbook series, highlighting content from the science and social studies course, or adding a piece of related literature. Some PTs focused on making small modifications to word problems from textbooks such as including of names of children from field experiences.

I felt frustrated with results of the learner analysis. Informal methods of assessment remained virtually undiscussed. Also, the mathematics teaching practices PTs described were not always consistent with practices encouraged by local, state, and national mathematics groups. Other instructors reported similar trends in the science course. The social studies instructor, however, reported more emphasis on informal observations, perhaps because no standardized assessment existed for them to refer to. I began to consider how I might support PTs to recognize that formalized assessments did not always paint an accurate picture of children's experiences.

Field Supports to Move Thinking

Although I was initially unsure why PTs were reliant on formalized assessment data, I gained insight during my visits as a field-based supervisor. During planning sessions at school sites, much of the discourse revolved around test data from prior years or prior chapter tests. Emphasis was on maintaining a certain instructional pace and timeline, regardless of children's backgrounds. This signified a lack of attention to the unique and diverse experiences that children were having outside of school. My experiences as an elementary classroom teacher had shown me the power of having conversations with children and looking at their work samples, but

I grappled with how I might convey this to my PTs without imposing my viewpoint upon them.

Because I worked with PTs in the methods course and the field, when I was at the school site, I coached them in order to model conversations with children about their out-of-school experiences with mathematics. These conversations centered on how children saw their families using mathematics in their homes and where they themselves felt they used mathematics outside of school. In addition, as I saw PTs working with children during their initial field visits, I would probe them to more closely examine classwork and classroom discourse. An example of this approach began when a PT in a first-grade classroom wondered how she might use her calendar time to gain insight into children's mathematical capabilities. The PT wanted to include events happening in the community, home, and school environments of children in her class as a springboard to review counting (e.g., how many days until/beyond) and addition and subtraction strategies (using the dates to determine how many days unit/beyond).

Small Shifts-Iteration 2

PTs' work products informed my next revision. In addition to pressing them to use informal assessment methods, I wanted PTs to attend to the overall mathematics environment in the classroom, including the availability of types of supporting materials, both textual and manipulatives. I also wanted PTs to examine how (if at all) pedagogies for instruction we discussed in the methods course, such as using open-ended problems, were used in practice.

As my understanding of a sociopolitical perspective evolved, I wanted to ensure that the assignment helped PTs become aware of and draw upon knowledge of children's experiences with mathematics in various settings and recognize that children's experiences may match or differ from their own. I included reference to children's home experiences in the assignment's extended description, directing PTs to reexamine prior discussions during fieldwork seminars and course instruction related to funds of knowledge. I hoped this would prompt PTs to talk with children and families about their mathematical experiences in school and how mathematics was used outside the classroom and in the home. I knew conversations with PTs and scaffolding around these goals could occur during my field visits, but there was a need to make these ideas more explicit to provoke class discussions.

Course Accompaniments

I added supporting class activities to provoke insights about different experiences with mathematics. PTs reflected on their own mathematics experiences and described them as a dirt road, paved road, highway, or yellow brick road. These descriptions symbolized varying experiences with mathematics from struggle to progressing through the content with ease, respectively. I asked PTs to consider classroom practices and tools that may have impacted their perspectives about mathematics. These questions helped them engage in dialogue about effective instructional practices. PTs voiced a passion for mathematics content they perceived to be connected to the real world and to their lives. Some recalled having common classroom experiences, sports they played, or locations in the community included in word problems, which motivated them to determine solutions. For me, this addition helped to highlight that not everyone has the same experiences in mathematics over the course of their educational experiences or even in one year. My understanding of a sociopolitical perspective began to include the idea that children's out-of-school experiences were critical in helping make connections between mathematics and their lives, enabling them to recognize mathematical experiences beyond textbooks or assessments. I wanted my PTs to see that they had felt empowered when these connections were made by their teachers as well.

I also focused our discussion of classroom practices on areas where there was disagreement among PTs, such as the use of timed tests and solving problems by applying algorithms. PTs who were stronger in mathematics reported enjoying timed tests and applying algorithms, whereas others did not. I probed with questions that helped to problematize these experiences such as asking how PTs felt during these classroom experiences. For example, I asked them to think about the reaction they had when the test was placed in front of them, the feeling when the timer started and stopped, and what strategies they typically employed to complete such assessments. I also shared my own experiences using alternative strategies while doing mathematics in high school and how, although my strategy was sound and the answer correct, my methods did not match my mathematics teacher's and my answer was therefore marked incorrect. Anecdotes and narratives from classmates appeared to support PTs seeing such practices from multiple perspectives. They appreciated that practices they enjoyed may have been uncomfortable and anxiety ridden for others. This discussion provoked PTs to describe the importance of knowing more about the instructional tasks that children in the class connected with and why.

During initial course meetings we discussed the norms of our classroom space by first examining those of PTs' field sites and how they worked for children in the classroom. We then developed norms for our work together, such as having a willingness to explore and challenge varying strategies and ideas presented by others without directly challenging the person. Additionally, for whole-class lessons that emphasized PTs' understanding of content, I used more problem-solving situations that mirrored the ways in which PTs indicated they used or could use mathematics outside of the school setting. This included ideas such as cooking, paying for hotel rooms to an upcoming conference they were attending, and exploring fractional concepts with parking spaces on campus.

A second activity involved PTs bringing a mathematics lesson to class prior to their immersion in field experience. A requirement was that the lesson should be aligned to the content they were planning to address in their field experience during the upcoming month. PTs were asked to explore the problem posed in the lesson and critically examine whether children in their classroom would have the background and experiences to be successful with the content.

Field Support to Launch the Assignment

To support PTs' use of informal assessment in the learner analysis, I drew attention to these approaches during field visits. I worked with PTs to explore what they saw children producing. I encouraged them to take photos we could discuss during weekly field meetings, during which I asked them to examine these work samples for children's proficiency with concepts. The PTs were able to use these informal assessment pieces during the learner analysis. For example, one PT monitored children as they worked to solve word problems using concepts from prior learning. She shared one child's mathematics journal that included approaches to indirect measurement comparisons. In her narrative, she wrote:

> This student had no problem recognizing the given facts in the word problem and the question. However, she did not know how to begin solving the problem. When I asked her if she had to solve for a number she said no and said she did not want to use a ruler to solve the problem. I told her there are other ways to measure other than using a ruler and asked her how she would be able to tell who was taller than Bobby if the people were standing in front of her. After I asked that question she immediately started drawing lines of different lengths and labeling them according to what the word problem said and she figure out that Samantha was the only person taller than Bobby. Other students used different strategies to solve the same problem. (PT Assignment, September 2015)

She began to reflect on strategies children could use to solve word problems and the underlying issue of children's conceptualization of facts, extra information, and main idea of the problem.

Assignment Outcomes

After these revisions were made, many PTs shared information related to children's interests in and outside of school. Several created interest inventories, asking children to identify their favorite subject in school and share

TABLE 14.1 PT Sample Inventory Data, September 2015	
Student	Math Information Obtained
1	• Needs visuals to learn • Heavy assistance needed
2	• Confident in math • Would like computers and groups to help him learn • Understands greater/lesser • Can use manipulatives to solve simple problems • Needed assistance with money

what they liked and did not like about mathematics. One such PT included a table (see Table 14.1).

This PT shares children's views of mathematics and interpretations from classroom observations. She later used this evidence to describe ways she could engage children in mathematics learning, such as with technology tools or small-group experiences. These ideas served as a springboard as she designed her lesson plans, focusing on using the informal data gathered. Many PTs went beyond formal assessments to describe their observations of children in the mathematics classroom. They provided samples of children's problem solving and included questions they had used to facilitate student thinking about concepts being taught. Some PTs scripted children's responses to questions during calendar time to capture counting strategies, for example, but still lacked emphasis on children's out-of-school experiences.

Connections to children's home experiences were still scant. Only one PT included information related to family structure and home language. Many PTs interpreted the assignment to mean they should include children's ethnicities and make tables or graphs of this information. Some PTs asked children how they used mathematics at home but then stated that children were unsure of the connections between the content they were learning and real-world situations. Consistent with findings of Garii and Appova (2013), as PTs began to discuss lesson ideas, many identified using children's names and cultural holidays in word problems to engage learners.

After feedback on the assignment and in-the-field coaching, PTs included information about children's out-of-school experiences. For example, one PT reported that a child helped prepare family dinners each Sunday following church. The PT reflected on this particular child and the information gained from their conversations together:

> When I learned about <his> family dinners, that made such a difference in my lesson planning. We talked about being able to cook and bake and I suddenly realized wow, there is a lot of background on measuring here. If <he> is using these measuring tools every week with his family, <he> must know how

to measure. I wonder why he missed those items on the beginning of the year test then? Making sure that when I focus on measurement during the full immersion I bring in some of the tools and experiences <he> has cooking. I wonder if others in the class do this too? [sic] (PT assignment revision reflection, October 16, 2015)

This PT reflected on the valuable information she gained from her insight into this child's home experience. She later used this information to ask the child questions on the test orally and one-on-one. The PT discovered that much of his struggle on the initial assessment was not due to lacking understanding of the content but rather due to their placement at the end of the test (he indicated that he felt like he was running out of time and picked answers quickly) and the English terms being used on the assessment. As a result, when it came time for more preassessment in the classroom and lesson planning, the PT worked with her CT to use a cooking context, some measuring devices (e.g., measuring cups, spoons), literature related to families cooking together, and pictures of the measuring devices in both English and his native language (Creole).

Moving Forward

My revision of the learner analysis has helped me to more closely approach my goal that PTs gather evidence of children's experiences with mathematics both inside and outside of the school setting. Through my revisions and learning more about the sociopolitical perspective, I can see how my initial iterations, which focused on assessments, resulted in inaccurate pictures of children. This focus did not draw on and reflect the experiences of the child inside and outside of school, which was a goal of having PTs engage in the assignment.

I have learned that by shifting the view of PTs to include information gathered informally and the experiences of children outside the school setting, more in-depth connections to the lives of children can be made when planning. Supporting PTs in the field as they engage in these conversations has been a powerful tool to help use the information gathered to inform instruction. Furthermore, when looking at these elements, PTs can begin to explore if children feel represented by the lessons and methods being presented in the classroom. When this is not the case, PTs have a clear rationale, the voices and experiences of the children, as to why instructional decisions are being made.

As I reflect on my own understanding of the sociopolitical perspective and continue to learn and grow as a methods instructor, this activity is one that I envision continually changing.

I want to ensure I am modeling asset-based discourses rather than those that paint populations in a negative light. The next iteration will include the removal of the terms *below level, on level, above level,* as this language allows PTs to view children in the classroom through a deficit perspective from the onset of the assignment. I want PTs to begin to have an active voice in reframing deficit perspectives they encounter in schools in relation to children's capabilities in mathematics.

The PTs I work with are beginning to question practices in the classroom and district. My own learning is driving me to question former practices as a classroom teacher. Because of this, I will continue to revise the assignment to have PTs critically examine and question the practices they are seeing and learning. Do these approaches work and benefit all children? Do these approaches allow all children to have voice in and access to the mathematics being presented?

With the diverse nature of the schools in which PTs work, I need to make sure I am increasingly aware of the home and community connections PTs could make. This involves me exploring the community more in depth through community walks and attending community events. By demonstrating my engagement in these tasks, I can provide PTs with an explicit example of how to begin to look at mathematics beyond test scores and observations. Having PTs learn about children during course and field experiences may help them maintain these practices in induction.

APPENDIX A
Learner Analysis Syllabus Description

You will gather information on every student in your class and use these data to develop both an individual and holistic sense for "where children are now" on their journey toward mastery of math concepts. In addition to gathering data on "where children are now" through previous assessment data, you will also consider "where children need to go," as a part of the overall learner analysis. To accomplish this, you will consult the standards and school district curriculum guide for mathematics (at your grade level) and speak with your collaborating teacher about his/her forecast for student needs. You will talk with your collaborating teacher about his/her plans for the four-week period of your full-time internship. This will help you develop a sense of "where children need to go."

Additionally, you will need to critically examine the mathematics environment within your internship classroom. What materials are present? What materials may be needed? Describe the strategies/pedagogies used to instruct mathematics. Are there modifications to the environment or pedagogies which may be made based on course readings? How are children assessed on mathematics content?

Lastly, as a part of the learner analysis, you will describe some specific elements related to "how children will get there." These elements include activity sequencing; children's out-of-school experiences and cultural, linguistic, and family backgrounds; and special learning needs. You will discuss how these factors will inform your instruction. A rubric will be provided for the learner analysis.

APPENDIX B
Learner Analysis Bulleted List of Ideas

In essence, you will be exploring the strengths and needs of learners within your classroom through the following means:

- *Exploring the community context in which your children live*
  - *Where mathematics is located and used within the community*
  - *Making observations vs. inferences about the community in which children live.*
  - *What types of learning are available for children/ how do children learn within the community? Who is responsible for this learning?*
- *Exploration of classroom operations in mathematics*
  - *How is mathematics instructed within the classroom? What routines are used?*

- *Who is seen as having high levels of knowledge in mathematics? How is this determined?*
- *Do children see themselves in the mathematics being taught (e.g., are the problems they are given related to their lives and experiences?)*
- *How is success determined with the classroom?*

- Interest inventories
- Personal communication
- Child-written narratives *or autobiographies*
- **Exploration of student funds of knowledge** (Moll, Amanti, Neff, & Gonzalez, 1992) (what they bring from their home or outside school).
  - *With which types of out-of-school activities do children and their families engage?*
  - *What values are important within the community or context of the children?*
  - *What types of mathematics is done or valued within the student's home and community?*
- Background information on children: (*Family composition*, language, interests, SES, gender, race, etc.)
- Sociocultural norms of your children
  - Use of communication means, dispositions, prior experiences with content, real-world connections to content present in their communities
- Assessment Information
  - Formal Assessments: SAT-10 Scores, End-of-Year or Beginning-of-Year Tests, Chapter Tests, District Formative Testing
  - Informal Assessments: Anecdotal notes, observations, journal writing
  - ~~Student reading levels (above level, on level, below level)~~
- Methods of engagement and closure that are beneficial to children in your placement

REFERENCES

Frankenstein, M. (1983). Critical mathematics education: An application of Paulo Freire's epistemology. *Journal of Education, 165*(4), 315–339.

Garii, B., & Appova, A. (2013). Crossing the great divide: Teacher candidates, mathematics, and social justice. *Teaching and Teacher Education, 34*, 198–213.

Gutiérrez, R. (2002). Enabling the practice of mathematics teachers in context: Towards a new equity research agenda. *Mathematical Thinking and Learning, 4*(2–3), 145–187.

Gutiérrez, R. (2013). The sociopolitical turn in mathematics education. *Journal for Research in Mathematics Education, 44*(1), 37–68.

Herbel-Eisenmann, B. (2007). From intended curriculum to written curriculum: Examining the "voice" of a mathematics textbook. *Journal for Research in Mathematics Education, 38*(4), 344–369.

Moll, L., Amanti, C., Neff, D., & Gonzalez, N. (1992). Funds of knowledge for teaching: Using a qualitative approach to connect homes and classrooms. *Theory Into Practice, 31*(2), 132–141.

Rogoff, B. (2003). *The cultural nature of human development.* New York, NY: Oxford University Press.

Vomvoridi-Ivanović, E. (2012). Using culture as a resource in mathematics: The case of four Mexican-American prospective teachers in a bilingual after-school program. *Journal of Mathematics Teacher Education, 15*, 53–66.

SECTION V

ACTIVITIES AND IMPLEMENTATIONS

CHAPTER 15

BRINGING MATHEMATICS METHODS INTO CLASSROOMS

Rajeev Virmani
Sonoma State University

Megan W. Taylor
Trellis Education

Chepina Rumsey
University of Northern Iowa

Co-authoring with: Tabatha Box, Elham Kazemi, Melinda Knapp, Sararose Lynch, Catherine Schwartz, Barbara Swartz, Tracy Weston, Dawn Woods

The Blue Ribbon Panel Report of the National Council for Accreditation of Teacher Education begins with a dire statement and call for action to teacher education programs across the nation:

> The education of teachers in the United States needs to be turned upside down. To prepare effective teachers for 21st century classrooms, teacher edu-

Building Support for Scholarly Practices in Mathematics Methods, pages 233–248
Copyright © 2018 by Information Age Publishing
All rights of reproduction in any form reserved.

cation must shift away from a norm that emphasizes academic preparation and coursework loosely linked to school-based experiences. Rather, it must move to programs that are fully grounded in clinical practice and interwoven with academic content and professional courses. (Blue Ribbon Panel on Clinical Preparation and Partnerships for Improved Student Learning, 2010, p. ii)

To address this charge, we suggest that mathematics teacher educators (MTEs) situate prospective teacher (PT) learning within the P–12 context to purposefully observe, plan, enact, and reflect on their experiences in ways that support the development of ambitious teaching (Kazemi, Franke, & Lampert, 2009). To help PTs develop the knowledge, skills, and professional dispositions to positively impact P–12 students, Grossman, Hammerness, and McDonald (2009) suggest that teacher preparation "move from discussing what one might do as a teacher to actually taking on the role of the teacher...while receiving feedback on their early efforts to enact a practice" (p. 283). Integrating mathematics methods courses in the P–12 setting affords teacher educators unparalleled opportunities to engage their PTs in authentic instructional activities—bounded tasks and participation structures with clear goals for both teacher and student learning (Lampert, Beasley, Ghousseini, Kazemi, & Franke, 2010)—while providing real-time feedback and debriefing on experiences. Central to these experiences are the opportunities to develop meaningful relationships with students and practicing teachers and learn from and with young people in the classroom.

This chapter explores possibilities about how MTEs can bring their mathematics methods courses into local schools by creating a community of practice in which PTs can learn ambitious mathematics teaching and develop a stance that learning together from and with students is central to high-quality teaching (Kazemi, Ghousseini, Cunard, & Turrou, 2016). We share three exemplars highlighting variations in the integration of methods courses into P–12 classrooms. By describing the exemplars, which are on a continuum of varying degrees of integration and authenticity, we hope to offer MTEs flexibility and options for integrating their own teacher preparation courses into local schools. The first exemplar highlights the integration of a teacher preparation course into an after-school program. The second exemplar describes a modified lesson study cycle, and the third exemplar details a completely integrated course in a public school setting.

LITERATURE REVIEW

Many researchers have noted the divide between teacher preparation coursework and the experiences teaching candidates have in the field (Grossman et al., 2009; Korthagen, Kessels, Koster, Lagerwerf, & Wubbels, 2001; Lampert & Graziani, 2009; McDonald, Kazemi, & Kavanagh, 2013; Weston &

Henderson, 2015) that leaves PTs with the burden of figuring out how to enact what they are learning in their teacher education programs in their clinical placement classrooms. In a now classic study, Borko et al. (1992) discussed how PTs struggled with enacting complex teaching practices and brought attention to the need for teacher education programs to strengthen PTs' pedagogical content knowledge by providing them with opportunities to practice and reflect on their teaching. More recent scholarship continues to challenge the field about how to create approaches in methods instruction that enable PTs to make sense of the various bodies of knowledge that are important in teaching (e.g., subject matter teaching, differentiated instruction, learning and using community funds of knowledge, supporting language development) (Cochran-Smith et al., 2016; Ensor, 2001; Turner et al., 2012). A practice-based teacher education curriculum uses "practice as a site of inquiry in order to center professional learning in practice" (Ball & Cohen, 1999, p. 19) and creates opportunities for PTs to examine and develop their praxis within a supportive community of practice.

Our work is grounded in carefully selected instructional activities that require PTs in university-based teacher preparation programs to develop knowledge and experience in enacting ambitious teaching practices. Instructional activities support PTs' learning of important pedagogical moves while maintaining flexibility in working with students by attending to the specific structures of how teachers and students interact with content and with one another (Lampert & Graziani, 2009). In addition, emerging research indicates that the promotion of ambitious teaching practices with PTs is facilitated by the use of rehearsals of practice within cycles of enactment and investigation (Kazemi et al., 2016; Lampert et al., 2013). The cycle of enactment and investigation draws upon the Grossman et al. (2009) "pedagogies of practice in professional education" (p. 2058) framework that includes representation, decomposition, and approximation of practice. Approximations such as rehearsals are considered simulations of different aspects of a profession that "may fall along a continuum, from less complete and authentic to more complete and authentic" (Grossman et al., 2009, p. 2078). McDonald and colleagues (2013) conceptualized the preparation of PTs by developing types of pedagogies for the cycle of enactment and investigation wherein PTs engage in an authentic and ambitious instructional activity. This cycle for learning to enact core practices (Core Practices Consortium, 2013) is strongly grounded in a situated perspective on learning that views learning as a "process that occurs over time in interaction with the particular settings in which and students with whom teachers learn to teach" (McDonald et al., 2013, p. 381).

The opportunity for PTs to learn alongside MTEs, practicing teachers, and students within a P–12 setting bridges the expectation of preparing teachers to teach mathematics in an intellectually and socially ambitious manner. As PTs engage with P–12 students with explicit concurrent feedback and reflection, the experience is authentic because PTs are asked to simultaneously evaluate and reflect on their teaching, as they teach.

> This cycle intends to offer guided assistance to candidates to learn particular practices by introducing them to the practices as they come to life in meaningful units of instruction, preparing them to actually enact those practices, requiring them to enact the practices with real students in real classrooms, and then returning to their enactment through analysis. (McDonald et al., 2013, p. 382)

When teacher preparation courses are integrated into local P–12 classrooms, PTs are given opportunities to respond in the moment to student ideas and conceptions about mathematics while simultaneously connecting their experience to what they are learning in their coursework as they engage in a shared experience with peers and MTEs.

CONCEPTUAL FRAMEWORK

In this section, we describe how the situative perspective shapes our conceptual framing for bringing PTs into classrooms. This perspective on learning asserts that people learn as they deepen their participation within a community of practice (Lave & Wenger, 1991). Situating learning within practice moves away from decontextualized talk of knowledge to understanding the specific knowledge, practices, and dispositions that are being developed within the structure of the community of practice. The collaborative relationships between PTs, MTEs, P–12 students, and practicing teachers within the cycle of enactment and investigation (Kazemi et al., 2016) allow authentic learning opportunities as the participants engage with core teaching practices in a P–12 classroom. As collaborative relationships are developed between the many participants, our aim is to position the MTEs and PTs in structures of interaction that are designed to generate knowledge and to make learning visible through representations of practice that are common to those who are more knowledgeable within the community (Greeno, 2006). This commitment to participation within a community of practice enables members to learn from and with each other as they work to generate meaning and develop shared resources (e.g., experiences, tools, language). The learning environment that the MTEs and practicing teachers design is critical for PTs as they develop knowledge, skills, and identities (Putnam & Borko, 2000). Within the pedagogies of the cycle of enactment and investigation, PTs develop professional knowledge of teaching in an authentic context when situated in the classroom.

When developing learning environments for PTs, using a situative perspective focuses MTEs on key understandings that can "result in learners increasing their capabilities for participation in ways that are valued" (Greeno, 2006, p. 80). By highlighting specific aspects of teaching, PTs learn to exercise professional judgment that prepares them for the in-the-moment decision-making that teaching requires (McDonald et al., 2013). Furthermore, situating PTs' learning in P–12 classrooms enables MTEs to make authentic connections to the sociopolitical and cognitive perspectives. For example, we have the opportunity to work on developing PTs' political knowledge through activities that foster solidarity and commitment among PTs and students (Gutiérrez, 2013), as well as providing authentic work on key pedagogical aspects of mathematics teaching that would empower PTs to understand connections between teaching and learning (Simon, 2008).

THREE EXEMPLARS OF METHODS INTEGRATION

We selected three exemplars to highlight the ways MTEs integrated mathematics teacher preparation courses into school classrooms. The activities used in each of the exemplars aimed to provide PTs with authentic experiences to develop their knowledge about eliciting, interpreting, and responding to student thinking. The core teaching practice of eliciting, interpreting, and responding to students is an essential skill for teaching mathematics and can be challenging for PTs to learn. The three exemplars are on a continuum where the time spent in the classroom increases and the nature of the enactments are increasingly complex. During the school-based methods courses, the MTEs and PTs engaged in at least one activity from the enactment and investigation learning cycle (McDonald et al., 2013). In some cases, only the enactment of the instructional activity took place in the classroom setting, whereas in other cases, the complete cycle including introducing, rehearsing, enacting, and reflecting upon the instructional activity took place in the schools (McDonald et al., 2013).

The first exemplar is exploratory, where the MTE began to build relationships with schools, teachers, and classrooms of students. This initial immersion consists of a university course held in the school a few times during a semester where PTs observe practicing teachers and interact with students using a particular course assignment. For example, PTs may meet with students one-on-one to learn about student thinking of a particular mathematics concept. The second exemplar in this chapter involves MTEs modeling lessons and creating opportunities for PTs to make sense of the core practices by engaging with students. In particular, the PTs and the MTEs collectively engage with students to learn about facilitating student discourse through eliciting and responding to student thinking. The third exemplar includes extended approximations as PTs work with students in methods courses that are fully integrated in P–12 classrooms. MTEs work closely with practicing teachers and school administrators to develop detailed plans for weekly integration. Within each exemplar, we describe how MTEs structured activities for PTs to learn about eliciting, interpreting, and responding to student thinking within a variety of school contexts. Our goal is to share our experiences with the field of mathematics education and to encourage other teacher educators to explore ways to bring mathematics methods into schools that support pedagogies to enact and closely investigate teaching practice.

Exemplar 1: Exploratory Integration of Methods Work in Classrooms

During the fall 2015 semester, Sararose Lynch (Westminster College) implemented an after-school second-grade mathematics club as an on-site component of her elementary mathematics methods course for 20 PTs. The MTE integrated this experience into the course to engage PTs in anticipating, eliciting, and responding to student thinking. A focus of this elementary mathematics methods course was to purposefully promote meaningful discourse and include assignments focused on content from *Intentional Talk* (Kazemi & Hintz, 2014) and *5 Practices for Orchestrating Productive Mathematics Discussions* (Smith & Stein, 2011). The school was selected because of the proximity to the college and its status as a Title I school.

The MTE created an 8-week course that supported learning for both the second-grade students and the PTs. The first 3 weeks of the methods course was held on the university campus where the PTs and MTE engaged in content and investigated core teaching practices. The last 5 weeks of the methods course were held both at the university and at the school-based mathematics club. All second-grade students from the local elementary school were invited to participate, and approximately 65% of the second graders in the school (30 students) registered for the weekly program. In the 1-hour mathematics club sessions, the PTs and students played basic-skill card games focused on second-grade students' thinking and state standards, followed by a mathematics or coding task. PTs worked with the same students each week in small groups to build personal connections as they worked together on the mathematics activities.

To prepare for the weeks at mathematics club, the MTE designed an instructional activity for PTs to investigate student thinking while teaching tasks that addressed second-grade mathematics standards. The MTE facilitated learning about the instructional activity through a cycle that included the following stages: (a) PTs engaging in the task as learners, (b) identifying the learning goal(s) of the task, (c) observing a rehearsal of the task, (d) coplanning instructional adaptations for the task, and (e) enacting the task with second graders, followed by (f) debriefing in a large-group setting with PTs, MTEs, and the practicing teacher. During the first four stages of the instructional activity, the PTs were supported by the MTE in the methods class held at the university. The fifth and sixth stages took place in the school-based mathematics club, and both the MTE and a practicing teacher provided in-the-moment support to the PTs. "Time-outs" were used to make PTs' thinking visible and allowed for peers, second graders, the MTE, and the practicing teacher to give feedback on the PTs' instructional moves. During the debrief, PTs were asked two specific questions to reflect upon from their enactments: (a) "What did you learn about your students today?" and (b) "What did you learn about yourself as a teacher?"

Within the university-based methods class, the MTE first engaged the PTs in the mathematics task by having them act as learners to solve the task. The PTs then identified the learning goals of the task and any possible misconceptions their students might have. These misconceptions were made explicit during a whole-class discussion. For the majority of the classes, the MTE used Lesson*Sketch* (Herbst, Aaron, & Chieu, 2013) to create a hypothetical classroom scenario for the PTs to collectively discuss and reflect on instructional moves and student thinking related to the mathematics task. The MTE created a comic-based depiction using Lesson*Sketch* to convey possible questions and comments the second graders might pose, and the PTs planned and rehearsed how they would respond to these anticipated student ideas. During the small-group reflection of the rehearsal, the PTs were asked to respond to student-posed questions, justify why they responded in the manner they did, and identify how they expected the hypothetical second graders to respond to their prompts. Following this cycle (later that same afternoon), the PTs attended mathematics club and enacted the instructional activity with their student(s). During the instructional time, the MTE and practicing teacher circulated the room to coach PTs and employed time-outs during small-group instruction when necessary. Time-outs involved a purposeful pause in the lesson for educators to debrief and discuss next steps. Following the mathematics club session, the MTE facilitated a debriefing session where PTs, the practicing teacher, and the MTE provided feedback on what they had learned and noticed during their interactions with students.

An affordance of integrating mathematics club into the methods coursework, noted by both the MTE and PTs, was that PTs could collaboratively plan purposeful instructional tasks for second graders while anticipating student thinking, and then immediately enact the task with students. The proximity of the school provided the MTE enough time for PTs to do the planning, enacting, and debriefing all in one class session. A constraint of this integration was that the majority of PTs' mathematics content knowledge for teaching instructional tasks focused on second-grade standards throughout the semester. Given this constraint, consideration for future iterations of a similarly structured methods course could be to purposefully have the PT connect what they are teaching to standards in previous and later grades. Continuing to facilitate the one-on-one relationships developed between PTs and students while helping PTs connect the content they are teaching to prior and later grades will strengthen this integrated experience to benefit both PTs and elementary students. As a PT noted in a course evaluation,

> Math club taught me that not everything you plan will always go as planned. Most of the time you have to change what you're doing according to how the students are thinking and/or acting. It has taught me to be on my toes constantly and to try to figure out the students' next move before they make that move. I liked having the opportunity to teach with Dr. Lynch [MTE] or Ms. Taylor [practicing teacher] as a "coach" supporting my teaching.

This quote, as well as other similar responses, indicated that through PTs' site-based experiences during mathematics club they learned the importance of and developed skills in anticipating and responding to student thinking, which were goals of the course that were focused on through this integration.

Exemplar 2: Consistent Integration of Methods Work in Classrooms

During the fall 2015 semester, Melinda Knapp (Oregon State University, Cascades) implemented a series of half-day lesson study cycles in a local elementary school with a small group of six PTs, taken from a larger group of 25 who were enrolled in the elementary mathematics methods course. This particular elementary school was chosen because of the existing strong professional relationship between the MTE and the principal and practicing teachers. Because of their prior work together, the MTE was aware of the ambitious teaching practices of one of the fourth-grade teachers and knew the classroom would provide a vision of mathematics teaching she wanted her PTs to experience firsthand. The practicing teacher held the belief and expectation that students were mathematical sense makers, and she planned for and routinely utilized ambitious teaching practices that focused on eliciting, interpreting, and responding to students' mathematical thinking.

The lesson study structure provided an opportunity for the MTEs, PTs, and practicing teachers to collaborate authentically around the work of mathematics teaching. They coplanned a mathematics lesson, observed the enactment of that lesson by the practicing teacher, and debriefed the lesson using student discourse data collected by PTs during the classroom enactment. The MTE and practicing teacher selected intentional mathematical and pedagogical goals that connected to what the PTs were studying in their methods course work.

After the collaborative planning session, the PTs and MTE observed the practicing teacher as she taught the lesson to her students. During this time, PTs collected student discourse data by carefully listening for and noting specific student discourse and situations in which the practicing teacher elicited and responded to student thinking. During the lesson enactment, the MTE called multiple time-outs in order to highlight important mathematical ideas, allow the PTs to ask questions, or to have PTs engage with students in an intentional way. During one particular time-out, the MTE paused the class to allow each PT to ask a small group of students questions that would uncover their current understanding of the mathematical model being investigated. The purpose of this time-out was to have PTs rehearse in-the-moment question asking and to practice listening to students' mathematical thinking.

The debrief session included time for the group of educators to collectively analyze student discourse data and written work, review decisions the practicing teacher made during instruction, and discuss evidence of student learning. The debrief session also allowed time to better understand mathematical ideas that were shared by students in whole-group and small-group discussions. Time was also given for reflection and goal setting.

The PTs' reflections provided evidence of learning new ways to position students as sense makers and beginning to understand the role of the teacher in positioning students competently; one PT wrote:

> I need to remember to plan for and use structured talk moves in my classroom. I habitually answer questions for students and I need to elicit their thinking because *they* can make sense of the problems and answer the questions.

A challenge mentioned by another PT was the idea of bringing a wrong answer into a whole-class discussion "because I'm unsure if I'll be able to lead a discussion with the class which will help students revise their thinking." As a possible solution to this challenge, the practicing teacher discussed the importance of anticipating the range of understandings students bring when planning lessons. This solution directly connected with what the PTs were learning in their university coursework, thus supporting this work in practice.

During these half-day lesson study cycles, many opportunities arose to support PTs in applying the conceptual and practical tools presented in their coursework in more authentic ways. Through the collaborative work, MTEs, practicing teachers, and PTs began developing and articulating a common language for specifying practice, identifying and specifying pedagogies referred to in teacher education and breaking down the perennial and persistent divide between university coursework and clinical experiences.

The PTs were brought into the larger education community by participating with the methods class in a school setting in a lesson study cycle structure, as they had the opportunity to observe the lesson being taught by the practicing teacher, have strategic time-outs to discuss the lesson as it was happening, and debrief and reflect after the lesson. Purposeful planning was a key component because the PTs were engaged in researching a concept and developing a lesson as part of a team considering the mathematics, questioning strategies, and possible student partial conceptions and responses. These purposely planned elements were highlighted during time-outs and during the debrief session. The PTs were able to study student thinking in a way that would not be possible in the university classroom and could make observations specific to the lesson and student thinking. The in-the-moment work provided PTs the opportunity to understand students' mathematical thinking. At the end, the PTs reflected on the experience through a debrief discussion. We explicitly highlighted how student thinking was elicited and made sense of, and how it guided modifications to the lesson. As a result of these learning experiences, the PTs set personal development goals that focused on eliciting and responding to student thinking as they returned to their student teaching placement.

Exemplar 3: Integration of Majority or All of Methods Work in Classrooms

In the fall 2015 semester, Megan W. Taylor (Trellis Education) and Jake Disston (Masters and Credential Program in Mathematics and Science Education [MACSME] at U.C.–Berkeley), in partnership with Sonoma State University, piloted a collaborative, site-based methods course model that brought university faculty, PTs, and practicing teachers together weekly at middle and high school sites in San Mateo, San Francisco, Oakland, Marin, Petaluma, Santa Rosa, and Sonoma. These collaborative learning events included both *pedagogies of investigation* and *pedagogies of enactment* (Grossman & McDonald, 2008): Participants had the opportunity to observe and debrief expert teaching; engage in planning for, enacting, and reflecting on teaching practice with each other and with 6–12 students; and track and reflect on changes in their teaching practice over time. The goal was to ensure PTs developed measurable expertise enacting two critical teaching practices for secondary mathematics and science: *facilitating productive discourse* and *eliciting, responding to, and capitalizing on students' ideas.*

Each week two MTEs, 40 PTs, and a variable numbers of practicing teachers on release from their sites congregated at a school and were divided among practicing teachers' classrooms at that site. Schools were chosen based on practicing teachers and student teaching placements, and it was the responsibility of the practicing and PTs at each focal site to develop the schedule for the visiting group and provide context about the school and visited classrooms. MTEs worked with practicing teachers to choose or design instructional activities that would allow ample opportunity for practicing teachers to model at least one of the two core teaching practices and then for PTs to rehearse and practice enacting them with students. As homework, the PTs were expected to engage with the focal task(s) and do the anticipatory work of predicting student thinking.

A typical visit began with visitors observing one or more teachers implementing the instructional activity, with an eye on one of the two focal teaching practices, followed by a structured debrief led by an MTE. After the debrief, small groups of PTs planned a bounded enactment of part or all of the same instructional activity and rehearsed it with one another away from students, then reentered a classroom to practice enactment with students. Sometimes this work happened over multiple back-to-back periods (40–70 minute period), and sometimes within one long block period (90–120 minute period), depending on the practicing teachers' schedules and school-day schedule. Who rehearsed and practiced with students varied from visit to visit, but over the course of the semester, all PTs had at least six opportunities to plan for, rehearse, enact, and reflect on an instructional activity. All visits ended with a whole-group debrief led by at least one MTE.

One of the primary assignments of the methods course was for PTs to gather evidence of, make sense of, and improve their enactment of one or both focal teaching practices. PTs collected evidence throughout the semester of how their practice was changing in the learning cycle and presented analyses of these data to each other and their practicing teachers. For example, a large number of PTs focused on their use of questioning in facilitating discourse and eliciting student ideas. In approximations of teaching practice with one another, these PTs collected audio or video records of their questioning that they transcribed and examined afterward. In looking across these data, PTs could make claims about how their questioning practice was shifting toward more effective practice, before practicing more effective questioning practice with actual students. Some PTs showed evidence of how their use of questions shifted away from a peppering of yes/no questions to a more strategic use of different question types for a variety of purposes. Some PTs showed evidence of learning to wait and allow for silence in classroom discussions, a notoriously difficult, uncomfortable, yet critical thing for new teachers to do. One PT reflected early on that he wanted "to avoid asking so many 'what' questions and [instead ask] more 'how' and 'why' questions." In this PT's public demonstration of change in practice months later, he shared how his frequency of "what" questions decreased over time and that he was asking fewer and more effective questions. In his final reflection, he wrote: "I am asking better questions, and I know this because I can say so much more about what my students know. I'm more comfortable waiting for students to think, and I'm asking questions that encourage them to do so."

One of the affordances of holding methods coursework in classrooms is the rapid, visible improvement in PTs' enactment of targeted teaching practices as they engage in consistent, iterative cycles of planning, teaching, and reflection. By focusing on a few key, learnable practices over time, early career teachers are able to learn to enact these practices with an initial level of competency before taking responsibility for student learning in their own classrooms. A second affordance of classroom-integrated methods instruction is that PTs can start small and move towards more complex practice over the course, which enables PTs and their teacher educators to build and cultivate a culture of risk-taking and ideally trust in making one's practice public. The MTEs continually learn about what PTs do, what questions they have, and how feedback is or is not helpful in developing PTs' practice. Finally, in models of site-based methods work in which PTs visit the same school or classroom each week (Kazemi & Wæge, 2015; Weston & Henderson, 2015), a hybrid space (Bhabba, 1990; Zeichner, 2010) can be created that benefits the learning of all participants, both adults and students. The activity system created by consistent, mentored rehearsals and enactments of teaching in schools and with students enables PTs to build relationships with students and come to know them intellectually and socially, a vital component of effective teaching. The planning that happens week to week can be done with specific students in mind and with emerging understandings about children's academic and social strengths and needs.

DISCUSSION

By integrating mathematics methods courses in the P–12 schools, MTEs provide PTs with multiple opportunities to engage with students in authentic learning activities as they learn about core teaching practices. The PTs in the work described in this chapter were able to teach in response to students' ideas and actions and learn about ambitious teaching practices. Additionally, the PTs had the opportunity to develop meaningful relationships with students and practicing teachers as all the participants learned within the community of practice. Below, we summarize commonalities between the exemplars into five themes, which we argue can serve as an initial set of guidelines for future integration of methods courses in schools.

1: *PTs facilitate lessons with P–12 students with at least one mathematics teacher educator present for coaching and support.*

Having the MTE present in the classroom allows for coaching and support for the PTs as they interact with students and enacted activities. Further, the participants' shared experiences allow for opportunities to engage in meaningful and specific debriefs following the activity. In addition, the shared experiences allow a strongly connected university and clinical experience.

2: *Integrated classroom visits focus on observing, studying, and/or trying something related to eliciting, interpreting, and responding to student thinking, a specific core teaching practice.*

By being in the classroom, PTs work on eliciting, interpreting, and responding to student thinking, an essential core teaching practice that could not be approximated as authentically without student interaction and contextual features of the classroom.

3: *When PTs rehearse or when they teach P–12 students, MTEs use time-outs or planned/unplanned pauses in ways that (a) provide opportunities for all teachers to think in the moment (e.g., make sense of student thinking, discuss a next teacher move), and (b) position students as powerful and productive thinkers.*

The time-outs and pauses allow for MTEs to coach PTs during activities and for PTs to receive support from MTEs. In particular, the MTE helps PTs focus on (and rehearse again) certain aspects of practice that need additional work. Similarly, the PT can call a time-out to ask for support if unsure what to do in a situation. These opportunities to interact during teaching enable the community to work together on the relational aspects of teaching and learning and for the PTs to experiment with instructional decisions in response to student thinking.

4: *Participants' lessons are designed to elicit student thinking, and they approach rehearsals and enactments with students with a shared curiosity and/ or plan for learning something particular about teaching and learning.*

Providing a shared experience for the PTs and MTE can allow for more productive discussions about practice before, during, and after the classroom enactment.

5: *All participants debrief, analyze, reflect on the experience in some way (including in conversation with a teacher educator), and maintain a focus on what was learned about specific students' thinking.*

The debrief following the enactments provides MTEs and PTs the opportunity to make sense of and reflect upon the classroom interactions. The debriefing is essential in having PTs reflect upon learning goals of the activity and shaping their practice.

One of the primary goals in sharing our work of integrating teacher preparation courses into school settings is to encourage other MTEs to find ways within the structure of their university to hold courses in the P–12 classroom. The five themes could act as guidelines to develop relationships and design school-integrated pedagogies. To begin to integrate a teacher preparation course, we suggest understanding the role of the school experience in the course as well as the needs of the local schools. By addressing the common needs between the school and university, the MTE will have a solid base to begin to build relationships with schools, principals, and teachers. We also suggest starting small, first trying a less intensive (Exemplar 1) approach before moving to fully integrate a course in the school (Exemplar 3). As the partnership grows between university and schools, we encourage MTEs to continue to explore different pedagogical approaches to supporting PTs' development of their teaching practice.

REFERENCES

Ball, D. L., & Cohen, D. K. (1999). Developing practice, developing practitioners: Toward a practice-based theory of professional education. In G. Sykes & L. Darling-Hammond (Eds.), *Teaching as the learning profession: Handbook of policy and practice* (pp. 3–32). San Francisco, CA: Jossey Bass.

Bhabba, H. (1990). The third space. In J. Rutherford (Ed.), *Identity, community, culture and difference* (pp. 207–221). London, England: Lawrence and Wishart.

Blue Ribbon Panel on Clinical Preparation and Partnerships for Improved Student Learning. (2010). *Transforming teacher education through clinical practice: A national strategy to prepare effective teachers.* Washington, DC: National Council for Accreditation of Teacher Education.

Borko, H., Eisenhart, M., Brown, C., Underhill, R., Jones, D., & Agard, P. (1992). Learning to teach hard mathematics: Do novice teachers and their instructors give up too easily? *Journal for Research in Mathematics Education, 23,* 194–222.

Cochran-Smith, M., Villegas, A. M., Abrams, L., Chavez Moreno, L., Mills, T., & Stern, R. (2016). Research on teacher preparation: Charting the landscape of a sprawling field. In D. Gitomer & C. Bell (Eds.), *Handbook of research on teaching* (5th ed., pp. 439–547). Washington, DC: American Educational Research Association.

Core Practices Consortium. (2013, April). *Building a shared understanding for designing and studying practice-based teacher education.* Symposium presented at the annual meeting of the American Educational Research Association, San Francisco, CA.

Ensor, P. (2001). From preservice mathematics teacher education to beginning teaching: A study in recontexualizing. *Journal for Research in Mathematics Education, 32*, 296-320.

Greeno, J. G. (2006). Learning in activity. In R. K. Sawyer (Ed.), *The Cambridge handbook of the learning sciences* (pp. 79–96). New York, NY: Cambridge University Press.

Grossman, P., Hammerness, K., & McDonald, M. (2009). Redefining teaching, reimagining teacher education. *Teachers and Teaching: Theory and Practice, 15*(2), 273–289.

Grossman, P., & McDonald, M. (2008). Back to the future: Directions for research in teaching and teacher education. *American Educational Research Journal, 45*(1), 184–205.

Gutiérrez, R. (2013). The sociopolitical turn in mathematics education. *Journal for Research in Mathematics Education, 44*, 37–68.

Herbst, P., Aaron, W., & Chieu, V. M. (2013). LessonSketch: An environment for teachers to examine mathematical practice and learn about its standards. In D. Polly (Ed.), *Common core mathematics standards and implementing digital technologies* (pp. 281–294). Hershey, PA: Information Science Reference. doi:10.4018/978-1-4666-4086-3.ch019

Kazemi, E., Franke, M., & Lampert, M. (2009, July). Developing pedagogies in teacher education to support novice teachers' ability to enact ambitious instruction. In R. Hunter, B. Bicknell, & T. Burgess (Eds.), *Crossing divides: Proceedings of the 32nd annual conference of the Mathematics Education Research Group of Australasia* (Vol. 1, pp. 12–30). Palmerston North, NZ: MERGA.

Kazemi, E., Ghousseini, H., Cunard, A., & Turrou, A. C. (2016). Getting inside rehearsals: Insights from teacher educators support work on complex practice. *Journal of Teacher Education, 67*(1), 18–31. doi: 10.1177/0022487115615191

Kazemi, E., & Hintz, A. (2014). *Intentional talk: How to structure and lead productive mathematical discussions.* Portland, ME: Stenhouse.

Kazemi, E., & Wæge, K. (2015). Learning to teach within practice-based methods courses. *Mathematics Teacher Education and Development, 17*(2), 125–145.

Korthagen, F. A. J., Kessels, J., Koster, B., Lagerwerf, B., & Wubbels, T. (2001). *Linking practice and theory: The pedagogy of realistic teacher education.* Mahwah, NJ: Erlbaum.

Lampert, M., Beasley, H., Ghousseini, H., Kazemi, E., & Franke, M. (2010). Using designed instructional activities to enable novices to manage ambitious mathematics teaching. In M. K. Stein & L. Kucan (Eds.), *Instructional explanations in the discipline,* (pp. 129–141). New York, NY: Springer.

Lampert, M., Franke, M. L., Kazemi, E., Ghousseini, H., Turrou, A. C., Beasley, H., Cunard, A., & Crowe, K. (2013). *Journal of Teacher Education, 64*(3), 226–243. doi: 10.1177/0022487112473837

Lampert, M., & Graziani, F. (2009). Instructional activities as a tool for teachers' and teacher educators' learning. *The Elementary School Journal, 109*(5), 491–509. doi: 0013-5984/2009/10905-0005

Lave, J., & Wenger, E. (1991). *Situated learning: Legitimate peripheral participation.* Cambridge, England: Cambridge University Press.

McDonald, M., Kazemi, E., & Kavanagh, S. S. (2013). Core practices and pedagogies of teacher education: A call for a common language and collective activity. *Journal of Teacher Education, 64*(5), 378–386. doi: 10.1177/0022487113493807

Putnam, R. T., & Borko, H. (2000). What do new views of knowledge and thinking have to say about research on teacher learning? *Educational Researcher, 29*(1), 4–15.

Simon, M. (2008). The challenge of mathematics teacher education in an era of mathematics education reform. In B. Jaworski & T. Wood (Eds.), *International handbook of mathematics teacher education* (Vol. 4., pp. 17–29). Rotterdam, The Netherlands: Sense.

Smith, M. S., & Stein, M. K. (2011). *5 practices for orchestrating productive mathematical discussions.* Reston, VA: National Council of Teachers of Mathematics.

Turner, E. E., Drake, C., McDuffie, A. R., Aguirre, J., Bartell, T. G., & Foote, M. Q. (2012). Promoting equity in mathematics teacher preparation: A framework for advancing teacher learning of children's multiple mathematics knowledge bases. *Journal of Mathematics Teacher Education, 15*(1), 67–82.

Weston, T. L., & Henderson, S. C. (2015). Coherent experiences: The new missing paradigm in teacher education. *The Educational Forum, 79*(3), 321–335.

Zeichner, K. (2010). Rethinking the connections between campus courses and field experiences in college- and university-based teacher education. *Journal of Teacher Education, 61*, 89–99.

CHAPTER 16

PROSPECTIVE TEACHERS ANALYZING TRANSCRIPTS OF TEACHING

Laura M. Singletary
Lee University

Zandra de Araujo
University of Missouri

AnnaMarie Conner
University of Georgia

The main goal of teacher education is to prepare prospective teachers (PTs) for the complex work of teaching. Past approaches to teacher education have attempted to produce expert teachers upon graduation from a teacher education program. Current research on teacher education, however, suggests a different approach. Learning to teach mathematics well is a challenging and dynamic process that should not end with the completion of a teacher preparation program (Ball & Forzani, 2011; Morris, 2006). Morris (2006) claims that it is unrealistic to expect that PTs will be expert teachers at the

Building Support for Scholarly Practices in Mathematics Methods, pages 249–262
Copyright © 2018 by Information Age Publishing
All rights of reproduction in any form reserved.

completion of a teacher education program. Rather, she suggests that it is important for PTs to develop skills that allow them to learn from their teaching so they can continue to grow in their professional knowledge.

Kennedy (1999) argues that teacher education should consider the difficulties presented by what she describes as the *problem of enactment*. The problem of enactment is caused by two or more parties having different frames of reference. For example, we may apply different meanings to words we use or draw on our individual backgrounds when interpreting ideas. In mathematics education, it is likely that PTs and their mathematics teacher educators (MTEs) will draw from different frames of reference when discussing concepts related to teaching.

> Novice teachers often approach the formal study of teaching with a frame of reference they developed during their childhood, while their university faculty are likely to approach the formal study of teaching from the perspective of reformers.... Moreover, it is likely that all the sentences spoken by faculty will be interpreted in the light of novices' initial frames of reference, rather than in the light of a reform frame of reference. (Kennedy, 1999, p. 71)

To address the problem of enactment, teacher educators must not only provide PTs with a refined frame of reference but also an understanding of how pedagogical knowledge translates into the enactment of instructional practices. This kind of knowledge is situated knowledge; it is understood not only abstractly but also through particular practice-based situations.

To develop this type of situated knowledge for teaching mathematics, the literature strongly advocates situating PTs' learning about teaching through representations of practice so that they can learn about teaching from teaching (Lampert et al., 2013). In this chapter, we describe "analyzing transcripts of teaching," an activity developed to support each PT in developing and using a transcript of his or her teaching practice to analyze and learn from teaching. This activity provided an opportunity for the PT to engage in internal reflection on the activity and then provided an external manifestation of his or her internal learning. We examine the opportunities our PTs had to learn about teaching through this activity and our insights into their learning. Although exploratory in nature, this work provides evidence for the potential effectiveness of this activity in helping PTs learn about teaching through an analysis of their teaching actions.

THEORETICAL FRAMEWORK FOR ANALYZING TRANSCRIPTS OF TEACHING

The two methods courses described in this chapter were designed from a situative perspective on learning to teach mathematics (Peressini, Borko,

Romagnano, Knuth, & Willis, 2004). From a situative perspective, learning occurs as one participates more fully in a particular social practice (Lave, 2009). We consider teaching to be a social practice, where we do not consider an individual to be completely separable from the community in which he or she is situated. For this reason, the methods courses were structured to engage PTs in authentic activities as a means to situate their learning about the complexities of teaching. These courses were also designed so the main goal was to turn PTs not into expert teachers, but rather into competent novice teachers who adopt a mindset of continual reflection and growth so that they may develop the ability to learn from teaching (Hiebert, Morris, Berk, & Jansen, 2007). To emphasize this mindset, we coupled a representation of practice with structured analysis to refine their professional knowledge and teaching practices.

The theoretical perspective framing the design of these methods courses informed the development of the analyzing transcripts of teaching activity, whether PTs focused on eliciting student thinking or on asking questions to prompt discussion. The structure of this activity follows from McDonald, Kazemi, and Kavanagh's (2013) framework for PTs to learn from engaging in an authentic and ambitious instructional activity through the engagement of core teaching practices (see Table 16.1).

METHOD

We report on the implementation of two activities with similar goals and designs. The first implementation involved elementary PTs in a methods course focused on student thinking. The second centered on secondary

TABLE 16.1 Structure of Learning Cycle Including Analyzing Transcripts of Teaching

McDonald, Kazemi, and Kavanagh's (2013) Learning Cycle Quadrants	Learning Cycle Implementation
Introducing and learning about the activity	PTs read about, discuss, and analyze videos related to facilitating discourse
Preparing for and rehearsing the activity	PTs use concepts related to facilitating discourse to plan a task or a lesson, including anticipated questions and answers
Enacting the activity with students	PTs enact the lesson plan in field experience and capture the enactment for transcription and analysis
Analyzing enactment and moving forward	PTs use transcript of enactment to conduct analysis of enactment and make plans for improvement of practice

PTs in a mathematics methods course focused on developing pedagogical content knowledge for teaching. We examined how these activities with a focus on the development and use of a transcript informed PTs' learning experiences in different contexts.

We collected task/lesson plans, transcripts of the enactment, and reflections from each of the PTs enrolled in the courses. We analyzed these data using the constant comparative method (Corbin & Strauss, 2008). Codes were developed inductively with a focus on what the PTs attended to as they reflected on their enactment of their plans. In the following sections, we describe the two implementations of the analyzing transcripts of teaching activity by detailing important features from each context and highlighting examples of PTs' learning.

IMPLEMENTATION IN AN ELEMENTARY METHODS COURSE

At de Araujo's institution, prospective elementary teachers take two combined mathematics methods and content courses. Each of these three-credit-hour courses meets for 75 minutes, twice a week. The first course in the sequence focuses on children's thinking, fractions, and number and operation. The PTs in this study were enrolled in the first course and were concurrently enrolled in a weekly, full-day field experience. Part of the field experience involved each PT working one-on-one with a fifth-grade student. This field experience supported the overall goal of the methods course, allowing the PTs to explore students' mathematical thinking.

Description of Analyzing Transcripts of Teaching

In the first part of the learning cycle, PTs learned what it means to elicit and respond to student thinking. The MTE modeled and provided exemplars of this practice. She followed up on these activities with whole-class debriefings during which she highlighted particular instructional moves she enacted in trying to elicit and interpret the PTs' thinking. Other class activities were developed so PTs could begin to interpret student thinking. For example, PTs extensively examined student solutions, both correct and incorrect, and read articles and watched videos exemplifying this practice. The PTs also read articles related to the course's mathematical focus, fraction concepts (e.g., Empson, 1995). Together, these activities provided examples of teacher-questioning techniques PTs could use during subsequent parts of the learning cycle.

The learning cycle continued as each PT planned to enact a particular mathematics task during the initial meeting with his or her elementary student: "There are three children sharing eight candy bars. How should the children share the candy bars so that each child can get the same amount of candy bars?" The assignment required PTs to detail learning goals for students, possible student solution strategies, and extensions and supports for their student. In addition to these conventional lesson plan features, the instructor required each PT to create a script of the most likely scenario of how his or her meeting with his or her student would occur. These scripts served as rehearsals for the PTs to anticipate how to elicit and respond to student thinking.

Following the rehearsal of the task, the PTs enacted the task with elementary students. PTs were asked to provide an initial reflection on interactions with students. They were also asked to describe what surprised them about their meeting with the student, aspects of their practice they would like to improve, and the "big takeaway" from this activity.

After this initial reflection, the PTs conducted a more extensive analysis of their enactment. The PTs had recorded the meetings with students using Explain Everything (Explain Everything Inc., 2015), a screen-casting application for iPad that captures written work on the iPad and the accompanying audio as a single video file. The PTs transcribed the interactions with students (including screen shots of student work), analyzed the transcripts (e.g., marked questions asked, tracked wait time), and answered reflection questions. In this final and more structured analysis of their teaching, PTs compared their anticipated script with the actual transcript, described things they did and did not do well, identified and explained what they thought were the most and least effective questions they asked and why, highlighted instances in which questions or statements were too leading, reflected on who did the most talking, and made suggestions for improvement. The goal of the interaction with students was for the PT to elicit student thinking; the goal of the transcription and reflection was for PTs to identify instances in which they elicited or failed to elicit student thinking.

Illustration of Analyzing Transcripts of Teaching: The Case of Jada

Jada, an elementary PT, rehearsed for this task by creating a script of imagined dialogue with her student. The anticipated script included questions she might ask to elicit her student's thinking and how she might respond to the student's thinking. Jada began the anticipated script by asking the student to read the task and the student easily solving the task without assistance. The script concluded with Jada asking the child to explain her solution, which she did, and then Jada provided validation for her student's solution.

Jada's actual transcript differed greatly from her anticipated script. The most notable difference was in length. When asked to describe the differences between the two scripts, Jada wrote the following:

> The biggest difference between the two is that the original script is short and sweet. I predicted the student would get the answer with little trouble or help from me. What I didn't anticipate happening was what actually happened: that my student would struggle to get to the idea of equal distribution. . . . I ended up having to ask a lot more questions.

From this excerpt, we see that the elementary student required far more support than Jada had anticipated. In addition, the student required encouragement and motivation to persist with the task. Jada found eliciting her student's thinking more difficult than she had initially intended. Through transcribing her practice, Jada had a concrete artifact to compare to her anticipated script. This allowed her to analyze and better understand what she may have under- or overestimated in terms of her interaction with her student. This also allowed the MTE to suggest strategies Jada might try in subsequent interactions to elicit and respond to student thinking.

In terms of the mathematical content, the transcript and the corresponding screenshots of the student's work allowed Jada to provide an extremely detailed description of the student's solution strategy. Further, she was able to analyze the student's solution strategy and to communicate that analysis with her MTE. For example, the following is an excerpt from her description of the student's solution:

> She [the student] drew three cookies on the Explain Everything app (Figure 16.1). I didn't understand her thinking behind this, and when I asked her why she drew three cookies [the student used cookies in the place of the candy bars described in the task], she said that she was drawing 11 total cookies because "[the problem] says 'same' and when a math problem says same you know you have to add." So she had just added 3 + 8 together to get 11. . . . In an attempt to redirect her thinking, I reread the problem to her, and it seemed like she knew what to do after that so I left her with little guidance again to think through it. She then drew eight cookies on the Explain Everything app saying, "They each get 4 because 4 and 4 is 8." I then said, "There are 8 *candy bars*, but there are 3 *children*." . . . She began manipulating the pattern blocks again, with the yellow ones representing the children and the orange ones representing the cookies. She began distributing the cookies evenly, and in doing so came to realize that that wasn't going to work. I asked her what she was going to do with her last two cookies and she insisted on "throwing them away because that was the only way to make it fair."

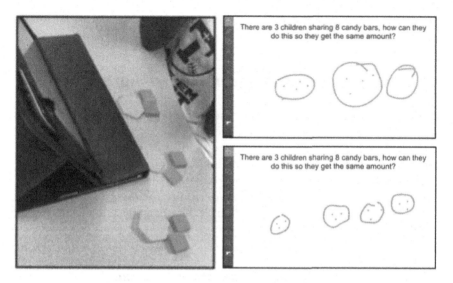

Figure 16.1 Student work Jada included in her analysis.

The in-depth analysis of the student's solution strategy allowed both Jada and her MTE to connect the strategies the student used to readings and concepts from class. For example, the MTE was able to follow-up with Jada to discuss the student's sharing strategy of first giving out as many whole cookies as possible and then attempting to deal with the leftover cookies. The MTE pointed out that this strategy was one discussed by Empson (1995) in an article the PTs read in class. The MTE also directed Jada to this article in thinking about other strategies she might try to help her deal with the leftover cookies.

The case of Jada illustrates the general outcome that analyzing the transcript led the elementary PTs to an increased awareness of the teacher's role in enacting tasks and thinking through ways to elicit and respond to student thinking in the moment. The power of the analyzing transcripts of teaching activity is demonstrated by the ways PTs were able to frame their responses and the depth of their analysis when they referred to the transcripts as compared to their initial reflections on the activity. In Jada's initial reflection before the transcription she merely wrote, "I do not think [the learning goal] was met.... The student did break the cookie into two equal halves with the first problem, which demonstrated to me that she had a basic understanding of partitioning." After analyzing the transcript, Jada was able to describe her student's work more fully, showing an increased understanding of her student's thinking, and the MTE was able to gain a much deeper understanding of how the student approached the task and use that knowledge to help Jada more fully understand the particular strategies the student used.

IMPLEMENTATION IN A SECONDARY METHODS COURSE

At Singletary's institution, prospective secondary mathematics teachers take a single mathematics methods course during the final semester of coursework prior to a semester-long student-teaching experience. The 15-week methods course met once each week for 2 hours, and the accompanying field experience required 30 hours of classroom observation and teaching. This field experience provided an opportunity for the PTs to reflect on teachers' instructional practices and enact the pedagogical concepts developed in the methods course.

Description of Analyzing Transcripts of Teaching

To begin the learning cycle, explicit instruction on facilitating discourse lasted one class session during the methods course with several additional discussions or activities throughout the semester integrating concepts of facilitating discourse with additional pedagogical methods, such as selecting and orchestrating mathematical tasks. Prior to this class period, PTs were assigned readings (e.g., Chapin, O'Connor, & Anderson, 2009; Singletary & Conner, 2015) to provide conceptual frameworks about questioning, argumentation, and discussions to inform their thinking about facilitating discourse. In class, PTs discussed the readings and then analyzed video exemplars using each framework. These video cases provided models of productive discourse and allowed the PTs to discuss key features of the discourse.

This learning cycle continued as the PTs planned to teach mathematics lessons in their field experiences that focused on facilitating productive discourse. Each PT planned a lesson in which a high-cognitive-demand task (Stein, Smith, Henningsen, & Silver, 2009) was used as a foundation for the classroom discourse. PTs included several questions in their lesson plans to prompt student thinking, and the lesson plans included possible student solutions and misconceptions. Singletary viewed the combination of questions and anticipated student solutions as a modified rehearsal because of the inclusion of potential student–teacher contributions to the classroom discourse. PTs then enacted the lesson plans in their field experiences. The lesson was recorded, and each PT transcribed 15 minutes of whole-class instruction. Each PT used the transcript as evidence to analyze the enactment of his or her plan and reflect on ways to improve instruction using the conceptual frameworks provided by the discourse readings. Following the

PTs' initial reflection and analysis of the lesson, they each met with the MTE to collaboratively analyze the enactment of the lesson, focusing on how the PT facilitated discourse using the frameworks provided in the readings as a frame for the analysis.

Illustration of Analyzing Transcripts of Teaching: The Case of Naomi

To illustrate the analyzing transcripts of teaching activity in a secondary context, we examine the case of Naomi. She planned to teach a lesson on writing equations to model linear functions to an introductory algebra class. She used the following task to begin the classroom discussion.

> Let's say we have a snake that starts out as four units long and grows by one unit each time it eats. How long is the snake after it eats five dots? How long is the snake after it eats 100 dots? How long is the snake after it eats n dots? That is, can we create an equation for the length of the snake? (modified from Snake on a Plane, n.d.)

In her lesson plan, Naomi planned to have students work in groups to solve the problem, allowing time to conclude the task with a whole-class discussion.

Naomi focused her class on exploring the task she planned. The following excerpt details the classroom discourse following the introduction of the task.

> **Naomi:** Ok, so what if we ate 5 dots?
> **Class:** [Pause] Nine.
> **Naomi:** Nine. What if we ate 100 dots?
> **Class:** 104.
> **Naomi:** 104. These are really easy questions that I'm asking you guys. Can we write an equation for the nth dot? Can we take the pattern that we are noticing? Are you seeing a pattern here?
> **Class:** Mhm. [affirmative]
> **Naomi:** Ok. So can we take the pattern that we are noticing, and can we create a formula out of it? Can we create an equation?
> **Class:** Yes.
> **Naomi:** I see head nodding and a yes. You are on the right track. What would my equation be? We're measuring length so we are going say length equals, and then how can I figure that out?
> **Student :** $D + 4$

The transcript provided Naomi with evidence to analyze her role in facilitating classroom discourse. She examined her transcript and noticed that many of her questions were "very computational." In her written analysis, she applied her understanding from reading Chapin and colleagues (2009), and she recognized that the majority of the classroom discourse during her lesson enactment followed the Initiation-Response-Evaluation (IRE) format, which she characterized as a weakness of the lesson.

> I found the simple, computational nature of the questions I asked produced an unexpected side effect and what I view to be a weakness. The majority of the questions I asked produced an IRE form of class discussion.... IRE is an undesirable from [sic] a classroom discussion because it normally leaves the "Why" out of classroom discussion, leaving students behind. A second weakness in my lesson is that the work of Chapin, O'Connor, and Anderson puts an emphasis on using the students as speakers in the math talk moves. I only talked or invited one student to comment at a time. The math talk moves would have been much more powerful if I had asked students to repeat or add on. So, in order to improve math talk moves in my teaching I need to move from rapid IRE to having more student discussions and from teacher dominated revoicing to asking students to add onto and repeat student contributions. (Chapin, O'Connor, & Anderson, 2009)

Naomi analyzed her classroom transcript for the kinds of questions she had asked throughout her lesson. She used Singletary and Conner's (2015) categories of questions as a framework and classified each of the questions she had asked the class. Through this detailed analysis, she noticed that the overwhelming majority of her questions had requested factual answers. She also noted that she did not "request ideas"; nor did she "request elaboration" of the mathematics (referring to specific aspects of Singletary and Conner's framework, p. 146). After collaborating with her MTE about how this artifact of her practice could be used to help her develop as an educator, Naomi's reflection and analysis included plans for improvement, suggesting modifications to support developing productive discourse in her future instruction. She concluded that her task needed to be "more challenging," and that she needed to specifically plan to ask students different types of questions to further prompt their thinking and reasoning about the mathematics. The case of Naomi was representative of the others in this methods class. Naomi, like her peers, used the transcript to analyze her teaching as a means to learn from her teaching.

BENEFITS OF ANALYZING TRANSCRIPTS OF TEACHING

Across both methods courses, using the transcript as an artifact of practice allowed PTs to systematically analyze their teaching. The use of transcripts enabled the PTs to refer specifically to moments in the enactment that might have been forgotten or unnoticed without the use of the transcript. This activity had flexibility to meet course goals and to work across contexts that differed by grade level, focus, and relative experience of the PTs. That is, in the elementary methods course, the PTs were able to focus on eliciting and responding to an individual student's thinking, whereas in the secondary methods course, the PTs were able to study how teacher moves supported whole-class discourse.

Use of Specific Transcript Evidence in Analysis

A main goal of the analyzing transcripts of teaching activity was for PTs to use the transcript to learn from their teaching (Hiebert et al., 2007). For example, as the PTs listened to and transcribed their interactions with students, they became more cognizant of several characteristics of discourse. One elementary PT described how different her actual language was from what she had anticipated:

> I used much more formal or mathematical language in the original (antici-
> pated) script compared to the actual script. It was difficult to be focused on
> using "proper" terminology when my student was struggling so much just
> understanding the basics of the problem at hand. . . . In the actual script I
> resorted to a casual register of language by saying, "What could you do dif-
> ferently to break these up so that it is even? What is a different way that you
> could break those two candy bars up?" . . . As soon as I started to work with my
> student I forgot completely about those terms simply because I was so focused
> on just getting my student to get his thinking going in the right direction.

The PTs observed how they used mathematics terms and supported students in using precise language. Analyzing transcripts allowed the PTs to identify mistakes, mathematically and linguistically, of which they were unaware in the moment.

In addition to the words used, the PTs discussed their wait time in detail. They were able to measure the time they provided students. This quantification was important because prior to transcribing their enactment, many of them thought they had given their students adequate time to think. However, after transcribing, they realized that their wait time was not as extensive as they had thought. In both classes, many of the PTs made statements similar to the following reflective comment made by one of the prospective elementary teachers:

> A couple of times during the problem, I cut off my student to ask anoth-
> er question. I should not do this. Instead, I should let my student finish his
> thought to make sure he is doing his own thinking.

With the transcript, the PTs noticed how many times they interrupted their students, thereby cutting off student thinking. Through this analysis, the PTs reiterated the importance of wait time to afford students the opportunity to contribute to the mathematical discussion.

Many of the PTs quantified additional aspects of their discourse beyond the wait time. Some PTs, like Naomi, analyzed the kinds of questions asked to elicit student thinking. In this analysis, many of the PTs recognized how quickly they resorted to asking rather simplistic questions. Others quantified the amount of student contributions to the discourse as they analyzed their teaching. For example, one prospective secondary mathematics teacher wrote,

> After looking at the transcript, I saw that the majority of student responses were one or two words. Most of my questions did not prompt the students to give warrants with their claims, and as a result, I was not encouraging them to think very deeply or relationally about the mathematics.

In this way, the transcripts allowed the PTs to quantify specific aspects of the discourse to learn from this teaching experience. Additionally, references to the readings allowed them to make informed recommendations about improving their teaching. By citing portions of their transcript, PTs provided evidence for the claims they made about what they learned from teaching. Without the transcript, this depth of analysis would not have been possible. For example, using the transcript, Naomi identified her use of the IRE instructional format and suggested math talk moves to increase student contributions to the discourse. This use of specific transcript evidence situated her abstract understandings of talk moves and contextualized them in her facilitation of a whole-class discussion.

Providing a Common Artifact for Collaborative Reflection

The specificity of transcripts and the PTs' detailed analyses gave us, as MTEs, a better understanding of the PTs' conceptions of effective teaching. It was evident that some PTs characterized aspects of the discourse differently than their methods instructor would have described it, mirroring what Kennedy (1999) described as the *problem of enactment*. The elementary PTs noted each question they asked and determined whether the questions helped or hindered student thinking. For example, Corey wrote the following with regard to his most effective question:

> The most effective question was, "what is it called when you split them into threes?" I think this is an effective question because I am asking him if he knows if you call a whole split into three pieces as thirds. Also, the question does nothing to reveal the answer to the student.

From Corey's response, it seemed he defined an effective question as prompting mathematical vocabulary. After de Araujo read his transcript, she thought there were other, more effective questions he could have chosen that focused on exploring the student's thinking. This difference between perspectives exemplifies Kennedy's (1999) problem of enactment. Using the transcript as a common artifact, it was possible for de Araujo to discuss effective questioning with Corey to refine his professional knowledge.

CONCLUSION

Looking across the two implementations of the analyzing transcripts of teaching activity reveals the potential of this activity for the development of reflective novice teachers. The activity serves as a clear connection of field experience to the content of the methods course. This connection emphasizes our situative perspective on learning to teach mathematics (Peressini et al., 2004) in that the activity was purposefully structured to engage PTs in an authentic activity as a means to situate their learning about the complexities of teaching through teaching. Additionally, the transcripts allowed more realistic critiques of the PTs' practice connected to course readings and discussions. Written analysis of the transcript helped the PTs to operationalize various pedagogical constructs and supported MTEs in understanding and responding to the analysis. Experiences of this type can also inform MTEs about their students' understandings of important aspects of classroom practice. Methods instructors rarely have access to records of instruction for all of their students, and the opportunity to interact with each PT about his or her practice is an important affordance.

Analyzing transcripts of teaching has potential for use in courses for elementary, middle, and secondary PTs. It can be useful for those at the beginning of their preparation programs as well as those who are approaching their student-teaching experience. We have used it to examine aspects of discourse and student thinking, but we see its potential to address many of the goals of methods courses, such as issues of equity or assessment. This kind of activity incorporates the best aspects of the literature on rehearsal (Lampert et al., 2013) and positions teaching as an explicit object of reflection for learning and growth.

REFERENCES

Ball, D. L., & Forzani, F. M. (2011). Building a common core for learning to teach and connecting professional learning to practice. *American Educator, 35*(2), 17–21, 38–39.

Chapin, S., O'Connor, C., & Anderson, N. (2009). *Classroom discussions: Using math talk to help students learn, grades K–6* (2nd ed.). Sausalito, CA: Math Solutions.

Corbin, J., & Strauss, A. (2008). *Basics of qualitative research: Techniques and procedures for developing grounded theory* (3rd ed.). Thousand Oaks, CA: SAGE.

Empson, S. (1995). Research into practice: Young children's fraction thinking and the curriculum. *Teaching Children Mathematics, 2,* 110–114.

Explain Everything Inc. (2015). *Explain everything.* Retrieved from http://explain everything.com/

Hiebert, J., Morris, A. K., Berk, D., & Jansen, A. (2007). Preparing teachers to learn from teaching. *Journal of Teacher Education, 58,* 47–61.

Kennedy, M. M. (1999). The role of preservice teacher education. In L. Darling-Hammond & G. Sykes (Eds.), *Teaching as the learning profession: Handbook of teaching and policy* (pp. 54–86). San Francisco, CA: Jossey Bass.

Lampert, M., Franke, M. L., Kazemi, E., Ghousseini, H., Turrou, A. C., Beasley, H., Cunard, A., & Crowe, K. (2013). Keeping it complex: Using rehearsals to support novice teaching learning of ambitious teaching. *Journal of Teacher Education, 64,* 226–243.

Lave, J. (2009). The practice of learning. In K. Illeris (Ed.), *Contemporary theories of learning: Learning theorists . . . in their own words* (pp. 200–208). London, England: Routledge.

McDonald, M., Kazemi, E., & Kavanagh, S. S. (2013). Core practices and pedagogies of teacher education: A call for a common language and collective activity. *Journal of Teacher Education, 64,* 378–386.

Morris, A. K. (2006). Assessing pre-service teachers' skills for analyzing teaching. *Journal of Mathematics Teacher Education, 9,* 471–505.

Peressini, D., Borko, H., Romagnano, L., Knuth, E., & Willis, C. (2004). A conceptual framework for learning to teach secondary mathematics: A situative perspective. *Educational Studies in Mathematics, 56,* 67–96.

Singletary, L. M., & Conner, A. (2015). Focusing on mathematical arguments. *Mathematics Teacher, 109*(2), 143–147.

Snake on a Plane. (n.d.). *Illustrative Mathematics.* Retrieved from https://www.illustrativemathematics.org/content-standards/tasks/1695

Stein, M. K., Smith, M. S., Henningsen, M. A., & Silver, E. A. (2009). *Implementing Standards-based mathematics instruction: A casebook for professional development* (2nd ed.). New York, NY: Teachers College Press.

CHAPTER 17

DOING MATHEMATICS ACROSS LANGUAGES

Exploring Possibilities for Supporting Emergent Bilinguals' Mathematical Communication and Engagement

Frances K. Harper
University of Tennessee

Wendy B. Sanchez
Kennesaw State University

Beth Herbel-Eisenmann
Michigan State University

Given the growing number of students learning English as an additional language in public schools across the United States (National Center for Education Statistics, 2012), all teachers must be well equipped for the complex task of teaching emergent bilinguals (see García, Kleifgen, & Falchi, 2008, for a discussion of using *emergent bilinguals* versus *English Language Learners*).

Building Support for Scholarly Practices in Mathematics Methods, pages 263–277
Copyright © 2018 by Information Age Publishing
All rights of reproduction in any form reserved.

Teachers often look at their subject *through* language but rarely *at* the language required to interact with the content (Lucas, Villegas, & Freedson-Gonzalez, 2008). Language is the primary mode by which teachers make sense of content and students' interactions with content, yet they rarely explicitly examine the essential role language plays in teaching and learning. Planning effective lessons for emergent bilinguals necessitates focusing on language, and teacher educators play an important role in guiding prospective teachers (PTs) to explore how language and other modes of communication mediate academic learning (Lucas et al., 2008). In this chapter, we, three mathematics teacher educators (MTEs), describe one of our efforts to look *at* language in our secondary mathematics methods courses.

Our emphasis on preparing PTs to work with emergent bilinguals was guided by a sociopolitical perspective that considers issues of power, privilege, and identity (Gutiérrez, 2013). This perspective rejects deficit discourses about emergent bilinguals, views differences as assets for learning, and challenges larger structural inequities that limit learning. We aim to draw attention to how students are positioned during mathematics discussions so that PTs might also consider the relationships among social, cultural, and political aspects of mathematics learning and mathematics classroom discourse (Herbel-Eisenmann, Choppin, Wagner, & Pimm, 2012). This goal extends beyond PTs' development of specific teaching practices for supporting emergent bilinguals and instead aims for deeper understanding of and empathy for emergent bilinguals' experiences in mathematics.

Although we recognize limitations of an isolated activity to accomplish broader goals for PTs' sociopolitical awareness, we describe our enactment of a single activity designed to address narrower goals (described in the Methods section) for developing equitable teaching for emergent bilinguals. In this activity, we limited PTs' English use and introduced additional languages while doing mathematics with the hope that PTs' experiences communicating mathematically would approximate emergent bilinguals' experiences and would make the role of language in mathematics teaching and learning more transparent. The following question guided our inquiry: How do PTs relate their experiences with limited English use while doing mathematics to mathematics teaching and learning for emergent bilinguals?

MATHEMATICS TEACHING AND LEARNING
WITH EMERGENT BILINGUALS

Emergent bilinguals face difficulties inherent in learning mathematics while also acquiring new vocabulary and learning multiple meanings of words in conversational and academic English. A focus on vocabulary acquisition and learning multiple meanings, however, can overemphasize obstacles emergent bilinguals face and encourage a deficit view of their abilities to learn mathematics (Moschkovich, 2002, 2007). Because developing grade-level appropriate academic language takes years, teachers must adapt instruction to productively engage emergent bilinguals with conceptually rich, grade-level appropriate mathematics (Lucas et al., 2008; Moschkovich, 2002, 2007). Such linguistically responsive pedagogy relies on teachers' strong familiarity with emergent bilinguals' linguistic and academic backgrounds and an understanding of language demands (Lucas et al., 2008).

Although emergent bilinguals' needs differ from their peers', these differences should not be framed as deficiencies (Moschkovich, 2007). Instead, learning about emergent bilinguals and language demands can help teachers recognize that mathematical communication involves more than language (Lucas et al., 2008; Moschkovich, 2002). When teachers and students value gestures and objects (e.g., drawings, manipulatives) in mathematical communication, emergent bilinguals can demonstrate competencies that might not be evident with language alone (Moschkovich, 2007). Moreover, when given opportunities to engage with others to negotiate meaning in a welcoming environment and, at times, through their primary languages, emergent bilinguals' learning can flourish (Lucas et al., 2008).

Group work crafted for genuine collaboration (e.g., complex instruction; Cohen, Lotan, Scarloss, & Arellano, 1999) can support emergent bilinguals to use resources and engage meaningfully in mathematical work. When group norms support delegation of authority to students, students can accomplish tasks however they choose but are accountable to each other and the teacher for final products (Cohen et al., 1999). This requires students to listen to each other and value diverse contributions (Zahner, 2012). Tasks with multiple entry points also reinforce accountability by fostering positive interdependence and leveraging abilities beyond language ability (Cohen et al., 1999; Zahner, 2012). Students' use of multiple abilities creates opportunities for teachers to promote equitable interactions within groups by assigning mathematical authority and competence to emergent bilinguals and valuing their primary language use (Cohen et al., 1999; Zahner, 2012). In designing our activity, we sought to make transparent some of the language demands inherent in learning mathematics, but we also wanted PTs to recognize how they (and emergent bilinguals) could use multiple modes of communication while doing mathematics collaboratively.

RESEARCH CONTEXT AND METHODS

To focus on opportunities created by restricting English use in a mathematics activity, we attempted to replicate our enactments in two methods courses at different universities. Using a multiple-case, embedded case study design (Yin, 2009), we compared findings from the two contexts and considered, holistically, how the activity enactments created opportunities for PTs to engage with ideas about equitable mathematics teaching and learning for emergent bilinguals.

Activity Description and Goals

The activity was adapted from professional development materials created within the *Mathematics Discourse in Secondary Classrooms* project[1] (Herbel-Eisenmann, Cirillo, Steele, Otten, & Johnson, in press; see Cirillo et al., 2014; Herbel-Eisenmann, Steele, & Cirillo, 2013). The original collaborative activity, *Hidden Triangles*, explores triangle congruence. During each round, the *director* provides information about sides and angles of a triangle (without showing it) to their group mates, the *drawers*. Each drawer works behind a divider to sketch a congruent copy of the triangle. The hidden nature of the work fosters positive interdependence and ensures accountability—everyone helps by explaining or questioning but everyone does the work for themselves. Students explore various combinations of sides and angles, towards the goal of recognizing that particular combinations (e.g., side-angle-side) are sufficient for determining triangle congruence.

PTs completed initial rounds in English. Then, for 1–3 rounds, the directors used a language other than English. Our goal for this adaptation was to create opportunities to unpack ideas from a course reading (Zahner, 2012) that highlights four concerns for group work with emergent bilinguals in mathematics: (a) selecting appropriate tasks, (b) assigning students to groups, (c) setting up group norms, and (d) assessing students' understanding (p. 156). We aimed for PTs to look more closely *at* language and other modes of communication in their own experiences with the activity in order to reflect, during a whole-class discussion and in writing, on supporting emergent bilinguals' mathematical engagement as described in the reading. These prompts guided the PTs' reflections: (a) What features of the activity helped you engage with the mathematics, even though you did not speak the same language as the director?; (b) Thinking about your own experience and the reading you did before this class, what strategies can you use as a teacher to give all students, regardless of language ability, access to the mathematics?; (c) How can language diversity be a resource?

Description of Methods Courses and PTs

We enacted the activity approximately halfway through the semester in two secondary mathematics methods courses, one co-taught by Harper and

Herbel-Eisenmann at Michigan State University (MSU) and one taught by Sanchez at Kennesaw State University (KSU) in Georgia. There were 24 (9 male, 15 female) PTs in their senior year at MSU and 20 (6 male, 14 female) PTs in their junior year at KSU. All PTs were traditional undergraduate students pursuing degrees in mathematics/mathematics education. PTs at MSU had more experience in field placements and in secondary mathematics methods and language/literacy courses. MSU PTs also had more experience with complex instruction (having completed a 6-week unit). Sanchez briefly introduced complex instruction using materials adapted from Harper (2016).

At MSU, PTs select a teaching minor, and five PTs were pursuing minors in foreign languages (1 French, 2 Spanish, 2 German). These PTs were proficient enough to speak their additional language as a director but were not necessarily fluent. KSU does not require a minor, but four students spoke additional languages well enough to serve as directors (1 German-as-an-additional-language; 1 Korean-as-primary-language; 2 Spanish-as-primary-language, 1 of whom also spoke Japanese as an additional language), and Sanchez's husband and son (Spanish-as-an-additional-language speakers) also served as directors.

Data Sources and Analysis

Data sources included transcripts of whole-class discussions focused on language/pedagogy and written reflections. Initially, we used iterative rounds of open coding to identify themes for the content of PTs' reflections on their experiences during whole-class discussion. We applied the same codes to written reflections. Next, we categorized all coded excerpts across the data using the four concerns for group work with emergent bilinguals from Zahner (2012). Finally, we compared themes across the contexts, looking for similarities and differences in the PTs' reflections on their own experiences and their ideas for teaching emergent bilinguals.

FINDINGS AND DISCUSSION

Despite differences in contexts, PTs in both courses emphasized the same two considerations for group work with emergent bilinguals from the reading (Zahner, 2012), namely *selecting tasks* and *establishing norms*. Although PTs in the different contexts made sense of readings in subtly different ways, we saw more similarities than differences in the overarching themes. Thus, we concluded that the activity itself afforded opportunities to consider various modes of communication in equitable mathematics teaching and learning. In this section, we discuss how PTs related their experiences during the activity to mathematics teaching and learning for emergent bilinguals. Although PTs' reflections mostly suggested an asset-based view of emergent bilinguals, potentially problematic ideas about mathematics and language, which could indicate deficit perspectives of emergent bilinguals, surfaced from some statements. We share both promising ways PTs related their experiences to ideas from the reading and the potentially problematic statements that arose so other MTEs might use this activity to foster asset-based views of emergent bilinguals in mathematics.

Selecting Tasks

Zahner (2012) emphasized the importance of selecting *group worthy* tasks to support emergent bilinguals' mathematical learning. Group worthy tasks for emergent bilinguals have four characteristics: They (a) focus on central mathematical concepts or connections, (b) invite collaboration, (c) allow for multiple entry points, and (d) avoid unnecessarily complex language (Zahner, 2012, p. 157). PTs in both courses discussed all four characteristics when reflecting on their experiences. Although our analysis of PTs' discussions and reflections suggested that the activity created opportunities for PTs to think differently about triangle congruence postulates (i.e., the focus on central mathematical concepts or connections), we felt such claims would require analysis of data sources beyond the scope of this chapter. Consequently, we limit our focus to the way PTs made sense of their experiences with the latter three characteristics of group worthy tasks and how they related their experience with those characteristics to equitable mathematics teaching and learning for emergent bilinguals.

Invite Collaboration

PTs at KSU, who had limited experience with complex instruction, tended to focus on their experiences with collaboration more broadly. For example, some PTs described how the structure of Hidden Triangles pushed them outside their comfort zone for working in groups either by letting other people lead as a drawer (Mara, KSU; all names are pseudonyms) or by taking more of a lead as the director (Angel, KSU). At MSU, where PTs had engaged in more extended work with complex instruction, PTs' reflections on collaboration focused on variations in group interactions between English rounds to limited-English rounds. During English rounds, PTs described collaboration as interactions among the director and individual drawers who would ask clarifying questions. Kim (MSU) explained, "The director was the person that we would look to for advice on how to solve the problem. Even when other people in the group knew what was happening, we often automatically looked at the director for guidance." During the limited-English rounds, however, drawers relied more on other drawers. Helena (MSU) offered her perspective as a director speaking a foreign language. "Since [none of the drawers knew] what I was saying, it took multiple people to be like, 'Oh I think she means this' or 'I think she means that.' It wasn't an individual activity. It required multiple people to contribute to the group in order to understand what the [director] was saying." Chloe (MSU) corroborated, saying, "There was a lot more working together trying to figure it out.... There was a lot more teamwork versus figuring it out on your own." MSU PTs' reflections suggest that limiting the use of English more authentically approximated how high school students might rely on positive interdependence to work through a collaborative task. We see this recognition of the importance of true collaboration as supporting an asset-based view of emergent bilinguals as students who are capable of engaging in challenging group work when the task design encourages positive interdependence.

Allow for Multiple Entry Points

PTs utilized the activity's multiple entry points to draw on multiple abilities as resources (e.g., relying on gestures and physical objects), some of which are typically undervalued in mathematics (Moschkovich, 2007). Leveraging multiple abilities during limited-English rounds was the most common theme across the data. In general, PTs noted how the limited-English round required more creativity, both in using "other means [besides language] to get the point across" (Jessica, KSU) and in using language. For example, Diana (MSU) described a creative use of language: "When we were trying to figure out the angle, like 45...we knew the numbers one through ten [in French], so we said what number's in the tens position and what number's in the ones position." PTs drew on gestures and their abilities to decipher foreign languages to communicate mathematically. We view these experiences with multiple entry points as supporting an asset-based perspective of emergent bilinguals as students who can creatively communicate with others and engage in mathematics.

Using gestures. PTs at KSU emphasized hand gestures. "The shape of a triangle is easy to make, so that was something [directors] could make with their hands...to give information about the triangle" (Natalie, KSU). Directors used "hand motions to get their point across such as angles, lines, points, and numbers" (Jessica, KSU). PTs at MSU mainly used hand gestures to communicate numbers, and PTs in both courses found other ways to gesture to relate words from conversational English to information about triangles. For example, Chloe (MSU) communicated "side (of the triangle)" by "pointing to the side of [her] body." Directors used objects to support gesturing, such as "pointing at the ruler for the measurements"

(Angel, KSU) and using a pen to represent an orthogonal line (Frank, MSU). Because directors' heavy reliance on gestures approximates ways emergent bilinguals might communicate mathematically, PTs' experiences highlighted the importance of creating entry points for emergent bilinguals, other students, and the teacher to use gesture to communicate mathematical ideas across languages.

Deciphering foreign language. Directors who spoke diverse languages gave PTs opportunities to reflect on the ease or difficulty of moving between languages in mathematics. Korean was most challenging, and communication relied mostly on gestures (Natalie, KSU). In contrast, drawers generally found Spanish fairly easy to decipher. "For Spanish, I feel like there's a lot more similarities [with English], with at least this context, there's like centimeters, *centimetros*" (Frank, MSU). Many PTs had some background in Spanish, at least with numbers, "which made it not too difficult" (Tammy, KSU). Drawers primarily relied on peers able to "pick up on certain words that might have sounded similar to [familiar words] . . . and then relay that in the other language to the rest of the group" (Grace, MSU). Once drawers deciphered some of the language they "were able to draw the triangle fairly quickly and accurately" (Mara, KSU). Drawers' attention to familiar words and use of collaboration to decipher language approximates ways emergent bilinguals might make sense of English alongside mathematics when tasks support such forms of mathematical communication.

Avoid Unnecessarily Complex Language

Zahner (2012) focused on avoiding unnecessarily complex language in *written* tasks, but PTs recognized the need to also adjust *spoken* language when working with emergent bilinguals. PTs highlighted two particular aspects of language when reflecting on their interactions with the director during limited-English rounds: (a) explicitly attending to the language of mathematics and (b) simplifying and slowing the pace of language. Again, this recognition of the role of language in teaching and learning fosters asset-based views of emergent bilinguals as capable mathematical learners when language supports are present.

Explicitly attending to the language of mathematics. PTs recognized that all students are learning the language of mathematics. Several MSU directors explained they learned numbers but no other mathematical vocabulary in their additional language, and they struggled with the academic language of mathematics. For example, Helena (MSU) said she did not know the word for *angle* in Spanish, so she just said "angle," approximating emergent bilinguals' struggles to communicate mathematics in an additional language. KSU PTs did not struggle as much with academic language because most of the directors were primary speakers of their language. Nonetheless, KSU PTs also recognized challenges of learning the language of mathematics. Nate (KSU) explained, "We shouldn't discredit mathematics itself as a language . . . [students] don't innately know what [the word] function means even though it is English."

Adding the English restriction to a mathematical task sensitized PTs to consider the development of mathematical language, mathematical understanding, and the English language all simultaneously. PTs considered how a student's primary language could be a resource (i.e., an asset) in making mathematical connections and developing mathematical language. Diana (MSU) explained the importance of allowing students to reason and make sense of mathematics in their primary language: "Obviously you want [students] to be using English . . . but in a math classroom you want them to understand the content. . . . If they're making connections and reasoning through it, then allow them to do that in both languages." As emergent bilinguals learn the language of mathematics, they need to be able to connect it to the mathematical structures they have already developed in their primary language. Angel (KSU) suggested employing "notation used in the language of mathematics. By doing this, you are giving the words a symbol that connects it to the lessons. The English learner [can] associate the word to a physical thing." Diana (MSU) recognized the need for teachers to know their emergent bilinguals' linguistic histories. For example, she pointed out that notation can be different in different languages. She explained, "Word order in Spanish is different, so specifically Spanish speakers will struggle with word order." Students in both methods courses recognized that mathematics is itself a language and that emergent bilinguals' primary language should be used as a resource in developing the language and concepts of mathematics.

Simplifying and slowing the pace of language. PTs recognized the importance of using simplified spoken language during limited-English rounds. Chloe (MSU) explained: "I wasn't saying, 'vertices,' in Spanish. I was saying, 'point.' I wasn't saying, 'orientation.' I was saying, 'the side goes here.'" Similarly, Nate (KSU) said, "In English, we say, 'the angle is so and so [degrees]', but the only three words . . . are *angle, so and so,* [*degrees*]. That gets the point across, good enough." Through limited-English rounds, drawers recognized what helped *them* understand an unfamiliar language, and directors found ways to communicate. PTs at KSU were hearing mostly primary speakers during the foreign language rounds and therefore had to ask the directors to slow the pace of their speech. Jason (KSU), for example, discovered, "I will need to slow down when I am trying to explain something [as a teacher]. I had to say, 'repeat that please' or 'slow down please' several different times. I will need to ask if [students] understand what I am explaining." The experience of communicating with limited English helped PTs understand the importance of hearing an unfamiliar language at a slower than typical pace. Consequently, PTs recognized a need to adjust their own communication practices to better support emergent bilinguals' mathematics learning. PTs may also be better positioned to help their future students who will work with emergent bilinguals in groups understand the importance of using simpler language for group communication.

Establishing Norms

Zahner (2012) emphasized establishing norms for how students work together with their peers. The author identified two norms that are particularly crucial for group work in classrooms with emergent bilinguals: (a) all students listen to one another, value contributions, and respond to ideas, and (b) students' languages are valued. During their discussions and in their reflections, PTs connected the three considerations for selecting tasks, discussed previously, to these two norms that foster productive collaboration and an asset-based view of emergent bilinguals, and they extended ideas from the readings to identify a third norm: Challenging mathematics is fun.

All Students Listen to One Another, Value Contributions, and Respond to Ideas

PTs' own successes, despite English restrictions, were important for helping them see possibilities for collaboration, multiple entry points, and multiple abilities to support emergent bilinguals' mathematics learning (i.e., an asset-based view). Reflecting on their own experiences reinforced how "using hand gestures and other visual tools helps give all students access to mathematics" (Jason, KSU). Moreover, positive interdependence was essential to each group's success, positioning diverse contributions as valuable and highlighting listening and responding to each group member. PTs also extended beyond their own experiences to imagine how pictures/diagrams or manipulatives "are a good start to give all students access to mathematics" (Jayden, KSU). Reflecting on another assigned reading from the MSU course (Shulman, Lotan, & Whitcomb, 1998), Becky (MSU) noted, "Miguel [an emergent bilingual]...was being picked on...and students weren't listening to him, but [the group] had a diagram [that Miguel drew], and [everyone] was following the diagram because everyone could see it...so that could help." PTs' experiences encouraged a deeper appreciation for genuine collaboration and the use of pedagogical practices highlighted in various course activities and readings, such as leveraging multiple abilities.

Students' Languages Are Valued

PTs expressed both asset- and deficit-based views of emergent bilinguals and linguistic diversity in the mathematics classroom. Here we focus on the four themes related to valuing students' languages as assets that emerged from PTs' reflections on their experiences: (a) linguistic diversity as an asset for *all* students, (b) emergent bilinguals as leaders, (c) emergent bilinguals' comfort, and (d) respect for emergent bilinguals. Then, we conclude the findings section by attending to the potentially problematic statements that suggested deficit views of emergent bilinguals.

Linguistic diversity as an asset for *all* students. Linguistic and cultural diversity brings unfamiliar ideas, representations, and mathematical approaches into classrooms that can potentially expand learners' ways of thinking mathematically. For example, Kim (MSU) described a Chinese student in her Complex Analysis class who solved a problem in a unique way. The professor said, "That's how they learn in China. They deal with geometric [approaches] more than algebra[ic]." Kim suggested that all of the students in the class benefited from the different perspective offered by the Chinese student. Emergent bilinguals also bring vocabulary that does not translate directly into English. For example, Grace (MSU) explained, "[In French] there's a word that means . . . mediocre . . . and we don't have a word for it in English." Learning about words in different languages can help all students expand their understanding of concepts. PTs at both universities (Natalie, KSU and Grace, MSU) commented on learning the first 10 numbers in Korean and French, respectively, by engaging in Hidden Triangles. Natalie said, "When they're talking, we get what they're talking about. We're learning their language so that we're able to more easily communicate with them as well." PTs seemed to value what they, and emergent bilinguals, could learn from each other by attending to language in mathematics class.

Emergent bilinguals as leaders. Positioning emergent bilinguals as competent leaders was a more prevalent theme at KSU than MSU. We posit this is because there were KSU PTs who were emergent bilinguals themselves who highlighted features of the activity that others may not have recognized. For example, Angel (KSU) said, "We had control . . . on speed, on how we talked, on everything. . . . When you don't have the control, it's more difficult." After Angel's suggestion, other PTs gravitated towards its importance. Mara (KSU) explained, "Another strategy is to use activities such as this that include using a student's [primary] language as part of the lesson. First, it allows an English learner . . . to be in control of a class activity, something they probably do not get to do often." Similarly, Sandy (KSU) said, "Allowing the students to teach each other using their [primary] languages allows them to feel like they have some control over their learning." When emergent bilinguals are positioned as competent leaders, it helps all students recognize assets that emergent bilinguals bring to mathematical discussions and allows emergent bilinguals to gain confidence in both communication and mathematics.

Emergent bilinguals' comfort. PTs recognized emergent bilinguals' potential discomfort in mathematics classrooms. MSU PTs who were foreign language minors felt uncomfortable when using their language to speak mathematically. Their education had not prepared them to use the academic language of mathematics. This may sensitize them to their emergent bilingual students' discomfort when trying to speak mathematically in English. Diana (MSU) discussed an Asian student in microteaching lab, which took place in a college mathematics course: "She didn't know how to communicate with anybody, and it was her turn to go and she sat there, she didn't feel comfortable speaking." At KSU, Angel explained, "When one is not from the country they live in and they are just starting to learn the language and the customs, they feel blind, deaf, and mute." PTs at both universities expressed a desire to create comfortable learning environments for emergent bilinguals and suggested ways to do so. For example, Sandy (KSU) suggested, "Visual representations...assist students regardless of the language they use and perhaps even working to identify language-specific words that will help them understand...will help the students not feel so out of their comfort zone." Similarly, Helena (MSU) explained, "You don't have to learn [their] whole language but...showing that you're willing to take time away to get to know [emergent bilinguals] and put forth that effort shows that you care and could form trust between the two of you." Through the experience of Hidden Triangles, PTs developed empathy for emergent bilinguals and reflected on ways to make all students feel comfortable in their learning environment.

Respect for emergent bilinguals. Beyond empathy for emergent bilinguals, PTs' level of respect for emergent bilinguals increased when they were asked to engage with mathematics with limited English use. Kim (MSU) expressed her frustration: "I was trying to listen to both languages and I got really frustrated. Then I gave up." Leslie (KSU) realized, "I think some of the biggest things I could give to [emergent bilinguals] are understanding and grace. I now understand...what it is like to not know what the teacher is saying." The activity approximates the experience of being an emergent bilingual for PTs. Although the experience alone will not prepare PTs for equitable teaching, restricting English added a new level of complexity and provided new opportunities for reflection.

Challenging Mathematics Is Fun

PTs extended the norms discussed by Zahner (2012) and emphasized how Hidden Triangles was "more fun in the foreign language" (Diana, MSU). Leslie (KSU) explained, "I enjoyed the challenges of [the limited-English round] more than the easy [English] round. I wasn't 100% sure if my triangle was correct, but that made it all the better when I found out it was." Their experience drew attention to how mathematics can be both challenging and fun for emergent bilinguals. This norm suggests an asset-based view of emergent bilinguals as capable of engaging in and enjoying challenging mathematics, even though their academic language use might be limited. Such a norm supports the idea that emergent bilinguals should not be limited to remedial vocabulary acquisition and should be given opportunities to do challenging mathematical work.

Potentially Problematic Ideas About Mathematics and Language

PTs expressed an overall asset-based view of emergent bilinguals; however, some statements implied potentially problematic ideas about mathematics and language. For example, Olivia (MSU) described how she "found it more beneficial to dumb down . . . mathematical language as the director in order to help the drawers." Olivia's statement suggests she recognized the potential for emergent bilinguals to engage fully in mathematics through simplified language (an asset-based view), but her choice of the words "dumb down" suggests a deficit-based perspective. Mara (KSU) noted, "Because mathematics is a universal language, students can discover mathematics in their own language if they are given the freedom to explore it." Mara focuses on emergent bilinguals' use of language as a resource for learning, but her reference to "universal language" suggests a fixed view of the language of mathematics, which might encourage emphasis on memorized vocabulary or rules rather than fostering "the communicative competence necessary and sufficient for successful participation in mathematical discourse" (Moschkovich, 2012, p. 94). Additionally, Mara's view minimizes the language demands required for communicating about mathematics (Schleppegrell, 2007). Similarly, PTs questioned the feasibility of the pedagogical practices they highlighted as most beneficial to emergent bilinguals because of time constraints. Their dismissal of these instructional adaptations for emergent bilinguals may stem from lingering deficit-based perspectives that assume the responsibility for learning rests more heavily with the student than the teacher.

CONCLUSION

The design of the activity paired with the restriction of English use approximated the experiences of emergent bilinguals in mathematics classrooms. Firsthand experience with the activity from the student perspective allowed PTs insight into the complex pedagogical practices for supporting emergent bilinguals' mathematics learning. Yet PTs had fewer opportunities to consider two ideas from the assigned reading: assigning emergent bilinguals to groups and assessing student understanding. By framing the activity from the student perspective, PTs had limited opportunities to consider some "behind-the-scenes" work of teaching. Nevertheless, this activity was effective in creating opportunities for PTs to look *at* language in relation to equitable teaching and learning.

A single activity is insufficient for supporting PTs' understanding of sociopolitical dimensions of mathematics teaching and learning and to adopt, fully, asset-based views. The complexity of these issues played out in the ways that PTs expressed both asset- and deficit-based views of emergent bilinguals. We hope that other MTEs will view this activity as one possible starting point for developing PTs' sociopolitical awareness and asset-based perspectives. Our reporting might give MTEs a sense of what to expect if they enact this activity in their secondary mathematics methods courses and highlight ideas that might emerge as PTs reflect on this activity in relation to teaching and learning for emergent bilinguals. PTs' reflections provided us, and hopefully other MTEs, insights into supporting PTs' evolving understanding of mathematics teaching from an asset-based perspective of emergent bilinguals and into working to disrupt lingering deficit-perspectives about emergent bilinguals. Our analysis also illuminated some pedagogical considerations for emergent bilinguals (e.g., assigning emergent bilinguals to groups) that must be emphasized in our methods courses in other ways. We encourage other MTEs to consider adapting mathematics tasks so that PTs' experiences more closely approximate the experiences of future students, while at the same time balancing the emphasis on student perspective with the behind-the-scenes work of teaching.

NOTE

1. The research reported in this chapter was supported with funding from the National Science Foundation (NSF, Award #0918117). Any opinions, findings, and conclusions or recommendations expressed in this material are those of the authors and do not necessarily reflect the views of the NSF.

REFERENCES

Cirillo, M., Steele, M. D., Otten, S., Herbel-Eisenmann, B., McAneny, K., & Riser, J. Q. (2014). Teacher discourse moves: Supporting productive and powerful discourse. In K. Karp (Ed.), *2015 Annual perspectives on mathematics education: Using research to improve instruction* (pp. 141–149). Reston, VA: National Council of Teachers of Mathematics.

Cohen, E. G., Lotan, R. A., Scarloss, B. A., & Arellano, A. R. (1999). Complex instruction: Equity in cooperative learning classrooms. *Theory Into Practice, 38*(2), 80–86.

García, O., Kleifgen, J. A., & Falchi, L. (2008). *From English language learners to emergent bilinguals. Equity Matters. Research Review, 1.* New York, NY: Campaign for Educational Equity, Teachers College, Columbia University.

Gutiérrez, R. (2013). The sociopolitical turn in mathematics education. *Journal for Research in Mathematics Education, 44,* 37–68.

Harper, F. K. (2016). Complex instruction. *Solving world problems.* Retrieved from http://francesharper.com

Herbel-Eisenmann, B., Choppin, J., Wagner, D., & Pimm, D. (Eds.). (2012). *Equity in discourse for mathematics education: Theories, practices, and policies.* New York, NY: Springer.

Herbel-Eisenmann, B., Cirillo, M., Steele, M. D., Otten, S., & Johnson, K. R. (in press). *Mathematics discourse in secondary classrooms: A case based professional development curriculum.* Sausalito, CA: Math Solutions.

Herbel-Eisenmann, B., Steele, M., & Cirillo, M. (2013). (Developing) Teacher discourse moves: A framework for professional development. *Mathematics Teacher Educator, 1*(2), 181–196.

Lucas, T., Villegas, A. M., & Freedson-Gonzalez, M. (2008). Linguistically responsive teacher education: Preparing classroom teachers to teach English language learners. *Journal of Teacher Education, 59*(4), 361–373.

Moschkovich, J. (2002). A situated and sociocultural perspective on bilingual mathematics learners. *Mathematical Thinking and Learning, 4,* 189–212.

Moschkovich, J. (2007) Bilingual mathematics learners: How views of language, bilingual learners, and mathematics communication affect instruction. In N. S. Nasir & P. Cobb (Eds.), *Improving access to mathematics: Diversity and equity in the classroom* (pp. 89–104). New York, NY: Teachers College Press.

Moschkovich, J. N. (2012). How equity concerns lead to attention to mathematical discourse. In B. Herbel-Eisenmann, J. Choppin, D. Wagner, & D. Pimm (Eds.), *Equity in Discourse for Mathematics Education* (pp. 89–105). Dordrecht, The Netherlands: Springer Netherlands.

National Center for Education Statistics. (2012). *The condition of education 2012. Indicator 8: English language learners in public schools.* Washington, DC: U.S. Department of Education. Retrieved from http://nces.ed.gov/pubs2012/2012045_2.pdf

Schleppegrell, M. J. (2007). The linguistic challenges of mathematics teaching and learning: A research review. *Reading and Writing Quarterly, 23,* 139–159.

Shulman, J., Lotan, R. A., & Whitcomb, J. A. (1998). Case 13: The chance I had been waiting for. In *Groupwork in diverse classrooms: A casebook for educators* (pp. 69–71). New York, NY: Teachers College Press.

Yin, R. K. (2009). *Case study research: design and methods* (4th ed.). Los Angeles, CA: SAGE.

Zahner, W. C. (2012). ELLs and group work: It can be done well. *Mathematics Teaching in the Middle School, 18*(3), 156–164.

CHAPTER 18

USING MATHEMATICS AUTOBIOGRAPHY STORIES TO SUPPORT EMERGING ELEMENTARY MATHEMATICS TEACHERS' SOCIOPOLITICAL CONSCIOUSNESS AND IDENTITY

Anne Marie Marshall
Lehman College

Theodore Chao
The Ohio State University

A common activity in an elementary mathematics methods course for prospective teachers (PTs) to confront their beliefs, dispositions, and histories surrounding mathematics is the mathematics story or mathematics autobiography. In this chapter, we analyze the literature on the use of mathematics autobiographies while reflecting on how we have used this activity in

Building Support for Scholarly Practices in Mathematics Methods, pages 279–293
Copyright © 2018 by Information Age Publishing

our own teaching. Anne Marie currently works at Lehman College in the Bronx, NY, an urban setting where many of her PTs are students of color and first-generation Americans. Theodore currently works at the Ohio State University in Columbus, OH, a Midwestern setting where many of his PTs self-identify as White and grew up in rural or suburban regions of Ohio.

We bring to this work an understanding of our role as mathematics teacher educators (MTEs) in helping to develop our PTs' sociopolitical awareness. We draw on Gutiérrez's (2013) and others' (de Freitas, 2008; Felton-Koestler, 2015) construct of sociopolitical mathematics teacher education within a context of schooling that often fosters "unexamined ideologies and myths that shape commonly accepted ideas and values in a society" (Nieto & Bode, 2008, p. 7). We believe that teacher identity (Drake, Empson, Dominguez, Junk, & LoPresto, 2003; Drake, Spillane, & Hufferd-Ackles, 2001; de Freitas, 2008) is strongly related to preparing socially and culturally competent teachers. We also believe the mathematics methods course is a rich space for PTs to learn not only about content and pedagogy related to teaching mathematics but also about becoming a community of learners working to develop a critical understanding of mathematics and what it means to be a teacher of mathematics. The mathematics autobiography story serves as a powerful assignment in mathematics teacher identity development in ways that are consistent with what Gutiérrez (2013) referred to as identity as a verb rather than a noun; PTs see their mathematics teacher identity changing as they realize their own power.

Autobiography/Stories

The modern mathematics story assignment descends directly from McAdams' (1993) story protocols for teachers to reflect on their own learning experiences, adapted specifically for mathematics teachers by Drake and colleagues (Drake, 2006; Drake et al., 2001). This mathematics story assignment focuses on PTs identifying highs, lows, and turning points in mathematics (LoPresto & Drake, 2005). The purpose of the mathematics story work is to help MTEs understand PTs' relationships to mathematics, which can then be used to design instruction (methods courses) to best meet the needs of those PTs (Drake, 2006; Drake et al., 2001; Empson, Drake, & Junk, 2002; McAdams, 1993). In fact, LoPresto and Drake (2005) published guidelines detailing exactly how to use the mathematics story when working with PTs and inservice teachers, complete with a ready-to-use protocol for MTEs. PTs often use this autobiography activity to share stories about discouraging experiences as learners of mathematics (Drake et al., 2001), pinpointing events that led to declines in their interest, efficacy, and confidence at some point in their elementary or early secondary school years.

This mathematics story work has had a major impact on preparing PTs, particularly within research-based activities for preparing mathematics teachers through the use of narrative and reflection (Chapman, 2008; Jong & Hodges, 2013; Phelps, 2010). More particularly, the mathematics story becomes a vehicle for opening up dialogue for PTs to confront issues of race, ethnicity, gender, sexual orientation, and other often-marginalized spaces within mathematics education (Chazan, Brantlinger, Clark, & Edwards; 2013; Oslund, 2012).

Photovoice

Theodore's mathematics methods teaching extends Drake's (2006) mathematics story activity by incorporating a photovoice component (Wang & Burris, 1997). Photovoice is a relatively new education research method with roots in healthcare and sociology in which the research participant introduces personal photographs to anchor the stories he or she tells (Clark-Ibáñez, 2004; Gauntlett & Holzwarth, 2006; Harper, 2002). When using photovoice to seed the mathematics autobiography story, the photographs recall past memories freshly into discussion, allowing participants to share about other worlds they access beyond their world as a PT (Harper, 2002). Additionally, just the process of finding and choosing a photograph requires a substantial amount of thought and reflection on who they are in relationship to mathematics teaching (Gauntlett & Holzwarth, 2006).

For the remainder of this chapter, we collectively refer to various activities described above (e.g., mathematics stories, mathematics autobiographies, and mathematics photovoice interviews) as the *mathematics autobiography story*. We consider the mathematics autobiography story as an activity in which PTs reflect on who they are as mathematics learners, envision who they want to be as mathematics teachers, and confront issues of power and privilege in their own past and future mathematics teaching experiences. We start with Anne Marie's experience using a journal-based mathematics autobiography story, then share Theodore's experience incorporating photovoice into his mathematics autobiography story activity. We end with several considerations and caution signs for using mathematical story assignments from a sociopolitical perspective in mathematics methods courses.

ANNE MARIE'S EXPERIENCE WITH THE MATHEMATICS AUTOBIOGRAPHY STORY IN THE BRONX

Anne Marie: I have taught elementary mathematics methods for several years and in many different contexts: Midwest state universities, large private universities in the Midwest and East Coast, and now Lehman College, a senior liberal arts public institution serving the Bronx and a historically Hispanic population.

In the Bronx, PTs' schooling experiences are widely varied. As in many classroom settings around the country, a cultural mismatch (Banks, 2001; Ladson-Billings, 1994) exists in my courses at Lehman. I am a White woman, whereas almost all of my PTs are students of color. Some of my PTs spent most of their grade school experiences in other countries, many in Puerto Rico or the Dominican Republic. Some come to the United States as older students, in high school, or as adults. Many of my PTs attended local Bronx public schools, are first-generation college attendees, and are being prepared to teach in schools where they experienced their own education. Their rich experiences parallel those of students PTs will eventually teach in the Bronx.

On the first day of class, I ask PTs to reflect on their favorite and least favorite mathematics topics. Their discussions start the reflection process on their experiences as learners of mathematics. Their homework assignment for that day is to write a mathematical autobiography story detailing their prior experiences and feelings towards mathematics learning. The next week, we spend time as a community sharing our experiences with each other. Their stories tend to fall into similar categories: (a) always loved mathematics, (b) used to love mathematics, or (c) always hated mathematics. Below, I share excerpts from my PTs' mathematics autobiography stories to illustrate these categories.

Some PTs claim they love and have always loved mathematics. Some PTs have no bad memories with mathematics and later seem surprised by classmates with conflicting stories.

> As far as I can remember I have always loved math class. It always came easy to me and I loved when the teacher called on me. My favorite activity was when we played a game called "Around the world" and I got to beat my classmates at recalling all of the multiplication facts.

Some PTs claim to have once loved mathematics until a particularly bad experience, and subsequently hated mathematics.

> I was really good at adding, subtracting, multiplying, and dividing. But those good feelings about math didn't last long. Once I hit middle school, a lot changed. There were math topics which made me feel very anxious. The words fractions, percentages, and probability were some [topics] which constantly made me doubt myself and made me feel frustrated. I no longer looked forward to doing math or learning new math. I would shut down and space out during math class. My grades began to deteriorate, as well as my confidence in math. Math no longer brought me joy.

Some PTs had early negative experiences with mathematics and their fear or dislike of mathematics continued throughout their schooling, creating strong mathematical anxieties.

> Frustration and fear—Those are the words that come to mind. When I think about my own experiences in math class, I remember always wondering why math was so hard. I felt defeated in math class. I figured there was something was wrong with me. Reminiscing on my experiences... makes me wonder. WHY? Why would children be placed in a position which could lead to really hindering their self-confidence? I think that I was traumatized in most math classes and now I'm nervous to teach math.

These three categories of mathematics stories are typical of the experiences and dispositions PTs hold towards mathematics in general (Drake, 2006; LoPresto & Drake, 2005). However, in my context of teaching in the Bronx, PTs' stories surface the intersection of race and culture with mathematical experiences much more directly than my previous contexts. The stories I hear go beyond personal mathematical learning experiences, connecting to navigate race, culture, and language. PTs write about cultural congruency (Ladson-Billings, 1994) through the rare but life-changing experience of meeting mathematics teachers who looked like themselves, or drawing upon prior mathematical learning histories from other countries and in other languages. I add these two categories and share snippets to illustrate these stories:

(d) Cultural congruency

> I don't remember anything about math before 4th-grade. All that changed when I walked into class and found out that my teacher was an African-American woman. This was a life-changing experience. This woman changed my whole world about learning—what I believed a hard and confusing subject. Up until that moment, I didn't care about math. But she taught me. She made me like the math that I was learning. She looked like me.

(e) Drawing upon prior mathematics learning, culture, and experiences from other countries.

> My favorite memories of being in math class are from my childhood in the Dominican Republic. I lived there until I was in 6th grade when my family moved to the U.S. In grade school, we would often have class outside because the weather was so nice. I can remember doing math in 3rd grade while sitting under a mango tree.

Using the mathematics autobiography story activity affords me an opportunity to understand my PTs' lived experiences as diverse learners of mathematics, particularly through the intersection of race, culture, immigration, and mathematics learning. My PTs' stories reveal the complex relationship with mathematics and identity that serves as a starting point for our work together.

THEODORE'S EXPERIENCE INCORPORATING PHOTOVOICE INTO MATHEMATICS AUTOBIOGRAPHY STORIES

Theodore: My context is similar to Anne Marie's, except that I started teaching in a multicultural setting and now teach in a more ethnically homogenous environment in the Midwest. I have been teaching an elementary mathematics methods course for over 5 years at two different institutions, starting as a doctoral student at the University of Texas at Austin, a large Southwestern state university. Currently, I teach at the Ohio State University, also a large state university. The majority of my current PTs share similar schooling experiences: matriculating through public or private K–12 schools in suburban or rural areas of Ohio and doing well enough in their coursework to gain admission into our selective teacher education program. Most PTs in our teacher education program express a desire to return to teach in their community or a community similar to the one they grew up in and can identify immediate family members who are also teachers. As with Anne Marie, a cultural mismatch (Banks, 2001; Ladson-Billings, 1994) exists between my PTs and me. I self-identify as a Chinese American man, whereas the majority of my PTs self-identify as White women. Most of my teaching experience was in Brooklyn, NY, whereas many of my PTs will choose to teach in Ohio.

In the first week of my elementary mathematics methods course, we kick off the mathematics autobiography activity by each bringing in at least two photographs that represent our identity as a mathematics student. I start the class by sharing personal photographs of my own mathematical journey. I share stories about my childhood growing up in a religious household that led to my continual questioning of authority, my struggles with how the model-minority myth pigeonholed me as someone naturally "good" at mathematics (Hartlep, 2013; Lee, 1996), and my current struggles as a tenure-track faculty member navigating the world of mathematics teacher education. After observing how I model this photograph-sharing activity, my PTs enact photovoice interviews with each other in pairs, sharing their photographs and stories that emerge from these photographs with each other. This assignment uses photographs to anchor the mathematics autobiography story activity, extending from my prior work exploring mathematics teacher identity (Chao, 2014). Although this photovoice activity elicits many of the same categories of stories Anne Marie described—PTs reflecting that they (a) always loved mathematics, (b) used to love mathematics, or (c) always hated mathematics—sharing the photographs with each other also encourages PTs to reflect on who they are in relationship to mathematics teaching.

For example, in the following reflection, Kerry considered that maybe she was not as good at mathematics as she thought but rather was a good listener and follower. This reflection accompanied a photograph of flowers and butterflies painted on a window.

> At an early age, I was labeled as a student who excelled in math. As I grew up, I continued to enjoy learning and participating in math classes. It wasn't until this past year that I realized I might not be great at math, but rather great at doing what the teacher wants me to do. Now, I realize that being "good" at math is actually being able to problem solve and think critically without following guidelines given by the adult.

Kerry reflected she always thought she was strong in mathematics, but realized (as a student in a master's in education teacher preparation program) it was not mathematics she was "good" at but rather following the teacher's directions. Note that this reflection came from Kerry at the beginning of the semester, after she had completed two semesters worth of mathematics for elementary teachers content courses but before she had taken most of the elementary mathematics methods course.

Additionally, even though the majority of my PTs come from ethnically homogenous backgrounds, photographs allow my PTs to delve into how their racial, ethnic, gender, sexual, religious, and socioeconomic identities intersect with their emerging mathematics teacher identities. The photographs themselves do not always address these aspects, but the stories that PTs tell surrounding the photographs do. For instance, consider Jordan's reflection accompanying a photograph of her creating bracelets with several children.

> I enjoyed math when it became relevant to my life. As I grew older math became more and more challenging when I was unable to solve equations or problems in ways teachers wanted me to. As I struggled I became insecure about my ability to do math. I would express my struggles to family and friends, and they further ridiculed my inability to quickly provide a mental answer. Now that I am older, when working with children I try to incorporate experiences they can relate to real life. For example, I had children ages 2–6 from my church construct a Cheerios bracelet with the purpose of choosing a number of people they wanted us to pray for. I provided an edible manipulative that could be used for a spiritual lesson (and counting practice for the little ones) but also be eaten for fun. Now, I feel that along the way I lost interest for math and became insecure when I started to believe it was irrelevant to my daily life.

In this reflection, Jordan invoked her religious identity in connection to her mathematics teaching identity, reflecting that the lesson she describes was more connected to her own identity as a teacher than mathematics lessons she worked with that involved hackneyed story problems or forced real-world connections.

WHY THE MATHEMATICS AUTOBIOGRAPHY STORY?

Twenty years ago, Bruner (1996) warned that education must embrace the use of a narrative construal if we wanted to connect learning with how children actually think and grow—through stories and narratives. Narrative-based inquiry and research has been embraced by many educators outside of mathematics (Doyle, 1997) and within the field of mathematics teacher education, extending heavily from work previously mentioned (Drake, 2006; Drake et al., 2001). Hence, we consider the use of the mathematics autobiography story as an exemplary activity that can be utilized in connecting stories and narratives from PTs to views of themselves as mathematics teachers. Specifically, the mathematics autobiography story connects mathematics teaching with sociopolitical issues of inequity, privilege, stereotype, and power. This work develops mathematics teacher identity not only on the knowledge and lived experiences of teachers, but on how these experiences, dispositions, and practices intersect to interrogate issues of race, culture, power, and privilege (Aguirre, Mayfield-Ingram, & Martin, 2013).

For the Prospective Teacher

The mathematics autobiography story helps PTs reflect upon aspects of their various identities and the ways they know themselves and understand their previous mathematics experiences (Drake et al., 2001; McAdams, 1993). This activity also provides PTs with a tool to make sense of their experiences and how those experiences shape their beliefs and dispositions about the discipline, teaching, and learning. This reflection is particularly important in mathematics because MTEs are actively trying to break the cycle of mathematics teachers using their past experiences to inform how they should teach in the future (Ball, 1997). Many of our PTs experienced traditional models of mathematics teaching. Autobiography stories help PTs understand how their past experiences shaped their understanding of what mathematics is and how that might conflict with child-centered or inquiry-based mathematics teaching. Furthermore, the mathematics autobiography story allows PTs to confront connections between personal identities surrounding race, culture, power, and privilege with their mathematics experiences and how they will disrupt these existing hegemonic structures as future mathematics teachers.

For the Mathematics Teacher Educator

The mathematics autobiography story provides an opportunity for MTEs to learn about their PTs' mathematical experiences. Knowledge of PTs' prior experiences helps the MTE create personalized instruction. Additionally, sharing stories brings PTs together as a community (LoPresto & Drake, 2005; McAdams, 1993). PTs compare similarities and differences in their mathematics autobiography stories. The revelation that other PTs struggle with mathematics is empowering for those who share feeling discouraged or "dumb" when it comes to mathematics. Finally, the mathematics autobiography story activity brings forth PTs' sociopolitical identities for MTEs to build upon in instruction. For instance, when one PT raised her transformative experience of having an African American fourth-grade mathematics teacher, Anne Marie incorporated ideas of cultural congruence and culturally relevant pedagogy (Ladson-Billings, 1994) into the course.

MATHEMATICS STORY ASSIGNMENT EXAMPLE: THE MATHEMATICAL AUTOBIOGRAPHY

The autobiography assignment can take on many forms. Here is how Anne Marie frames the mathematics story autobiography in her mathematics methods course:

> Write your mathematical autobiography. Focus on key moments that you have experienced as a student of mathematics. Why do these mathematics moments stand out? What is it about these moments that continue to give you joy or discomfort? What role did your parents, teachers, siblings, and friends play in shaping your current disposition about mathematics? In the 4–5 page paper, be specific about the role of particular teachers who contributed to your development as a doer of mathematics.

After the PTs write their autobiographies, PTs spend time sharing their experiences with each other to further build a classroom community. The following prompts help guide PTs' reflective conversations: What emotions emerged during your writing? What, if anything, surprised you? What are some similarities and differences you hear across our stories? What did you enjoy writing about in your own autobiography? How might engaging in this writing and talking be useful preparation for becoming a teacher of mathematics?

Identifying how race, culture, power, and privilege played a role in a PT's story is important for PTs to note and continually refer to as the semester continues. This can be done by connecting critical mathematics research and literature to themes that emerge in the autobiography assignment and incorporating or connecting these themes to social justice mathematics activities. Returning to the autobiography at the end of the semester as a reflective prompt for writing a "mathematics future" or "letter to my future self as a mathematics teacher" helps highlight how PTs have learned from and with their own experiences.

MATHEMATICS STORY ASSIGNMENT EXAMPLE: PHOTOVOICE INTERVIEWS

Incorporating photographs into the mathematics autobiography story adds a visual element to this activity. Theodore uses this assignment in his mathematics methods course:

> Digitally prepare two photographs to share in class that represent your mathematics identity when you were a student. In class, you will partner up and share your photographs with each other in a 20-minute interview (10 minutes each). During these interviews, the sharer can talk about whatever s/he wants to. The sharer does not have to only stick with what the photograph shows. The listener is to listen and is only allowed to ask two things: (1) Specific details of the photograph (e.g., Where was this photograph taken?, Who is in this photograph?) and (2) "Tell me more about that..." to extend or get more detail about what the sharer might be talking about. After the interview, you will reflect on the experience by writing down your mathematics teacher identity story. This will be less than a page long and encapsulate the story you told in the interview as well as reflections on the interview itself.

Like the previous activity, the critical reflection comes not from the creation of the mathematics autobiography story, but in the sharing of this story among peers. The photographs create very vulnerable and personal spaces, which help PTs immediately connect their mathematics experiences with intersections from their personal identities.

CONSIDERATIONS AND CAUTION SIGNS

We end with considerations and caution signs for fellow MTEs using the mathematics autobiography story. First, take caution when assigning, debriefing, and reflecting on mathematics stories with PTs. Because these activities bring up a myriad of emotions, stereotypes, and dispositions, we must honor our PTs' words and engage with their encounters carefully. Be open to the range of experiences and emotions that may emerge and ready to support the stories that your PTs will tell. We have found that our framing the mathematics autobiography story as a reflective identity activity reveals stories of identity that help PTs see how mathematics connects to issues of race, culture, power, and privilege much more than a traditional "icebreaker" activity would.

Second, share your story. From our experience, when we write and share own story first, we model how our experiences affected our dispositions towards mathematics and reveal our humanness. PTs may assume that because we teach mathematics methods courses, our relationship with mathematics has always been positive. This is somewhat true: As MTEs, we have collectively experienced success in mathematics and likely feel confident about teaching mathematics. However, our relationship to mathematics is not one-dimensional; focusing only on our positive mathematics experiences ignores the deep social inequities embedded within mathematics instruction (Weissglass, 2002). Many of us share experiences similar to those of our PTs (e.g., fear, history of failure, gender or ethnic discrimination). We must be honest with our PTs to begin to support their understanding of how to teach mathematics as connected to culture and identity—a sociopolitical framework (Weissglass, 1990).

For instance, here are ways in which we tell stories that reveal our vulnerability to our PTs. I, Anne Marie, share that I started my career as a third-grade teacher at a school using a mathematics curriculum very different from what I knew. My deep interest in mathematics only began because of my third-grade students' mathematical thinking. It is surprising to my PTs that I wasn't always interested in or an accomplished student of mathematics. Additionally, I, Theodore, often share with my PTs about how I felt that my Chinese American identity often invoked the myth of the model minority (Hartlep, 2013) from fellow students and teachers. For instance, I felt that my own mathematics teachers often assumed my mathematics abilities were far superior than I thought and promoted me to honors mathematics courses. But because I disliked this tokenizing identity, I deliberately tried to do poorly in my mathematics classes, not wanting to be pigeonholed as an Asian mathematics whiz kid. Sharing with our PTs shows them how to share more intricate personal stories about their own mathematics experiences.

Third, check yourself and your assumptions about your PTs. The mathematics autobiography uncovers stereotypes that everyone in the class might hold, including the MTE. Examining our experiences reveals and encourages us to confront our biases about our PTs. Understanding the complexity of our growing identities as critical MTEs has meant checking our assumptions or relying on critical friends to do so. For example, we have found ourselves generalizing about PTs who wear sorority letters to class. We question the roots of our stereotypes of PTs who are members of Greek organizations. Where do these stereotypes come from? Why have we internalized these stereotypes? What does it reveal about our comfort with the institution of higher education? Additionally, we have found ourselves speaking to PTs of color in ways that assume they understand the complexities in mathematics education for equity and social justice, and being surprised when these PTs hold deficit perspectives. Why do we assume all students of color share similar stories? Why do we even dichotomize students of color against other students?

Fourth, be prepared to thoughtfully respond to stereotyping comments and help move conversations along in a productive way. Your PTs' stories might reify stereotypes of each other, which should be seen as an opportunity for growth rather than judgment. For example, in Anne Marie's class, a PT shared surprise that a fellow classmate was "Dominican like me" because "you don't look Dominican . . . I thought you were White." Because stories surface potential deficit perspectives and experiences, it is important for MTEs and PTs to work to understand how these perspectives developed and how to challenge them. MTEs can help PTs understand the connection between their identity, perspective, and role as a mathematics teacher in working towards equity. If the MTE intends to use the story assignment from a sociopolitical perspective with goals of preparing socially and culturally competent teachers, then the MTE must be prepared to use this assignment to address conditions of schools and society, which can produce and perpetuate biases and stereotypes. Gorski (2011) has provided several strategies for helping to challenge and defeat deficit ideologies.

Finally, plan to support PTs to reflect upon and update their stories. By the end of the semester, our PTs are especially busy. Our collective capacity to engage in critical reflection about PTs' evolving mathematics teacher identity feels compromised. Yet revisiting the themes emerging from PTs' stories in nonsuperficial ways can help PTs build insights about the role of their experiences as mathematics learners in their identities as mathematics teachers.

FINAL THOUGHTS ABOUT THE MATHEMATICS AUTOBIOGRAPHY SUPPORTING SOCIOPOLITICAL IDENTITY

We end with thoughts and advice about situating the mathematics autobiography story in a methods class to help PTs embrace a sociocultural mathematics teacher identity, revisiting our definition of sociocultural mathematics teacher identity as a verb, not a noun (Gutiérrez, 2013). First, we as MTEs must learn to immediately recognize deficit ideologies when they surface including stereotypes or disregard for the sociopolitical context of schools (Gorski, 2011). We find these deficit ideologies appear when PTs compare their own stories with their current experiences working in schools. For example, we both often hear PTs' comment that "the parents of my students just don't *care* in the same way my parents cared." When these statements surface, we must name this as deficit language and offer a counternarrative. In class, we explore the word *caring* and how it relates to the sociopolitical context of low-income families, then guide PTs in reframing many ways parents care beyond dominant cultural norms of doing homework with or for students, providing tutoring, and generally spending dedicated time with the student. For me, Anne Marie, as I work predominantly with students of color and first-generation college students, I was surprised to hear many of my students using deficit language to characterize parents they work with. I found PTs of color had picked up and learned how to appropriate this deficit-based language of care, applying it to their own students (Freire, 1970).

Second, we must help our PTs use their mathematics autobiography stories to critically reflect upon their own class socialization and identity (Gorski, 2011). We can ask reflective questions about what the stories reveal, such as "How do I, intentionally and unintentionally, support systems of oppression or beliefs and stereotypes of students?" However, although we want to showcase a diverse set of experiences in class, we must be careful not to tokenize our PTs of color or to ask our PTs to speak for their ethnicity, culture, or sexual identity.

Finally, we must remember the critical importance of connecting mathematics education to social issues of economic injustice and poverty, helping our PTs develop critical lenses to reframe the images they hold of teaching and their students (Gorski, 2011). We have found the mathematics methods course to be a rich space in which we connect mathematics teaching to issues of injustice and oppression, helping our PTs (and their future students) learn to read and write the world around them with mathematics (Gutstein, 2006).

We encourage further conversations (such as in this book) about what MTEs would want to see as evidence of PTs' critical growth. How do we know our PTs are entering the classroom with critical lenses and capacities to teach mathematics for equity and social justice? We encourage readers to reflect on how this intersects with their own practice and research. We know that the mathematics autobiography story, as an activity, has immediate impact within the methods course. We hope the field of mathematics teacher education research will continue to focus on how critical mathematics teachers develop and grow, particularly in regards to activities and experiences they encounter through the mathematics methods course.

REFERENCES

Aguirre, J. M., Mayfield-Ingram, K., & Martin, D. B. (2013). *The impact of identity in K–8 mathematics: Rethinking equity-based practices.* Reston, VA: National Council of Teachers of Mathematics. Retrieved from http://www.nctm.org/catalog/product.aspx?ID=14119

Ball, D. (1997). From the general to the particular: Knowing our own students as learners of mathematics. *The Mathematics Teacher, 90,* 732–737.

Banks, J. A. (2001). Citizenship education and diversity implications for teacher education. *Journal of Teacher Education, 52*(1), 5–16.

Bruner, J. (1996). *The culture of education.* Cambridge, MA: Harvard University Press.

Chao, T. (2014). Photo-Elicitation/Photovoice interviews to study mathematics teacher identity. In J. Cai & J. Middleton (Eds.), *Current research in mathematics teacher education: Contributions by PME-NA researchers.* New York, NY: Springer.

Chapman, O. (2008). Narratives in mathematics teacher education. In D. Tirosh & T. Woods (Eds.), *The international handbook of mathematics teacher education. Volume 2: Tools and processes in mathematics teacher education* (pp. 15–38). Rotterdam, The Netherlands: Sense.

Chazan, D., Brantlinger, A., Clark, L., & Edwards, A. (2013). What mathematics education might learn from the work of well-respected African American mathematics teachers in urban schools. *Teachers College Record, 115*(2), 1–40.

Clark-Ibáñez, M. (2004). Framing the social world through photo-elicitation interviews. *American Behavioral Scientist, 47,* 1507–1527.

de Freitas, E. (2008). Enacting identity through narrative: Interrupting the procedural discourse in mathematics classrooms. In T. Brown (Ed.), *The psychology of mathematics education: A psychoanalytic displacement* (pp. 139–158). Rotterdam, The Netherlands: Sense.

Doyle, W. (1997). Heard any really good stories lately? A critique of the critics of narrative in educational research. *Teaching and Teacher Education, 13*(1), 93–99.

Drake, C. (2006). Turning points: Using teachers' mathematics life stories to understand the implementation of mathematics education reform. *Journal of Mathematics Teacher Education, 9*(6), 579–608.

Drake, C., Empson, S. B., Dominguez, H., Junk, D. L., & LoPresto, K. (2003, April). *The role of teacher knowledge and identity in classroom interactions in elementary mathematics.* Paper presented at the annual meeting of the American Educational Research Association, Chicago, IL.

Drake, C., Spillane, J. P., & Hufford-Ackles, K. (2001). Storied identities: Teacher learning and subject-matter context. *Journal of Curriculum Studies, 33*(1), 1–23.

Empson, S. B., Drake, C., & Junk, D. L. (2002). Teachers' identity and knowledge in elementary mathematics. In D. Mewborn (Ed.), *Proceedings of the twenty fourth annual meeting of the North American Chapter of the International Group for the Psychology of Mathematics Education* (pp. 1867–1878). Columbus, OH: ERIC.

Felton-Koestler, M. D. (2015). Mathematics education as sociopolitical: Prospective teachers' views of the what, who, and how. *Journal of Mathematics Teacher Education, 19,* 1–26.

Freire, P. (1970). *Pedagogy of the oppressed* (30th anniversary ed.). New York, NY: Continuum.

Gauntlett, D., & Holzwarth, P. (2006). Creative and visual methods for exploring identities. *Visual Studies, 21*(1), 82–91.

Gorski, P. (2011). Unlearning deficit ideology and the scornful gaze: Thoughts on authenticating the class discourse in education. In R. Ahlquist, P. Gorski, & T. Montaño (Eds.), *Assault on kids: How hyper-accountability, corporatization, deficit ideology, and Ruby Payne are destroying our schools* (pp. 152–176). New York, NY: Peter Lang.

Gutiérrez, R. (2013). The sociopolitical turn in mathematics education. *Journal for Research in Mathematics Education, 44*(1), 37–68.

Gutstein, E. (2006). *Reading and writing the world with mathematics.* New York, NY: Routledge.

Harper, D. (2002). Talking about pictures: A case for photo-elicitation. *Visual Studies, 17*(1), 13–26.

Hartlep, N. D. (2013). *The model minority stereotype: Demystifying Asian American success.* Charlotte, NC: Information Age.

Jong, C., & Hodges, T. E. (2013). The influence of elementary preservice teachers' mathematics experiences on their attitudes towards teaching and learning mathematics. *International Electronic Journal of Mathematics Education, 8*(2–3), 100–122.

Ladson-Billings, G. (1994). *The dreamkeepers: Successful teachers of African American children.* San Francisco, CA: Jossey-Bass.

Lee, S. J. (1996). *Unraveling the "model minority" stereotype: Listening to Asian American youth.* New York, NY: Teachers College Press.

LoPresto, K. D., & Drake, C. (2005). What's your (mathematics) story? *Teaching Children's Mathematics, 11*(5), 266–271.

McAdams, D. P. (1993). *The stories we live by: Personal myths and the making of the self.* New York, NY: Guilford Press.

Nieto, S., & Bode, P. (2008). *Affirming diversity: The sociopolitical context of multicultural education.* Boston, MA: Pearson.

Oslund, J. A. (2012). Mathematics-for-teaching: what can be learned from the ethnopoetics of teachers' stories? *Educational Studies in Mathematics, 79*(2), 293–309.

Phelps, C. M. (2010). Factors that pre-service elementary teachers perceive as affecting their motivational profiles in mathematics. *Educational Studies in Mathematics, 75*(3), 293–309.

Wang, C., & Burris, M., A. (1997). Photovoice: Concepts, methodology, and use for participatory needs assessment. *Health Education & Behavior, 24*(3), 369–387.

Weissglass, J. (1990). Constructivist listening for empowerment and change. *The Educational Forum, 54*(4), 351–370.

Weissglass, J. (2002). Inequity in mathematics education: Questions for educators. *The Mathematics Educator, 12*(2), 34–39.

SECTION VI

LOOKING INWARD

CHAPTER 19

INTERPRETATIONS AND USES OF CLASSROOM VIDEO IN TEACHER EDUCATION

Comparisons Across Three Perspectives

Stephanie Casey
Eastern Michigan University

Ryan Fox
Belmont University

Alyson E. Lischka
Middle Tennessee State University

Teachers can and should learn about teaching from investigating the work of teaching (Ball & Forzani, 2009). In addition to face-to-face classroom experiences, teacher educators commonly turn to classroom video as a representation of the work of teaching. Roth McDuffie and colleagues (2014) found that K–8 classroom video successfully bridges school classrooms and university mathematics methods classrooms, providing a context for approximation

Building Support for Scholarly Practices in Mathematics Methods, pages 297–310
Copyright © 2018 by Information Age Publishing
All rights of reproduction in any form reserved.

and decomposition of the practice of teaching (Grossman et al., 2009). However, effective use of classroom video must be embedded in structured instructional activities that support prospective teachers (PTs) in meeting clearly defined learning objectives identified by the mathematics teacher educator (MTE) and include prompts that focus PTs' attention on specific features of mathematics teaching and learning that are represented in the video (Roth McDuffie et al., 2014; Star & Strickland, 2008). MTEs' choices of specific features on which to focus are influential to PTs' learning. Seidel, Blomberg, and Renkl (2013) asked two groups of PTs to view and analyze the same video, but with two different instructional approaches aligned with different learning goals. The researchers reported that differences in learning goals and instruction seemed to result in differences in PTs' learning. This highlights the fact that PTs' learning from classroom video case analysis is influenced by instructional activities MTEs design.

An MTE's theoretical perspective potentially influences ways in which classroom video cases are used with PTs. As MTEs who identify as situative researchers, we (the chapter authors) value the affordances that using video can provide in our university-based methods courses. When we cannot immerse our PTs in K–12 classrooms due to contextual constraints, video provides an alternative through which we can examine the relationships and interactions that take place in mathematics classrooms. However, we wonder whether MTEs of different perspectives might approach the use of classroom videos in different ways. We contend that a MTE's theoretical perspective influences his or her analysis of a classroom video and the subsequent design of associated instructional activities. In this chapter, we present results of a survey study in which we examined what MTEs with different theoretical perspectives noticed and described about a classroom video case and how they proposed to use the video in mathematics methods. The study was designed to answer two questions: (a) How do MTEs of different theoretical perspectives analyze the task, teaching, learning, and power and participation in a mathematics classroom as represented in a selected public video?; and (b) How do MTEs describe potential instructional activities centered around the use of this video? This study explores the possible variation present in MTE-designed activities and emphasizes the need for mathematics teacher education to undertake scholarly inquiry focused on the use of classroom video to support the design and development of MTEs' scholarly practice (Lee & Mewborn, 2009).

BACKGROUND

Brief descriptions of three theoretical perspectives (Table 19.1) illustrate how we understand participants from each perspective might approach their work in mathematics teacher education. The descriptions are drawn from presentations given at the Scholarly Inquiry and Practices Conference (Sanchez, Kastberg, Tyminski, & Lischka, 2015). Although these descriptions may not capture the thoughts and beliefs of all MTEs who identify with each perspective, they represent our operationalization of the perspectives as implemented in this study. Chapters by Gutiérrez (Chapter 2, this volume), Simon (Chapter 3, this volume), and Kazemi (Chapter 4, this volume) provide additional insights regarding these perspectives. Descriptions of the sociopolitical perspective come from an earlier work by Gutiérrez (2013).

In this study, we also use the work of the Teachers Empowered to Advance Change in Mathematics (TEACH MATH) project (Roth McDuffie et al., 2013; Roth McDuffie et al., 2014) to frame our investigation and analysis. The TEACH MATH framework is organized around four lenses: Task, Learning, Teaching, and Power and Participation. When using the framework, teachers analyze videos of classroom instruction using these lenses and related questions that promote analysis. Stated learning objectives for the use of the TEACH MATH framework relate to the perspectives (i.e., sociopolitical, cognitive, and situative) investigated in our study: "Capitalize on students' diverse cultural, linguistic, and community knowledge in ways that support students' mathematics learning... [and] access and build on children's multiple ways of understanding and solving mathematical problems" (Roth McDuffie et al., 2014, p. 108). The first objective is aligned with the sociopolitical perspective, whereas the second is aligned with the cognitive and situative perspectives.

TABLE 19.1 Descriptions of Three Perspectives in Mathematics Teacher Education

Sociopolitical	• Understanding how oppression in schooling operates not only at the individual level but also the systemic level is central to your work. • You deconstruct the deficit discourses about historically underserved and/or marginalized students. • Buffering oneself, reinventing, or subverting the system in order to be an advocate for one's students are tools you use.
Cognitive	• The root of your work is in developing models of thinking about a concept (either mathematical or pedagogical). • A goal of your work is to help teachers learn to attend to and respond to student thinking in ways that promote student learning.
Situative	• Learning is conceptualized as an aspect of social activity and therefore you research and work within activity systems. • Learning is in relation to interactions and within contexts. • Your focus is on teaching, which occurs in schools with children and experienced teachers. • Prospective teachers' learning situated in interactive settings is valued.

We selected the TEACH MATH framework for our study because it addressed aspects of supporting students' development of mathematical knowledge that corresponded to all three perspectives, and the framework was developed for analyzing classroom videos, the focus of our study. We drew from questions aligned with the four lenses to construct our survey questions and analyze the resulting data. The four lenses of the framework also structure the presentation of our findings.

METHODOLOGY

To answer our research questions, we selected a video and created a survey about the content and uses of the selected video. Requests to participate in the study were sent electronically to all participants from the Scholarly Inquiry and Practices Conference (Sanchez et al., 2015) as well as to early-career MTEs selected as STaR Fellows (Association of Mathematics Teacher Educators, n.d.). Participants were asked to complete the survey within a 4-week time span and were reminded to participate once during that time. Twenty-nine respondents completed the survey, including 12 who self-identified as holding a cognitive perspective, 11 as situative perspective, and 6 as sociopolitical perspective. In the remainder of this chapter, we refer to participants who self-selected sociopolitical as their primary perspective as *sociopolitical participants* with similar references to *cognitive participants* and *situative participants*, and we acknowledge that participants may also espouse ideas from the other perspectives but were asked to identify a primary perspective.

The video, *The Case of Jeffrey Ziegler and the S-Pattern Task* (National Council of Teachers of Mathematics [NCTM], 2015) was selected because it was freely accessible to participants, was under 5 minutes in length, contained an example of classroom practice, and provided a variety of classroom and lesson elements. The selected video shows a high school teacher facilitating small-group discussion on the S-Pattern Task, shown in Figure 19.1.

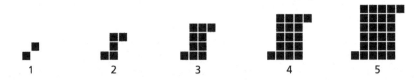

Figure 19.1 The S-pattern task pattern. (NCTM, 2015). *The Case of Jeffrey Ziegler and the S-Pattern Task*. (Retrieved from: http://www.nctm.org/Conferences-and-Professional-Development/Principles-to-Actions-Toolkit/The-Case-of-Jeffrey-Ziegler-and-the-S-Pattern-Task/)

Data were gathered through an online qualitative survey based on the TEACH MATH lenses for analyzing classroom video (Roth McDuffie et al., 2014). The lenses provided research-based focus questions about different aspects of the classroom and lesson. Participants first selected a perspective they identified most closely with from a list of descriptions (see Table 19.1). Next, they watched the video and responded to 10 open-ended prompts (see Table 19.2). In addition, participants were asked to discuss what they noticed in the video, the central mathematical ideas in the task, the discourse in the classroom, and aspects of culture. The first three questions were researcher generated in order to connect survey responses to participants' thinking concerning their self-selected theoretical perspective. The remaining questions were adapted from the TEACH MATH video analysis activity.

TABLE 19.2 Survey Questions and Sources		
Question Number	Survey Question	Source
1	From your perspective as a mathematics teacher educator, describe what you notice when watching this video.	Researcher created
2	Describe how you might utilize this video in your work with prospective mathematics teachers.	Researcher created
3	How does your perspective on mathematics education influence your response to the previous prompt?	Researcher created
4	What is/are the central mathematics idea(s) in the task that is the focus of this lesson (i.e., identify specific concepts, processes, skills, problem-solving strategies)?	TEACH MATH Task Lens
5	Do you consider the task that is the focus of the observed class session a good task? Why or why not?	TEACH MATH Task Lens
6	What specific math understandings and/or confusion are indicated in students' work, talk, and/or behavior?	TEACH MATH Learning Lens
7	How does the teacher elicit students' thinking?	TEACH MATH Teaching Lens
8	How does the teacher respond to student thinking?	TEACH MATH Teaching Lens
9	Who participates in the class session observed?	TEACH MATH Power and Participation Lens
10	Does the classroom culture value and encourage most students to speak, only a few, or only the teacher? Explain your response.	TEACH MATH Power and Participation Lens

Survey responses were first organized by question and open coded for themes (Merriam, 2009) within a question without attending to participants' perspectives. One author completed preliminary open coding and the grouping of open codes into themes for each question, and then a second author verified the initial codes and themes. For example, open codes for Question 1 included productive struggle, support of group talk, positioning of diverse students, mathematical representations, mathematics of the students, and others. The complete list of themes that emerged for Question 1 were: mathematical thinking, task description, discourse, productive struggle, teacher moves—discourse, authority and positioning, and mathematical representations. After themes were developed for each question, we identified trends within each perspective for each question, noting specifically cases of differences in the ways that participants of different perspectives responded to the questions. In the following section we present the findings, highlighting the prominent differences and similarities in responses across the three perspectives.

FINDINGS

We first present the findings according to the TEACH MATH lenses: Task, Learning, Teaching, and Power and Participation (Roth McDuffie et al., 2014). Then additional findings regarding MTEs' hypothesized uses of the video in methods classes are described. In each of these sections, an overview with general findings is presented first, followed by critical findings focused on similarities and differences in responses by MTEs from the three theoretical perspectives.

Task: Central Mathematical Ideas and Quality of the Implemented Task

Analysis of responses to questions related to the Task lens revealed no common theme across all three perspectives. However, participants from two of the three perspectives frequently commented in similar ways. Participants from the cognitive and situative perspectives tended to view the implementation of the task positively, whereas participants from the sociopolitical perspective saw the same implementation of the task as problematic.

When describing the good qualities of the task, participants in both the cognitive and the situative perspectives identified the multiple representations possible in the solution to the task. As illustration of the commonalities across the two perspectives, one participant from the cognitive perspective mentioned that the task "allows for students to demonstrate and see the connections between algebraic, tabular and pictorial representations and it also allows for students to see the connections between different expressions" (Participant 4). In a similar manner, one participant from the situative perspective noted that the task "has multiple entry points, multiple possible approaches, the sequence of figures can be 'seen' and generalized in multiple ways" (Participant 36).

In contrast, three of the five responding sociopolitical perspective participants offered a critique of the teaching or students' interactions in the lesson. One participant from the sociopolitical perspective critiqued the teacher's implementation of the task and desire to produce a single representation:

> The teacher has focused students on writing the symbolic form of a rule to count the number of tiles given the term in the sequence. But the task itself is more about generalization and thinking / problem solving with functional relationships—no particular representation is requested. (Participant 2)

Some sociopolitical participants noted other problematic aspects of the selected task in their responses to Question 1. Observing that the mathematical ideas in the task might be below grade level for the students in the video, one sociopolitical participant commented "this is a 7th or 8th grade task, too bad these grade 11–12 students have had poor math experiences to this point that it is a challenge" (Participant 2). Addressing problematic elements of the task, another sociopolitical participant noted that "the problem students were engaged with, while conceptually challenging, lacked a real-world context for many" (Participant 39). Only sociopolitical participants discussed problematic aspects of the selected task; no participants from the other two perspectives offered any critical feedback of the task in their responses.

Learning: Identifying Student Struggles

Participants from all three perspectives focused on confusion students appeared to have when completing the task. An equal number of participants from the three perspectives directly referenced students' confusion in their responses to Question 6. They identified that students struggled to connect the number of squares in each picture in Figure 19.1 to a closed formula for the number of squares in any picture. The following comment from a cognitive participant is representative of participants from all three perspectives:

> Several students showed difficulty moving beyond a recursive understanding of the pattern and going to an explicit form. Some also did not yet seem

to have a full grasp of the power of variables in these kinds of problems. Others were not necessarily confused but were somewhat stuck in thinking about the number patterns rather than the geometric structure of the pattern. (Participant 31)

A representative response from a situative participant discussed more specifically the students' inabilities to generate a single expression for their observation:

Since the change in the number of blocks were not constant, they had issues creating the equation. They had recognized a pattern in the change from one stage to the next, but had yet to figure out how to deal with that. They struggled discussing their thinking. (Participant 19)

A participant from the sociopolitical perspective, indicative of several participants in this perspective, additionally identified a lack of connections among representations available to the students:

[T]he first group...recognized that the pattern is growing in a predictable way, but they are not sure how to generalize the pattern to an equation because they only see the pattern recursively. The second group...found a general equation that they believe represents the pattern, but they have not yet connected, explicitly, the parts of their equation to the visual representation. (Participant 32)

Participants from the cognitive and sociopolitical perspectives tended to address mathematics understandings and confusions only in response to the learning question (Question 6). Yet, participants from the situative perspective addressed learning in their responses to other survey questions as well. For example, responding to Question 3 (addressing the influence of perspective on video use), a participant from the situative perspective discussed the learning of a particular student:

Male student 1 has discovered (and can articulate) several "patterns" in the diagram, describing how the top is "x" (the top row has the same number of blocks as the figure #).... Male student articulates patterns he sees—and has connected that to formula. He sees x times 2 as the top and bottom rows of blocks, then the middle is a square. (Participant 20)

Teaching: Eliciting and Responding to Student Thinking

The participating MTEs analyzed the pedagogical moves of the teacher, in particular the teacher's eliciting (Question 7) and responding to

(Question 8) student thinking. Participants found that the predominant move enacted by the teacher was active listening. They noted that the teacher was skilled at active listening, attentively considering what students had to say without interrupting them. Participants also noted the teacher followed up with questions that requested clarification about what the student said, made nonjudgmental statements that revoiced what the students said, or made requests for another student to revoice a statement.

Although many participants from all three perspectives noted active listening moves by the teacher to elicit student thinking, a greater proportion from the cognitive perspective noted the teacher's use of questions that sought to extend their understanding of mathematical ideas and relationships. For example, cognitive participants noted how the teacher "pressed to ask [if] there is no other way than to count it" (Participant 40) and "asks them to connect rules to original context" (Participant 43).

Although overall the participants' responses indicated the teacher did a good job of eliciting and responding to student thinking, seven participants criticized the teacher's teaching. Three of the five sociopolitical participants who responded to these survey questions raised concerns about the teaching, a larger percentage than did participants from each of the other perspectives. Criticisms focused on the teacher's use of funneling questions, with participants suggesting questions were used to encourage students to think in ways the teacher had in mind. For example, Participant 2 noted "He [the teacher] asks a loaded question, 'what do you have.' This is asked with an answer in mind: it is to get the student to say and then think as he wants."

Power and Participation: Classroom Culture and Involvement of Students and Teacher

MTEs' responses to the Power and Participation questions revealed differences in judgment regarding students' participation and role in the classroom. Approximately half of the respondents expressed that all students in the video participated and were valued; remaining respondents remarked that few students participated or were valued (with some participants noting inequities with respect to gender and race). All but one of the sociopolitical participants were in the latter group, noting that students who participated were called upon by the teacher. The remaining sociopolitical participant declined to answer the question, noting that a determination of participation could not be made with such a short video.

This group of sociopolitical participants further noted that the teacher's behavior reflected differing expectations regarding student competency. For example, in response to Question 9, one sociopolitical participant noted students were "not on equal grounds—there were status differences, or maybe better described as different expectations for competency" (Participant 2). In comparison, 75% of the cognitive participants and nearly all of the situative participants felt that most of the students were valued and encouraged to speak. Participant 35's response is representative of these participants: "It seems that many students feel comfortable speaking and the teacher encourages silent group members to join the conversation so that everyone is included."

General Noticings and Considerations for the Use of Video

Participants were asked what they noticed in the video, how they thought it related to their personal perspective (Questions 1 and 3), and to propose uses of the video in activities with PTs in methods courses (Question 2). Across all three perspectives, participants noted teacher moves that promoted discourse among the classroom learners in the video, with the situative group commenting more frequently concerning classroom discourse.

All three groups commented on mathematical ideas present in the video, but there were differences in the ways the participants discussed the mathematics by perspective. In particular, responses to Question 3 showed that nearly half of the cognitive participants prioritized the mathematics of the task (i.e., the specific mathematics it addresses, its cognitive demand, and representations it elicits) when considering how they would use this video in working with PTs. Participants from other perspectives mentioned mathematical aspects of the task at a lesser rate, and often in a generic sense. For example, a situative participant declared, "I would engage the prospective teachers in the mathematics task" (Participant 37) without elaborating the mathematical aspects of the task. The cognitive group discussed the cognitive demand of the task and highlighted connections between mathematical representations. For example, "[The teacher] does a nice job of re-focusing students on the visual representation and observing how the pattern relates to the figure number" (Participant 29). Situative participants' comments on mathematical ideas were often couched in the form of noticing student or teacher actions. For example, one situative participant reported:

> I noticed that one of the students has figured out an equation by using the shapes within the figures. Namely, the student realized that the top and bottom rows were length s and the rest of the rows formed a square with the number of blocks being $(s-1)^2$. (Participant 19)

Only two of the sociopolitical participants noted the mathematical aspects of the task in their responses to Question 1, 2, and 3. Mentions of mathematical aspects included reference to the task as "group-worthy" (Participant 2) and statements of the mathematical content they would ask PTs to observe (Participants 2 and 5).

In contrast, sociopolitical participants prioritized issues related to power and participation rather than mathematical aspects of the task when responding to Question 3. Presumably as a result of their educational perspective, four of the six sociopolitical participants noted they would use the video to raise issues of power and inequity, whereas issues of power and participation were mentioned by 3 of the 12 cognitive participants and not at all by the situative participants.

These similarities and variations continued in participants' suggestions of how they might implement the selected video as part of a methods-course activity. Participants in each perspective discussed potential uses that attended to the mathematical elements of the task in the video, with each perspective highlighting the mathematical content in different ways. In the cognitive group, the emphasis was on student thinking and mathematical representations. One cognitive participant identified a use of the video to highlight mathematical representations:

> Emphasizing the need for multiple representations in which I would have students do the task themselves and then compare their efforts to those that they see in the video. I would also ask them to identify knowledge that seems to be gained by the students in the video from looking at multiple representations. (Participant 35)

For the situative group, the focus was on opportunities for thinking about mathematical teaching practices. One situative participant focused on the practice of responding to student thinking:

> I could imagine using this video to ask teacher candidates to listen to each student's contribution and consider what each student knows and is working to understand. I would also ask teacher candidates to imagine how they would respond to the students' contributions. (Participant 13)

For the sociopolitical respondents, there was a theme of setting the mathematical work within overarching classroom goals. For example, a sociopolitical participant described using the video by asking "students to identify what mathematical (content or practice) goal they might use the task for? And what other classroom goals they might hope to achieve. Then return to critique the video and ask what this teacher's goals may have been" (Participant 2).

Regarding student thinking, 4 of the 12 cognitive participants and 2 of the 11 situative participants noted that their perspective influenced their decision to consider students' learning needs and how to make the classroom conducive to learning, particularly mentioning specific learning theories such as constructivism and learning through social processes. Participants from the sociopolitical perspective, however, did not identify these issues when responding to Questions 2 or 3.

The predominant use for the video suggested by participants from the cognitive group focused on mathematical thinking. The situative group suggested potential uses for identification of discourse moves and interactions within the classroom. The sociopolitical participants echoed the responses of the situative group and added elements that would encourage PTs to "problematize the ways in which students were engaged in the mathematical discourse" (Participant 39) and question "where the mathematical authority resides" (Participant 32).

DISCUSSION AND IMPLICATIONS

K–12 classroom videos are often used by MTEs to make mathematics methods practice-based. However, as MTEs draw on scholarly inquiry and practices of colleagues who implement activities involving video, it is important to recognize variations in implementation that may provide different affordances for PTs' learning. The findings of this study suggest that MTEs holding different theoretical perspectives will likely emphasize different elements of the same video. Our analysis suggests that MTEs holding a cognitive perspective were more likely to focus on mathematical elements and specific student thinking in a video and draw more clearly on those features when discussing the video with PTs. MTEs holding a situative perspective were more likely to focus on human interactions documented by the video, analyzing discourse moves of both the teacher and the students. For those holding a sociopolitical perspective, issues of power and authority were a more likely focus.

It is notable that the sociopolitical participants were more critical of aspects of the classroom interactions than either the situative or the cognitive participants. Participants from the cognitive and situative perspectives made laudatory comments about teacher moves that impacted group discourse. There were very few comments made by the six sociopolitical participants concerning teacher discourse moves. Instead, participants from the sociopolitical perspective critiqued the teacher's funneling of students' thinking and the lack of participation by all students. Attentiveness to specific aspects of teaching, such as sociopolitical researchers' critical stance toward teaching interactions or cognitive researchers' close attention to students' mathematical thinking, highlights each group's unique and necessary contribution toward building a comprehensive body of literature on the scholarly practice and inquiry of teaching.

We do note that there are variations within each perspective group, as MTEs prioritize various aspects of classroom practice differently. For example, some cognitive participants discussed focusing on the important mathematical ideas and structures, whereas others focused on building from student thinking. Within the situative perspective, some participants referred to their perspective as focusing on interactions within a classroom environment, whereas others discussed prioritizing practice-based learning opportunities in teaching. The sociopolitical participants focused on authority and status in classroom environments or on participation and engagement, with regards to who participates and in what ways. Variations in ways the perspectives are held by MTEs can potentially create additional variation in MTEs' uses of videos in methods classes. Taking up someone else's activity does not imply that PTs will have the same experience, even when MTEs' perspectives share the same name. Our findings suggest a need to look across the work of MTEs, within and among perspectives, on effective implementation of video in methods of teaching mathematics courses as an approximation of classroom practice (Grossman et. al, 2009).

We note three limitations of the study focused on population and design. First, the relatively small volunteer sample of 29 MTEs may not be representative of all MTEs. Second, every participant was asked to identify with one of three theoretical perspectives, yet our study did not account for the fact that MTEs may identify with multiple perspectives. Third, the length and interaction in the video used in this study may have influenced responses (e.g., a few participants noted they did not feel qualified to answer some of the questions because of the length of the video). A longer video, which more thoroughly represents the classroom environment (e.g., the student–teacher interactions are longer), may result in different responses from participants. We recommend that future studies use multiple videos documenting different approaches to mathematics teaching.

As MTEs work to develop academic practices through scholarly inquiry in methods classes (Lee & Mewborn, 2009), building on the work of other MTEs takes on great importance. There is much to be learned by considering the affordances and constraints for PTs' learning as viewed across multiple enactments of video use. Variations across MTEs' theoretical perspectives, among other factors, must be attended to as work is shared and built upon.

REFERENCES

Association of Mathematics Teacher Educators. (n.d.). Supporting the next generation of mathematics teacher educators. Retrieved from: https://amte.net/
star

Ball, D. L., & Forzani, F. M. (2009). The work of teaching and the challenge for
teacher education. *Journal of Teacher Education, 60,* 497–511.

Grossman, P., Compton, C., Igra, D., Ronfeldt, M., Shahan, E., & Williamson, P.
(2009). Teaching practice: A cross-professional perspective. *The Teachers College Record, 111,* 2055–2100.

Gutiérrez, R. (2013). Why (urban) mathematics teachers need political knowledge.
Journal of Urban Mathematics Education, 6(2), 7–19.

Lee, H., & Mewborn, D. (2009). Mathematics teacher educators engaging in scholarly practices and inquiry. In D. Mewborn & H. Lee (Eds.), *Scholarly practices
and inquiry in the preparation of mathematics teachers* (pp. 1–6). San Diego, CA:
Association of Mathematics Teacher Educators.

Merriam, S. B. (2009). *Qualitative research: A guide to design and implementation.* San
Francisco, CA: Jossey-Bass.

National Council of Teachers of Mathematics. (2015). *The case of Jeffrey Ziegler and
the S-pattern task.* Retrieved from: http://www.nctm.org/Conferences-and
-Professional-Development/Principles-to-Actions-Toolkit/The-Case-of
-Jeffrey-Ziegler-and-the-S-Pattern-Task/

Roth McDuffie, A., Foote, M. Q., Bolson, C., Turner, E. E., Aguirre, J. M., Bartell, T.
G., Drake, C., & Land, T. (2013). Use of video analysis to support prospective
K-8 teachers' noticing of students' multiple mathematical knowledge bases.
Journal of Mathematics Teacher Education, 10, 245–270.

Roth McDuffie, A., Foote, M. Q., Drake, C., Turner, E., Aquirre, J., Bartell, T. G., &
Bolson, C. (2014). Use of video analysis to support prospective K–8 teachers'
noticing of equitable practices. *Mathematics Teacher Educator, 2*(2), 108–140.

Sanchez, W., Kastberg, S., Tyminski, A., & Lischka, A. (2015). *Scholarly inquiry and
practices (SIP) conference for mathematics education methods.* Atlanta, GA: National
Science Foundation.

Seidel, T., Blomberg, G., & Renkl, A. (2013). Instructional strategies for using video
in teacher education. *Teaching and Teacher Education, 34,* 56–65.

Star, J., & Strickland, S. K. (2008). Learning to observe: Using video to improve
preservice teachers' ability to notice. *Journal of Mathematics Teacher Education,
11*(2), 107–125.

CHAPTER 20

THEORETICAL PERSPECTIVES, GOALS, AND ACTIVITIES FOR SECONDARY MATHEMATICS EDUCATION METHODS COURSES

Ryan C. Smith
Radford University

Cynthia E. Taylor
Millersville University of Pennsylvania

Dongjo Shin
University of Georgia

In the United States, there is a great deal of variability in mathematics teacher education programs (Darling-Hammond et al., 2000; Levine, 2006). This variability may be seen in programmatic features including the number and type of field experiences and the number and content of required mathematics and methods courses. Program differences may be related to the lack of

Building Support for Scholarly Practices in Mathematics Methods, pages 311–324
Copyright © 2018 by Information Age Publishing
All rights of reproduction in any form reserved.

shared standards and professional curriculum for the preparation of mathematics teachers (Ball, Sleep, Boerst, & Bass, 2009; Zaslavsky, 2007). Variability may also be related to mathematics teacher educators (MTEs) who teach and prepare prospective teachers (PTs). Moreover, we know very little about the practices of MTEs as these practices are not widely documented or disseminated (e.g., Bergsten & Grevholm, 2008; Floden & Philipp, 2003).

Researchers have made efforts to examine and define general practices of MTEs (e.g., Dixon, Andreasen, & Stephan, 2009; Kastberg, Tyminski, & Sanchez, in press). There have also been studies exploring MTEs' mathematics methods course syllabi (e.g., Harder & Talbot, 1997; Taylor & Ronau, 2006) and course topics MTEs value (Otten, Yee, & Taylor, 2015; Watanabe & Yarnevich, 1999). These studies highlight the variability in mathematics methods courses. Furthermore, working groups and conferences have focused on opportunities and challenges surrounding the support and work of MTEs (e.g., North American Chapter of the International Group for the Psychology of Mathematics Education [PME-NA], 2014; Scholarly Inquiry and Practices [SIP] in Mathematics Education Methods Conference, 2015 [Sanchez, Kastberg, Tyminski, & Lischka, 2015]).

Despite the efforts and progress exploring MTEs' practices, there remains a need to build descriptions of methods courses including goals, activities, and theoretical bases for the design and implementation of the courses. In addition, building descriptions of the extent of the alignment between and relationships among theoretical perspectives, goals, and activities will support understanding of variation described in prior research (Taylor & Ronau, 2006). To address this need, we sought to explore theoretical perspectives, goals, and activities MTEs draw upon when preparing for and instructing their secondary mathematics methods courses. We define a secondary mathematics methods course as a course PTs take designed to develop the mathematical and pedagogical knowledge, skills, and dispositions to teach mathematics at the secondary level. The research questions that guided the study were (a) What theoretical perspective(s) do MTEs draw upon when developing and teaching a secondary mathematics methods course? (b) What theoretical perspective(s) best align to the goals and activities MTEs deem most important for their secondary mathematics methods course? and (c) In what ways do the most important goals and activities discussed in (b) align with the MTEs' self-identified perspective(s) in (a)?

METHODS

To recruit participants for the study, we sent email invitations to MTEs who had taught secondary mathematics methods courses. We invited MTEs who were (a) Service, Teaching, and Research (STaR) fellows; (b) Association of

Mathematics Teacher Educators (AMTE) affiliate members; (c) members of a mathematics methods discussion group on Facebook; (d) members of a mathematics education researchers group on Facebook; (e) subscribers to Jerry Becker's mathematics education email listserv; and (f) SIP conference participants. Thirty-five MTEs who taught secondary mathematics methods courses responded, 17 male and 18 female, from 33 institutions in 23 states in the United States of America and 1 South African province. Participants had a mean of 10.77 years of experience (SD = 10.34 years) as an MTE prior to participating in the study.

Participants completed a 22-question online survey that included

- Briefly describe the perspective(s) (e.g., cognitive, situative, and sociopolitical) you draw upon when developing and teaching your methods course.
- What are the top three goals (e.g., knowledge, skills, and dispositions), in no particular order, that you have for your methods course?
- What are the three most important tasks/activities you use in your methods course?

They uploaded supporting documents for the activities they described and a copy of their course syllabus.

We began our analysis by creating a spreadsheet of the survey responses and examining the participants' descriptions of their theoretical perspectives. The first two authors coded the participants' self-identified perspectives through discussions and deliberations. Codes for the perspectives were based on the participants' descriptions of perspectives and not a predetermined list. The following perspectives were articulated: *situative, cognitive, sociopolitical, social-constructivist, sociocultural,* and *cognitive and constructivist.* We chose not to combine the participants who self-identified as drawing on a "cognitive and constructivist" perspective with those who self-identified as drawing solely from a cognitive perspective because our focus of this study was on their self-identification and not the researchers' presupposition. We created two additional codes to account for those who identified multiple perspectives (*multiple*; e.g., one participant wrote, "complex instruction; cognitive; constructivist; mathematics for problem solving"), and those that could not be determined (*cannot be determined*; e.g., one participant wrote, "Personal experience. The cognitive demands and clarity of concepts that will transfer from one year to the next is a major emphasis").

To examine the participants' goals and perspectives, the first two authors coded the participants' goals with a perspective (*cognitive, situative,* and *sociopolitical*) using the definitions of Casey, Fox, and Lischka (Chapter 19, this volume). The example goals listed in Table 20.1 illustrate our

TABLE 20.1 Perspectives as Defined in Casey et al. (this volume) and Examples From Participant Responses

Definition	Example Goal	Example Activity
Cognitive		
"The root of your work is in developing models of thinking about a concept (either mathematical or pedagogical). A goal of your work is to help teachers learn to attend to and respond to student thinking in ways that promote student learning."	"Develop conceptual knowledge across several mathematical strands."	"I pose the staircase problem where students work individually, then in groups to solve the task. As a class, we make student thinking public and get out the various ways PTs solved the task...discussing why the formula works and at the end, connecting Standards of Mathematical Practices to various portions of the task."
Situative		
"Learning is conceptualized as an aspect of social activity and therefore you research and work within activity systems. Learning is regarded in relation to interactions and within contexts. Teaching occurs in school with children and experienced teachers, and so prospective teachers' learning situated in interactive setting is valued."	"Knowledge of and skill with implementing a high-level task following a launch, explore, and discuss model."	"Teach a lesson in a secondary math classroom. Lesson plan is reviewed and edited multiple times before being taught. The cooperating teacher and I give feedback on the teaching. PTs write a reflection paper."
Sociopolitical		
"Development of teachers' political knowledge or *conocimiento* is your primary goal. Among other things, political *conocimiento* involves understanding how oppression in schooling operates not only at the individual level but also the systemic level; deconstructing the deficit discourses about historically underserved and/or marginalized students; and buffering oneself, reinventing, or subverting the system in order to be an advocate for one's students."	"Create a set of goals for student learning, including content, self-efficacy, and social/citizenship."	"PTs respond to 'mathematics autobiography' questions in order to begin the discussion about students' identities and the role of affect and identity in students' learning."

coding. In general, we coded goals as cognitive if the MTEs focused on PTs' learning and developing models of mathematical thinking. The first goal in Table 20.1 was coded as cognitive because the MTE wanted their PTs to build their own knowledge of mathematics. Situative goals were goals in which the MTEs wanted PTs to develop their pedagogical skills in the context of the mathematics classroom. The example situative goal centered around PTs developing their own skills in implementing a three-phase lesson plan model. Finally, we distinguished sociopolitical goals on the basis of the MTE's desire for PTs to learn that mathematics teaching influences more than just the learning of mathematics content (e.g., "understanding how oppression in schools operates at both the individual and systematic level"; Casey et al., this volume). For example, the goal coded as sociopolitical focused on developing PTs' ability to create learning goals for their future classroom that focus not only on mathematics but also on developing Grade 7–12 students' self-efficacy and social citizenship.

At times, one could make a strong argument that a goal would align well with multiple perspectives. For example, the goal "to gain knowledge in pedagogical skills" could be interpreted as situative or sociopolitical depending on the particular pedagogical skill being developed. In these cases, we coded the goals as *multiple*. Also, at times, we were unable to determine a perspective that best aligned with the goal (e.g., "Translate theory to practice"). In these cases, we coded the goals as *cannot be determined*. During coding, the first two authors discussed the codes, created arguments for each code, and came to an agreement.

We next coded activities based on the perspective (*cognitive, situative,* and *sociopolitical*) we believed most aligned with that activity. We examined both the MTEs' description in the survey and the supporting documents they uploaded. In general, we coded activities as cognitive if they could be used to develop PTs' understanding of mathematics and Grade 7–12 students' mathematical thinking. The cognitive example in Table 20.1 asked PTs to solve a mathematics problem to develop their own understanding of a particular formula. Activities coded as situative had the characteristic of PTs practicing or developing skills and abilities in the context of the mathematics classroom. The example provided in Table 20.1 engaged PTs in a task where they develop and teach a mathematics lesson to secondary mathematics students. Activities coded as sociopolitical were those that provided PTs the opportunity to develop their understanding of Grade 7–12 students' identity, how culture influences students' mathematical learning, and how to advocate for one's students. For example, the activity in Table 20.1 asked PTs to begin thinking about their own identity as a mathematics learner and the factors that could influence students' identity formation.

Similar to analysis of the goals, there were activities that aligned well with multiple perspectives (e.g., "Clinical Interview") and activities that we were unable to determine a perspective that best aligned with the activity (e.g., "Algebra Task"). Thus, we created two additional codes—*multiple* and *cannot be determined.* The two lead authors coded the activities simultaneously, discussing each activity and coming to an agreement for a particular code.

We recognize coding participants' goals and activities using only the situative, cognitive, and sociopolitical perspectives may not provide a holistic picture of the alignment between perspective and goal or activity. However, we wanted to gather evidence of participants' self-described perspectives to build insight into perspectives used in the design of methods courses. Then, we could explore and describe the alignment of perspectives, goals, and activities for those who self-identified as one of the three perspectives. In the following sections, we describe our findings from our analyses of the survey data focusing on MTEs' theoretical perspectives, goals, and activities and their alignment.

RESULTS

In the following sections, we discuss respondents' perspectives, goals, and activities (Table 20.2). We then examine alignment for 17 MTEs who exclusively drew upon the situative, cognitive, or sociopolitical perspective.

Perspectives

As shown in Table 20.2, the 35 participants draw on a variety of perspectives in the design and implementation of their secondary mathematics methods courses. Seventeen of the 35 MTEs stated that they draw exclusively from one of the three focal perspectives of the SIP conference (Sanchez et al., 2015): *situative, cognitive,* or *sociopolitical.* Of the eight MTEs who

TABLE 20.2 Classification of Secondary MTEs' Perspective, Goals, and Activities

	Perspective*		Goals		Activities	
Situative	11	(33%)	57	(55%)	57	(59%)
Cognitive	4	(11%)	21	(20%)	8	(9%)
Sociopolitical	2	(6%)	10	(10%)	2	(2%)
Multiple	8	(23%)	9	(9%)	15	(16%)
Cannot Be Determined	3	(9%)	6	(6%)	14	(15%)
Other	7	(19%)				
Total	35	(100%)	103	(100%)	96	(100%)

* The perspective was self-identified.

stated that they draw upon multiple perspectives, two indicated that they draw upon a cognitive and constructivist perspective and two from a cognitive and situative perspective. The remaining four MTEs who stated they draw upon multiple perspectives had unique sets of perspectives. For 9% of the participants (3/35), we could not determine a perspective they draw upon from their responses. The seven remaining participants stated they draw upon other perspectives—17% draw upon a constructivist perspective (6/35) and 3% draw upon a sociocultural perspective (1/35). Even though there was no majority perspective MTEs draw upon exclusively in the design of secondary mathematics methods courses, just under half of the participants (15/35) identified drawing upon the situative perspective (either solely or in conjunction with another perspective).

Goals

The 35 participants identified 103 goals for mathematics methods courses, as two participants listed two goals instead of the requested three. As illustrated in Table 20.2, we coded 85% of participants' goals using situative, cognitive, or sociopolitical perspectives. Ten participants had all three goals coded with the same perspective; nine of these were coded as situative and the remaining participants' goals were coded as sociopolitical. Interestingly, only five of those nine participants reported they draw exclusively from a situative perspective. Two others identified as drawing from a cognitive perspective, one stated drawing from a social-constructivism perspective, and one listed drawing from multiple perspectives. The participant whose three goals were coded as sociopolitical reported having a sociopolitical perspective.

Activities

Participants were asked to list their three most important activities (i.e., activities MTEs use in their methods classes) and upload one supporting document for each activity. Thirty-two participants described 96 activities—70% of which were coded as situative, cognitive, or sociopolitical. The majority of the activities, like the goals, were coded as situative. Comparing percentages of goals and activities for each of the three perspectives, the percentages of goals coded cognitive and sociopolitical are much greater than the percentages of corresponding activities. It seems that although MTEs report goals related to cognitive and sociopolitical perspective, the activities they use to address these goals are not among those they identify as the three most important they facilitate in their classroom.

Our analysis revealed that six MTEs' activities were all coded as aligning with the situative perspective. Of the six, three identified as drawing from a situative perspective, two stated they drew from a constructivist perspective, and one drew from multiple perspectives.

Drawing on Perspectives: A Closer Look

Of the 35 MTEs, 17 identified drawing exclusively from a situative, cognitive, or sociopolitical perspective. To understand how perspectives may influence secondary mathematics methods course design, we looked more closely at alignment for these respondents.

Cognitive Perspective

Of these 17 respondents, only 4 identified drawing from the cognitive perspective. Even though this number is low, we found it interesting that there was no noticeable pattern or alignment within and between the cognitive respondents. For example, one MTE had two goals coded as cognitive but all of the activities coded as situative. And the reverse was true for another MTE; none of the goals were coded as cognitive, but two of the activities were coded as cognitive. The lack of alignment in their activities and goals with their perspective may indicate the influence of the cognitive perspective is apparent in other aspects of the work of MTEs such as during the enactment of activities in the methods course.

Sociopolitical Perspective

Only two MTEs identified drawing exclusively from a sociopolitical perspective. The pair did share a common goal in developing PTs' understanding of the politics and inequities in mathematics education and how these influence students' mathematics learning and achievement. However, the influence of the sociopolitical perspective and their shared goal was not evident in the activities these MTEs submitted as those most important in their methods courses. Additional data are needed to draw any conclusions.

Situative Perspective—Goals

When designing, planning, and facilitating a secondary mathematics methods course, 11 of the 17 MTEs identified drawing from a situative perspective. At least two of the three goals identified by each of the 11 participants aligned with a situative perspective. As we examined their goals, two themes emerged. First, MTEs wanted PTs to develop their abilities and skills to plan engaging lessons. Lesson planning goals were coded as situative because this type of preparation is an integral part of teaching that takes into account the context and community of the classroom. When PTs design and prepare lessons, they learn about different components that may be preplanned in order to provide the opportunity for their future students to learn mathematical concepts through those preplanned activities. Eight of the 11 MTEs identified goals that would provide PTs with opportunities to develop and understand how to design lesson plans. These goals focused on PTs demonstrating creativity in lesson design, selecting and modifying activities to support student understanding, and using a variety of instructional strategies when preparing lessons. One MTE's lesson planning goal also involved PTs anticipating student thinking:

> Create lesson plans that involve meaningful tasks in the context of a given unit, that anticipate student thinking and how the lesson might build on that thinking, and that articulate how the [PTs] will orchestrate each phase of the lesson (launch, explore, discuss, unpack).

These types of goals were coded as situative as opposed to cognitive as they primarily focused on how the PTs could interact with the students and not the students' knowledge or actions. In the previous example, the MTE wanted PTs to understand lesson planning involved the selection and use of meaningful activities, anticipating student thinking, and considering implementation. MTEs also wanted PTs to consider the role of technology in lesson planning.

Second, 6 of the 11 MTEs included goals for PTs to develop knowledge of, disposition towards, and abilities to engage students in learning mathematics. One MTE wanted PTs to "develop a disposition toward building on student thinking through task enactment and questioning," while another MTE wanted PTs to "understand the need to actively engage [the PTs] in the mathematics they are studying through discussion, activities, problems, formative assessments, etc." Although these goals appear to involve different aspects of teaching, both MTEs wanted their PTs to develop mathematical knowledge used in teaching. These goals were coded as situative as they focused on the PTs' actions and reactions to students' learning rather than developing models of their thinking. In addition, the MTEs considered the PTs' learning of mathematics, which the MTEs would facilitate with fellow classmates by engaging them in meaningful mathematical discourse—similar to the "activity system" in which the PTs would eventually be working (i.e., the Grade 7–12 classroom).

Although the majority of the MTEs' goals were also coded as situative, four were coded as cognitive and two were coded both cognitive and situative. We coded goals as aligning to a cognitive perspective when MTEs stated they wanted PTs to have the opportunity to build on student thinking and focus on what Grade 7–12 students know, mathematical misconceptions they might have, and how to meet the students' mathematical needs to support their mathematical learning. In other words, they wanted PTs to build models of and attend to student thinking. These goals were not coded as situative because they primarily focused on the students' actions and not how the teacher could interact with the students. The final two goals the MTEs articulated seemed to align to situative and cognitive perspectives. For example, one MTE's goal was to develop "student engagement techniques," which we could have interpreted as developing students' thinking (cognitive) or the teacher's instructional strategies (situative). In all, 81% of the goals submitted by the 11 MTEs who identified drawing from a situative perspective were coded as situative.

Situative Perspective—Activities

The 11 MTEs who identified exclusively as having a situative perspective provided 20 of the 30 activities that were coded as situative (one MTE did not share any activities, so the total number of activities was 30 rather than 33). We coded seven of the remaining ten activities as situative and cognitive, two as cognitive, and one as cannot be determined. The most common activity MTEs shared centered around lesson planning including writing, teaching, and reflecting on lessons. Some activities involved PTs writing lessons and others included teaching written lessons to Grade 7–12 students or peers.

The majority of the seven activities aligning to both situative and cognitive perspectives engaged PTs in solving mathematical problems and considering ways to engage Grade 7–12 students in learning mathematics while also developing their pedagogical skills. In such activities, PTs built models of student thinking (cognitive) and planning lessons to "practice questioning." Some activities involved PTs peer teaching (situative) planned lessons or conducting student interviews. Other activities that were coded as both situative and cognitive involved writing papers, letter-writing activity, and culminating projects for the course.

Planning and teaching lessons were the most common activities identified by MTEs drawing on the situative perspective. For example, one activity an MTE described engaging her PTs in was, "Students design and implement mathematics lessons that are focused on specific content and [are] age appropriate." These planning and teaching activities align with the most prominent situative goal, to plan engaging lessons. The second most common activity involved PTs analyzing student thinking via student interviews. One provided activity was a "student interview task" in which PTs plan, implement, and analyze a task-based interview with three Grade 7–12 students. This activity aligns with the second most common goal, developing knowledge of how to engage students in learning mathematics. In conclusion, the two most popular goals and two most popular activities seem to align for the MTEs who identified as drawing from a situative perspective. Furthermore, for these 11 participants, the majority of their important goals and activities aligned with their perspective.

DISCUSSION AND IMPLICATIONS

Our findings indicate the most popular perspective identified by MTEs to design and implement their secondary mathematics methods courses was a situative perspective. Moreover, our findings suggest that the most important goals and activities for secondary mathematics methods courses are largely based on a situative perspective as well. A situative perspective may be the lens many MTEs draw upon because this is the perspective that centers on PTs' learning in the context of teaching and the interactions between teachers, among students, and within the classroom community. However, is the pervasiveness of a situative perspective a good thing? Should a situative perspective be the perspective MTEs draw upon when teaching methods courses? Because the work of teaching is extremely complex, drawing upon a single perspective may be limiting such that PTs would only see teaching with a particular lens. "Any one perspective is [not] better than any other. Different perspectives can allow us to gain insight and through acting on these insights, to change" (Brown & Coles, 2010, p. 380). By providing PTs with the opportunity to see, experience, and consider multiple perspectives, MTEs are able to provide PTs with a more robust and complete experience. Furthermore, it appears that many MTEs did that; only nine MTEs (26%) had all three goals from the same perspective and only five (16%) had all three activities from the same perspective. Thus, the majority of the MTEs seemed to draw upon multiple perspectives.

Evidence of Sociopolitical Perspectives

It is interesting to note that in the data there were few goals and activities designed from a sociopolitical perspective. Two MTEs indicated they draw exclusively upon a sociopolitical perspective. To gain further insight into goals MTEs might have but not identify as among the three most important goals for methods, we examined the MTEs' syllabi. Twenty-six MTEs uploaded syllabi as we requested, and only 10 (38%) included goals we coded as aligned with a sociopolitical perspective. In contrast, only 10% of the most important goals were based on this perspective. From these data, it seems MTEs do not identify sociopolitical goals for their methods course and even fewer consider them among the three most important in methods. In addition, only 2% of the most important activities were coded sociopolitical.

Perspective Will Influence Methods Courses

One could view the lack of alignment between our participating MTEs' perspectives, goals, and activities to mean that a perspective may not have a great deal of influence in the design of methods courses. Brown and Coles (2010) state, "There is a growing realization that what is offered with [mathematics teacher preparation programs] is often based on facilitators' (or administrators' or policy makers') perceptions of what teachers need as opposed to knowledge of what teachers actually do need (Ball, 2000)" (p. 378). The design and content of a methods course is likely not solely at the discretion of the MTE, which may limit the influence of an MTE's perspective on the design of their methods courses. Perhaps their perspective may not have as much influence on the design of the courses and the selection of activities. Rather, the influence may be seen in the implementation of the activities.

An MTE's perspective may have greater influence during the implementation of lessons rather than in the design of goals and activities. MTEs who have the same goals and use the same activity but draw upon different perspectives may focus on different ideas or aspects of an activity during its implementation. For example, one activity in a methods course is to have PTs complete a mathematical activity and then observe a video of students working on the same activity (e.g., Van Zoest & Stockero, 2008). MTEs from various perspectives may have the same goal for this activity, namely for PTs to develop their abilities to analyze student thinking and examine the reciprocal relationship between teaching and learning. However, an MTE with a situative perspective may focus on the teacher's actions and how they influenced student learning. An MTE with a cognitive perspective may focus on building models of student thinking in relation to the mathematical activity and have PTs consider what activity they would do next to further students' mathematical understanding. An MTE with a sociopolitical perspective may consider how the teacher's actions may have led to inequitable opportunities for all students to learn mathematics. In fact, the results of Casey et al. (Chapter 19, this volume) align well with these ideas; an MTE's perspective may be more influential in the implementation of activities than the selection of the activities.

Future Direction

The results of this study shed light on alignments between MTEs' perspectives, goals, and activities for their secondary mathematics methods courses. Furthermore, findings from this study build on research focused on what MTEs value in the field, the instructional approaches MTEs design for methods courses, and what MTEs emphasize with PTs in their interactions with them in those courses (e.g., Otten et al., 2015; Taylor & Ronau, 2006). However, additional research should be conducted that provides both a broader perspective as well as deeper insight into how an MTE's theoretical perspective, at all levels, influences his/her design and implementation of a mathematics methods course. Specifically, interviews with MTEs at all levels and observations of their teaching would allow for a deeper understanding of how their perspectives influence the design and implementation of mathematics methods courses.

REFERENCES

Ball, D. L., Sleep, L., Boerst, T. A., & Bass, H. (2009). Combining the development of practice and the practice of development in teacher education. *The Elementary School Journal, 109*(5), 458–474. doi: 10.1086/596996

Bergsten, C., & Grevholm, B. (2008). Knowledgeable teacher educators and linking practices. In B. Jaworski & T. Wood (Eds.), *The international handbook of mathematics teacher education* (Vol. 4, pp. 223–246). Rotterdam, The Netherlands: Sense.

Brown, L., & Coles, A. (2010). Mathematics teacher and mathematics teacher educator change—Insight through theoretical perspectives. *Journal of Mathematics Teacher Education, 13*(5), 375–382. doi: 10.1007/s10857-010-9159-3

Darling-Hammond, L., Macdonald, M. B., Snyder, J., Whitford, B. L., Rusco, G., & Fickel, L. (2000). *Studies of excellence in teacher education: Preparation at the graduate level.* New York, NY: AACTE.

Dixon, J. K., Andreasen, J. B., & Stephan, M. (2009). Establishing social and sociomathematical norms in an undergraduate mathematics content course for prospective teachers: The role of the instructor. In D. S. Mewborn & H. S. Lee (Eds.), *AMTE Monograph 6: Scholarly practices and inquiry in the preparation of mathematics teachers* (pp. 43–66). San Diego, CA: Association of Mathematics Teacher Educators.

Floden, R. E., & Philipp, R. A. (2003). Report of working group 7: Teacher preparation. In F. K. Lester & J. Ferrini-Mundy (Eds.), *Proceedings of the NCTM Research Catalyst Conference* (pp. 171–176). Reston, VA: National Council of Teachers of Mathematics.

Harder, V., & Talbot, L. (1997, February). *How are mathematics methods courses taught?* Paper presented at the annual meeting of the Association of Mathematics Teacher Educators, Washington, DC.

Kastberg, S. K., Tyminski, A. M., & Sanchez, W. (in press). Reframing research on methods courses in mathematics teacher education. *The Mathematics Educator.* Retrieved from http://edschools.org/pdf/Educating_Teachers_Report.pdf

Levine, A. (2006). *Educating school teachers.* Washington, DC: The Education Schools Project.

Otten, S., Yee, S. P., & Taylor, M. W. (2015). Secondary mathematics methods courses: What do we value? In T.G. Bartell, K. N. Beida, R. T. Putnam, K. Bradfield, & H. Dominguez (Eds.), *Proceedings of the 37th annual meeting of the North*

American Chapter of the International Group for the Psychology of Mathematics Education (pp. 510–517). East Lansing, MI: Michigan State University.

Sanchez, W., Kastberg, S., Tyminski, A., & Lischka, A. (2015). *Scholarly inquiry and practices (SIP) conference for mathematics education methods.* Atlanta, GA: National Science Foundation.

Taylor, P. M., & Ronau, R. (2006). Syllabus study: A structured look at mathematics methods courses. *AMTE Connections, 16*(1), 12–15.

Van Zoest, L. R., & Stockero, S. L. (2008). Using a video-case curriculum to develop preservice teachers' knowledge and skills. In M. S. Smith & S. Friel (Eds.), *AMTE monograph 4: Cases in mathematics teacher education: Tools for developing knowledge needed for teaching* (pp. 117–132). San Diego, CA: Association of Mathematics Teacher Educators.

Watanabe, T., & Yarnevich, M. (1999, January). *What really should be taught in the elementary methods course?* Paper presented at the annual meeting of the Association of Mathematics Teacher Educators, Chicago, IL.

Zaslavsky, O. (2007). Mathematics-related tasks, teacher education, and teacher educators. *Journal of Mathematics Teacher Education, 10*(4), 433–440. doi: 10.1007/s10857-007-9060-x

CHAPTER 21

THE "MIRROR TEST"

A Tool for Reflection on Our Sociopolitical Identities as Mathematics Teacher Educators

Andrea McCloskey
Pennsylvania State University

Brian R. Lawler
Kennesaw State University

Theodore Chao
The Ohio State University

This chapter builds upon sociopolitical perspectives to elucidate how the "mirror test" informs our approaches to mathematics methods instruction and might serve to shape the development of professional ethics for mathematics teacher educators (MTE). Sharing examples from our practice, we raise questions regarding intersections of sociopolitical perspectives and mathematics teacher education as a field of scholarship and practice. The three of us self-identify as early career mathematics educators who teach

Building Support for Scholarly Practices in Mathematics Methods, pages 325–340
Copyright © 2018 by Information Age Publishing

mathematics methods courses. We consider our histories as former public school teachers to be an important component of our professional identities. At the same time, we lay claim to differences in our professional and personal identities. Brian identifies with secondary education, whereas Theodore and Andrea work more in elementary education. We each lie at unique points in the intersection of gender, race, and other identifications. We self-identify as critical mathematics educators yet recognize the limitations of that label. That is, we are more interested in engaging MTEs in conversation than in definitional disagreements.

We begin the chapter by articulating our understanding of a sociopolitical perspective on mathematics teacher education, drawing from the Scholarly Inquiry and Practices Conference (Sanchez, Kastberg, Tyminski, & Lischka, 2015) work group discussion sessions. Our ideas are informed by the work of Rochelle Gutiérrez, who served as the keynote speaker representing a sociopolitical perspective at the Scholarly Inquiry and Practices Conference and participated in our work group. The remainder of the chapter is based on Gutiérrez's (2016) mirror test. We describe the mirror test, provide examples of how we have used it in our own methods instruction, and propose how and why the mirror test might be used within the MTE community.

The mirror test has been an essential tool for bringing integrity to our work as MTEs and empowering us to live out social justice commitments in mathematics methods instruction. We argue that MTEs as a community should also use the mirror test to live out social justice commitments. This stance raises difficult questions, such as *What is the relationship between the mirror test and professional organizations? Who can implicate such organizations (and us as members) in critiques of power and inequity? How should we individually act when we view a colleague through our own mirror and conclude that he/she is not aligning with the same ethics? Should there be a shared ethic or moral code for the MTE community? And if so, how we might we hold one another accountable?* We suggest responses to these questions and frame our discussion by proposing reasons few MTEs engage sociopolitical issues in their mathematics methods teaching. But first we articulate our conception of a sociopolitical perspective.

A SOCIOPOLITICAL PERSPECTIVE ON MATHEMATICS METHODS INSTRUCTION

We understand sociopolitical perspectives through historical considerations of mathematics education. Mathematics education established itself as a field of study with a focus on the individual as a learner of mathematics (Kilpatrick, 2012). During its foundational years, mathematics education focused on cognitive functioning and its development, especially in formal school settings. As the field matured, awareness of the structures and social nature of learning gained footing, so that thinking, reasoning, and meaning-making began to be regarded as products of social activity (Lerman, 2000). However, neither of these perspectives satisfied researchers who embraced equity, social justice, and antiracist orientations. Feminist, critical, and poststructural theoretical frameworks created new epistemologies and ontologies, allowing for an interrogation of the ways power and privilege impact and define what is learned and who is learning. Banks' (1993) words resonated for many in mathematics education: "All knowledge reflects the values and interests of its creators" (p. 4). What counts as mathematics knowledge became subject to interrogation. Mathematics education is primed and ready for sociopolitical orientations to enter the mainstream (Kilpatrick, 2012).

Just as the experiences of the powered and subaltern produce idiosyncratic realities, groups develop identifiable epistemologies, or ways of knowing (Collins, 1991). Following the recognition that knowledge and power are interrelated (Foucault, 1980), the identities of the learners—whether student, teacher, or any research subject—become an important consideration when naming goals, examining results, and measuring the impacts of mathematics education. By studying ways knowledge, power, and identity are taken up by members of oppressed communities, specifically related to mathematics education, MTEs can learn from the experiences of minoritized individuals and communities "how to rethink mathematics education" (Gutiérrez, 2013, p. 3) in order to improve conditions for every student.

Sociopolitical perspectives in mathematics education are grounded in recognition that mathematics is a human activity involving powered relationships and identities. Human interactions shape what becomes accepted as truth and knowledge (Lawler, 2012). Activities of teaching and learning mathematics are not neutral or apolitical (Martin, 2013). Ultimately, embracing a sociopolitical view recognizes that mathematics education is identity work (Gutiérrez, 2013), reflecting on "who am I" in relationship to mathematics (Lawler, 2012) and to others. Mathematics teaching builds upon a learner's prior identities toward an emerging mathematics identity, as learners recognize and use mathematics to understand and engage the world (Aguirre, Mayfield-Ingram, & Martin, 2013). Mathematics methods teaching plays a similar identity formation role. Learners are prospective teachers (PTs) with multiple and intersecting relevant identities—as former mathematics learners, current university students, and future teachers.

Goals for Mathematics Methods

The preparation of mathematics teachers must include the development of "skills to form deep connections to students and political knowledge"

(Gutiérrez, 2013, p. 17) that are necessary for negotiating political aspects of the profession, including standards and standardization; high-stakes testing and accountability; and connections with community, classical, and critical knowledge (Gutstein, 2007). PTs must learn how to advocate for students, self-examine for biases, and strategically subvert the system in which they teach to counteract student oppression. Our working group identified five goals toward helping PTs develop political knowledge. Students in a mathematics methods course that successfully attends to the intersection of mathematics teaching, learning, and social justice will

1. develop strategies for disrupting current mathematics education norms and agency for pushing back;
2. become aware of and draw on knowledge of context in which they do and might work, including families and communities;
3. develop a critical orientation to mathematics;
4. critique discourses of education (e.g., "Schools are failing," "The achievement gap is only about achievement"); and
5. critically analyze and intentionally develop their own mathematics teacher identity.

If these goals are charged to MTEs for their work in mathematics teacher preparation, what resources and what preparation would the MTE require in order to achieve such goals? In addition to resources and knowledge, there is a need to learn to negotiate the sociopolitical space—embracing and enacting this in our work is a tension all three of us have experienced. Instead of shifting responsibility to those who prepare MTEs to teach methods with sociopolitical goals in mind (i.e., MTE educators), we focus our discussion on a grassroots community effort for the development of MTEs by MTEs. In this effort, knowledge builds from within the community rather than being passed down from above by a more knowledgeable expert or bureaucratic agency. As members of the community of MTEs, we take ownership and responsibility for continuing this conversation among our peers. Our social justice orientation compels us to engage in deliberative, self-reflective discussion among colleagues and within ourselves. We begin by clarifying our notion of *identity*.

Identity

Within mathematics education, identity has often been used as a categorical device (e.g., gender, socioeconomic status, race/ethnicity, sexual orientation; Beauchamp & Thomas, 2009; Langer-Osuna, 2015). MTEs must help future mathematics teachers move beyond this checkbox view of identity to become aware of the ways they position themselves in relationship to power and how PTs can raise an identity consciousness within students that transcends categories. MTEs must help PTs see their mathematics teacher identities as negotiated between themselves and the worlds they inhabit and as continually evolving, recognizing identity as a verb (as opposed to a noun; Gutiérrez, 2013). Thereby, PTs may come to appreciate that students learning mathematics are cocreating a mathematics identity that connects to ways student see and position themselves (Gutiérrez, 2013).

MTEs may find eliciting overlapping narratives helpful for navigating emerging identities, in which MTEs confront their roles in oppression and privilege (Herbel-Eisenmann et al., 2013). These narratives invoke *intersectionality*, a construct with roots in both feminist and critical race theory that examines how combinations of marginalized spaces overlap in real-world settings (Crenshaw, 1991; Delgado & Stefancic, 2012). Intersectionality draws attention to uniquely marginalized spaces that exist within the status quo. For instance, the community of MTEs is typically positioned as composed of White women with backgrounds as former mathematics teachers. Positioning MTEs as a homogenous group takes power away from them, thereby diminishing their effectiveness in helping PTs develop sociopolitical mathematics teacher identities.

Using intersectionality brings marginalized spaces to attention and provides room for MTEs to model how to engage in identity work with their PTs. For instance, an MTE who self-identifies as a White woman who grew up in a non-English-speaking household may highlight the unique intersection point of being White, female, and an English language learner (ELL), a particular marginalized space with only some dimensions in common with those who identify as either White, female, or a second language learner. Using intersectionality to frame our MTE identities, we break the myth of MTEs as a homogenous group. This sort of *mirror test* allows MTEs to model how PTs can confront their emerging sociopolitical identities.

THE MIRROR TEST IN THE PREPARATION OF PTs

The mirror test has its origins in fields that consider the nature of consciousness: philosophy, animal studies, cognitive science, ethics, and others. Scholars in these fields ask: *What does it mean to have self-consciousness?* and *How can we define what it means for a being to know and recognize itself as an entity?* The mirror test posits that if a being recognizes itself in a mirror, then it has self-consciousness. Gutiérrez (2016) described the mirror test as a useful metaphor for her goals in her work with PTs:

I aim to prepare them with the ability to look themselves in the mirror every day and to be able to say "I'm doing what I said I was going to do when I entered the profession." And, if they are not, they need to be asking themselves "Why not?" and "What can I do about that?" (p. 260)

To prepare her students to engage in this orientation toward *critical vigilance,* Gutiérrez (2016) regarded the learning goals of mathematics methods courses as less about "correct views" or "best practices" and more about a way of being. Mathematics teacher educators must continually engage in critical reflection regarding the effectiveness of their instructional practices toward developing this critical vigilance. Speaking for ourselves, we intend for PTs' experiences in our mathematics methods courses to help them develop an understanding of mathematical proficiency in a broader and more critical sense than typically embraced in K–12 schooling (e.g., *Adding it Up,* National Research Council, 2001). Effectiveness of MTE instruction would also be reflected in PTs' more sophisticated notions of equity, for example, beyond simply closing "achievement gaps," in which achievement is measured by a test score and "group membership" is no more than a checkbox category. Likewise, we consider effective methods instruction to result in our PTs claiming explicit and informed respect for and appreciation of the roles of identities—their own, that of the students in classrooms, and beyond (de Freitas, 2008; Ma & Singer-Gabella, 2011; Walshaw, 2004). This understanding should inform PTs' responsibility for their identities as soon-to-be mathematics teachers, the temporal intersectionality of that identity, and the political nature of their position in their work as mathematics teachers. Such development of responsibility lays the groundwork toward critical vigilance as a way of doing—and being—a mathematics teacher.

USING THE MIRROR TEST TO INFORM OUR OWN METHODS INSTRUCTION

Our primary argument is that the mirror test can and should play a role for MTEs to reflect on their own practice. In particular, when MTEs teach mathematics methods courses, we should ask ourselves whether we are doing what we said we were going to do when we entered the profession. For the three of us, this includes a commitment to addressing the inequities and social injustices that persist in mathematics teaching and learning. Some questions we ask of ourselves include

- Does my methods instruction address the racial and cultural mismatch of the largely White teacher population and increasingly non-White student population in the U.S.?

- Does my methods instruction address the "achievement gap" rhetoric in ways that prepare teachers to engage in complex conversations in sophisticated and informed ways?
- Does my methods instruction put me in the shoes of my PTs (see Chapter 10, this volume, for Gutiérrez's description of the "In my shoes" activity), so that I do not create and reify generalizations of them, and I come to see my PTs as they see themselves?
- Are there ways my methods instruction contributes to the White supremacist capitalist patriarchy (hooks, 1994) or the White institutional space (Martin, 2013) of mathematics education? Even if I myself am neither White nor male?
- What are the ways in which my instructional practices further the hegemony of NCTM, AMTE, CAEP, NAEYC, AMLE, edTPA, and other status-granting institutions?
- Do I respond with integrity and courage when PTs in my class reveal deficit views and assumptions about students, especially students from marginalized groups?
- What action can I take when a professional organization or institution I am a part of takes a position ignoring issues of race as connected to mathematics teaching?
- Do I disrupt the veneration of a reified mathematics, reattributing it as the product of human activity?

We have found these questions helpful in reminding us of our commitment to social justice and providing correctives when our instructional and professional practices fall short.

In a later section, we argue that the mirror test should not only be used by individual MTEs, but also collectively by all who identify as a member of the MTE community. In doing so, we foreshadow a sort of professional ethic. This claim is manifested, for example, when we consider the role of ideas such as *evidence* and *impact,* positivist principles that have currency in establishing sociopolitical orientations as significant and deserving of a prominent voice in what should happen in mathematics classrooms. But first we provide some specific examples of how we have used the mirror test to critically reflect on our own practices.

Examples of the MTE Use of the Mirror Test

As examples of how we use the mirror test as MTEs ourselves, we present two illustrations. First, Theodore demonstrates how he puts into action the mirror test question: *Does my methods instruction put myself in the shoes of my PTs, so that I do not create and reify generalizations of them?* To provide context,

Theodore has been teaching an elementary mathematics methods course for the past 6 years at two institutions, both large public universities. At both institutions, he observed that the majority of his PTs presented themselves as White women from rural or suburban, middle- to upper-class socioeconomic backgrounds. These were quick generalizations he made at the beginning of every semester. In contrast, Theodore self-identifies as a Chinese American male from a middle-class socioeconomic background. He grew up in urban and suburban environments in Texas, New York, and Taiwan. He felt disconnected from the PTs because most of them self-identified as a different gender and ethnicity than him.

Theodore confronted the mirror test question noted above through an activity, the photovoice mathematics teacher autobiography, detailed in chapter 18 of this volume. In this activity, PTs brought photographs of themselves to represent how they felt as a mathematics student. They shared these photographs with each other in pairs at the beginning of class (Chao, 2014), then wrote reflections about what they had learned about themselves and each other in terms of their mathematics identities. Through these reflections, Theodore learned the complexity of who the students see themselves to be and the intersections they live within beyond his initial generalizations. For instance, he learned that although some PTs grew up in entirely White, rural areas, they might also have family members from other ethnic groups, care for siblings with special needs, volunteer at youth homeless centers, come from service and social-justice-oriented religious backgrounds, or self-identify as biracial or LGBTQ. The photographs opened up vulnerable spaces for PTs to share with each other and help Theodore know who his students are "in their own shoes." Most importantly, this activity helped him get past his own generalizations of his PTs as White women from a homogeneous culture and background.

A second example of utilizing the mirror test can be found in Brian's attempts through his methods course to disrupt the hegemony of institutions of power, in particular a corporate structure involved in the evaluation of teacher performance, the edTPA. Among criticisms of the edTPA[1] are the high-cost and high-stakes nature of the test, that hourly workers rather than educational professionals decide whether test-takers are prepared to teach, and the argument that edTPA "colonizes the curriculum of teacher education programs and narrows the focus on teaching as pre-determined and top down delivery of lessons" (Ayers, 2015, para. 1). Brian does not aspire to provide future mathematics teachers a narrow preparation, focused on properly filling out templates for planning, instruction, and assessment. Yet he is fully aware the edTPA serves as a barrier to PTs' personal and financial investments toward becoming mathematics teachers. To be able to resolve this dilemma yet maintain his core values and "pass" his mirror test, Brian took two specific actions.

First, Brian provided his PTs familiarity with the expectations and scoring mechanisms of edTPA. Brian slightly modified a lesson planning assignment to align with structures of the edTPA. He maintained former assignment features, specifically that students develop a lesson in pairs and then observe a practicing teacher implement this lesson while attending primarily to student reasoning. His modification was to utilize a peer assessment mechanism in which fellow PTs provided feedback utilizing the 15 edTPA rubrics. Using authentic artifacts of the edTPA provided the immediacy of experience for the PTs not only to begin to express reservations, concerns, or critiques of educational institutions and policy, but to also develop strategies as future teachers to push back (sociopolitical goal 1) against normalized yet inequitable practices, beginning with reconsidering assessment strategies for their own classrooms.

Brian's second action also involved a minor modification of a current activity to examine assessment and grading. Following readings, students discussed both positive and negative effects of evaluation and considered how feedback as well as authentic and multiple assessments are necessary for a supportive assessment model—one that must also fit within the constraints of their induction school site or future job. Through this classroom dialogue, PTs identified many classroom and educational practices that operate against principles of equity, fairness, or justice, adding edTPA as the newest one on their list.

We have shared how two of us—Theodore and Brian—have used the mirror test to enact our sociopolitical commitments. Andrea's enactment is shared in Chapter 7 of this volume. Next we consider how the mirror test is useful in engendering critical vigilance and reflection at a community level.

THE MIRROR TEST GUIDING THE COMMUNITY OF MTES

We write this chapter as MTEs who teach and study mathematics methods courses. Yet we are also researchers expected to and desiring to conduct and publish research on this domain in competitive, refereed academic journals. This intention raises the question of what counts as empirical, publishable research. In particular, for researchers who share our goals of teaching mathematics methods courses consistent with a sociopolitical orientation, how can we claim that the course and our teaching made an impact? What counts as "evidence" for such a claim? How can sociopolitically oriented researchers build a knowledge base about what works in mathematics methods courses? And how do we define or describe what "works" means, from a social justice perspective?

Multiple tools might be recognized as providing evidence that a mathematics methods course and the instruction therein has been effective at implementing goals; that is, they work: comparisons of a PT's beliefs about mathematics teaching and equity before and after the course, a case study of a PT several years after methods that points to ways the course influenced the PT's teaching toward more equitable and just ways, or comparisons of student achievement measures from participating PTs that show a lessening or nonexistent "achievement gap." All of these scenarios would help provide evidence for an argument that the goals—at least as implemented in a particular case—made an impact.

However, will these methods provide insight that we value? The challenge of what passes as research, what is acceptable as evidence, and what counts as having impact once again reflects the political nature of the work; the positivist science and quantitative methodologies reflected in these normalizing questions assume a particular epistemology (St. Pierre, 2002).

We argue that the most valuable evidence and insight of "what works" comes from an application of the mirror test within the community of MTEs. When MTEs openly share what we do in mathematics methods courses and the effects our teaching practices have on us and our students, and when we engage together in critical questioning and answering as an MTE community while being attentive to our individual moral compasses, we collectively come to some understanding of what works. Engaging in conversations with our colleagues at conferences and in journals moves mathematics education toward greater insight about MTEs' effectiveness at contributing to social justice goals.

We argue that the mirror test is a mechanism for building collective knowledge and change within our community of MTEs. We offer four suggestions for MTEs seeking to use the mirror test to help guard against unproductive applications of self-examination and reflection:

1. The mirror test should be applied in ways that resonate with you personally. Ask questions and consider principles that align with your own values. Be honest when you pose and answer questions, even (or especially!) to yourself.
2. Apply the mirror test as you are engaged with a community. You should be seeking and providing accountability to and for colleagues. Although the mirror test may be applied privately as an individual exercise, it is most powerful when engaged in more public and collaborative forums.
3. Regard shortcomings you find in your use of the mirror test not as failures but as opportunities for growth. Avoid creating a deficit view of your work and identity as a MTE.
4. Do not shy away from acting yourself into new ways of thinking. It may be a struggle to think yourself into new ways of acting. Follow your convictions, and you will find that they end up leading you.

The mirror test is primarily for self-examination and self-growth, but when applied by MTEs individually and collectively, we suspect it may lead to broader impact through the emergence of implicit professional ethics that strive for equity and justice.

BARRIERS TO BECOMING THE MTE YOU WISH TO SEE IN THE MIRROR

Institutional obstacles can impede MTEs' attention to sociopolitical goals in the preparation of PTs. We share several obstacles and offer MTEs possible resolutions. First, many MTEs have little opportunity to learn to navigate the sociopolitical space within a mathematics methods classroom. Many MTEs became professors by successfully navigating a system that may have been rigged in their favor (Oslund, 2012). They may be unlikely to critique such a system. The mirror test helps many of us MTEs recognize our privilege and power as mathematics learners and thereby realize inequities in mathematics teaching that worked well for us but was inaccessible and oppressive for others. Engaging in the mirror test as a community of MTEs (and not just as individuals) is important so that we recognize that ways we might have learned mathematics might not have been equitable for all students, helping us break the cycle of continuing inequitable mathematics teaching practices.

Second, MTEs might not feel empowered or protected sufficiently to engage students in these difficult conversations of power and identity. MTEs are incentivized by their universities to worry about student evaluations and to avoid upsetting colleagues. This is especially true for MTEs who are doctoral students or clinical or pretenure faculty members. This obstacle brings up systemic and hegemonic issues within the mathematics teacher education community that a mirror test could surface. Third, MTEs may be concerned about time constraints and the amount of content they need to cover in their methods course. In situations where PTs might only have a single combined mathematics methods content course or a combined mathematics and science methods course, MTEs might feel they do not have adequate time to attend to the sociopolitical goals set forth in the Scholarly Inquiry and Practices Conference (Sanchez et al., 2015), to delve into issues of equity, diversity, justice, and power. Although Lawler et al. (Chapter 13, this volume) suggest minor modifications to methods course activities that may offer some resolution, structural challenges such as consolidating courses or devaluing a dedicated mathematics content course demonstrate initiatives in which the collective MTE voice can be an important mechanism of pushback. Further, we argue that *any* amount of connection to issues of equity, diversity, social justice, and power is better than none at all.

A fourth reason MTEs may not embrace sociopolitical goals in mathematics methods instruction is they might feel that their PTs will never be in communities where issues of justice and power affect their teaching. Or MTEs may believe that issues of equity and diversity should be addressed in another course, outside of the mathematics education space. We find these perspectives to be naïve of the very real practice of teaching in today's schools. Inequity, injustice, and power dynamics operate in all social settings. No school or classroom is excluded, including mathematics methods. Perspectives that ignore these phenomena, like many obstacles to enacting sociopolitical goals, primarily reflect fear to acknowledge and confront the inequities existing within our educational system.

A final, and maybe most debilitating obstacle, is that many MTEs admittedly do not know enough about sociopolitical issues to adequately address them or make them a part of their class. For these MTEs, we offer this chapter and welcome their engagement with continued scholarship on how MTEs develop and enact their sociopolitical identities.

MOVING FORWARD TOWARD A JUST MATHEMATICS EDUCATION

We see much reason for hope. We are buoyed by the statements emerging in policy documents and position papers from our professional organizations. In *Principles to Actions*, the National Council of Teachers of Mathematics (NCTM, 2014) made strong statements about the detrimental effects of tracking, as well as bold and clear statements regarding unproductive beliefs about teaching and learning mathematics, access and equity, and other common practices that result in negative results for marginalized groups. In the recent draft for *Standards for Mathematics Teacher Preparation*, the Association of Mathematics Teacher Educators (AMTE, 2016) acknowledged the importance of teacher development and preparation to include a focus on equity, including an expectation that a well-prepared beginning teacher understands "the roles of power, privilege, and oppression in the history of mathematics education and [is] equipped to question existing educational systems that produce inequitable learning experiences and outcomes for students" (p. 18). Similarly, the recent joint position statement released by the National Council of Supervisors of Mathematics (NCSM) and TODOS (2016) ratified social justice as a key priority in mathematics education, stating, "A social justice stance interrogates and challenges the roles power, privilege, and oppression play in the current unjust system of mathematics education—and in society as a whole" (p. 1). These three documents have included calls for sophisticated discussions in mathematics education about equity, institutional systems of oppression, and professional accountability. This last charge—for collective commitments to sociopolitical perspectives and for a shared sense of responsibility—is where we think the mirror test might prove especially useful for initiating difficult but important conversations.

Let us return to the difficult questions we posed at the beginning of this chapter. *What is the relationship between the mirror test and formal professional organizations? Who can implicate such organizations (and us as members) in critiques of power and inequity? How should we act when we view a colleague through our own mirror and conclude that they are not aligning with the same ethics? Can there be a shared ethic or moral code for our community? And if so, how we might we hold one another accountable?* These are difficult questions, for which there are no easy answers. But we can identify an incorrect answer: "These questions are too difficult and too acrimonious! Best to avoid this territory altogether!" Each of us has experienced the temptation to keep our heads down, disengage from communities and colleagues who disagree, and close the doors to our methods classes and do what we think is right. Ultimately, our mirror test tells us that our methods classroom doors must remain open and our conversations must continue.

In the end, we must remind ourselves that we *are* our field, the field of mathematics teacher education. We *are* our organizations, our departments, our accrediting bodies, and our colleges and universities. These institutions, as powerful as they are, are made up of MTEs. It is up to MTEs to join together to look into our mirrors, to tell one another what we see in these mirrors, and to learn from and with one another. We must speak up for social justice in large, sanctioned platforms (such as plenary addresses) and in informal, one-on-one conversations with colleagues in the hallway. Tell others about your mirror and ask them about theirs. And then listen.

We regard the recent attention to social justice within mathematics education and mathematics teacher education professional organizations as reason for optimism, even as some proclamations fall short (Martin, 2015). We are especially energized by conversations about the intersections of mathematics teacher education and social justice happening in the "margins" of scholarly discourse: for example, the #EduColor hashtag in Twitter and the Creating Balance in an Unjust World Conference Facebook group. These "backchannel" communications have an important role to play in democratizing and broadening conversations. Indeed, these very conversations may be an important site for the mirror test to be rehearsed and applied. We look forward to these difficult but essential reflective dialogues about our methods courses and how they do—or do not—contribute to social justice commitments. With these tools available not just to mathematics teachers, but to MTEs, particularly ones who had previously been isolated at their institutions, we hope that MTEs will continue to enact mechanisms to question constructs of identity into their crucial work with beginning mathematics teachers. Until then, we look forward to seeing you in the mirror.

NOTE

1. Consider the position paper by the National Association for Multicultural Education at http://www.nameorg.org/docs/Statement-rr-edTPA-1-21-14.pdf

REFERENCES

Aguirre, J. M., Mayfield-Ingram, K., & Martin, D. B. (2013). *The impact of identity in K–8 mathematics: Rethinking equity-based practices.* Reston, VA: National Council of Teachers of Mathematics.

Association of Mathematics Teacher Educators. (2016, October 15). *AMTE standards for mathematics teacher preparation: Draft version for review.* Retrieved from https://amte.net/sites/default/files/AMTE_MTP_Standards_forReview_2016_10_16.pdf

Ayers, R. (2015). Smacking down the opposition: EdTPA advocacy in Illinois. *Huffington Post – The Blog.* Retrieved from http://www.huffingtonpost.com/rick-ayers-/smacking-down-the-opposit_b_8490892.html

Banks, J. (1993). The canon debate, knowledge construction, and multicultural education. *Teachers College Record, 100*(4), 4–14.

Beauchamp, C., & Thomas, L. (2009). Understanding teacher identity: An overview of issues in the literature and implications for teacher education. *Cambridge Journal of Education, 39*(2), 175–189.

Chao, T. (2014). Photo-Elicitation/Photovoice interviews to study mathematics teacher identity. In J. -J. Lo, K. R. Leatham, & L. R. Van Zoest (Eds.), *Research trends in mathematics teacher education* (pp. 93–113). New York, NY: Springer International.

Collins, P. H. (1991). *Black feminist thought: Knowledge, consciousness, and the politics of empowerment.* New York, NY: Routledge.

Crenshaw, K. (1991). Mapping the margins: Intersectionality, identity politics, and violence against women of color. *Stanford Law Review, 43,* 1241–1299.

de Freitas, E. (2008). Troubling teacher identity: Preparing mathematics teachers to teach for diversity. *Teaching Education, 19,* 43–55. https://doi.org/10.1080/10476210701860024

Delgado, R., & Stefancic, J. (2012). *Critical race theory: An introduction.* New York, NY: New York University Press.

Foucault, M. (1980). *Power/knowledge: Selected interviews and other writings 1972–1977* (C. Cordon, L. Marshall, J. Mepham, & K. Soper, Trans.) New York, NY: Pantheon.

Gutiérrez, R. (2013). The sociopolitical turn in mathematics education. *Journal for Research in Mathematics Education, 44*(1), 37–68. [appeared online first in 2010]

Gutiérrez, R. (2016). Nesting in nepantla: The importance of maintaining tensions in our work. In N. M. Joseph, C. Haynes, & F. Cobb (Eds.), *Interrogating whiteness and relinquishing power* (pp. 253–282). New York, NY: Peter Lang.

Gutstein, E. (2007). Connecting community, critical, and classical knowledge in teaching mathematics for social justice. *The Montana Mathematics Enthusiast, Monograph 1*, 109–118.

Herbel-Eisenmann, B., Bartell, T. G., Breyfogle, M. L., Bieda, K., Crespo, S., Dominguez, H., & Drake, C. (2013). Strong is the silence: Challenging interlocking systems of privilege and oppression in mathematics teacher education. *Journal of Urban Mathematics Education, 6*(1), 6–18.

hooks, b. (1994). *Teaching to transgress: Education as the practice of freedom.* New York, NY: Routledge.

Kilpatrick, J. (2012). Forward. In D. Stinson & A. Wager (Eds.), *Teaching mathematics for social justice: Conversations with educators* (pp. ix–x). Reston, VA: National Council of Teachers of Mathematics.

Langer-Osuna, J. M. (2015). From getting "fired" to becoming a collaborator: A case of the co-construction of identity and engagement in a project-based mathematics classroom. *Journal of the Learning Sciences, 24*(1), 53–92.

Lawler, B. R. (2012). The fabrication of knowledge in mathematics education: A postmodern ethic toward social justice. In A. Cotton (Ed.), *Towards an education for social justice: Ethics applied to education* (pp. 163–189). Oxford, England: Peter Lang.

Lerman, S. (2000). The social turn in mathematics education research. In J. Boaler (Ed.), *Multiple perspectives on mathematics teaching and learning* (pp. 19–44). Westport, CT: Ablex.

Ma, J. Y., & Singer-Gabella, M. (2011). Learning to teach in the figured world of reform mathematics: Negotiating new models of identity. *Journal of Teacher Education, 62*(1), 8–22. https://doi.org/10.1177/0022487110378851

Martin, D. B. (2013). Race, racial projects, and mathematics education. *Journal for Research in Mathematics Education, 44*(1), 316–333.

Martin, D. B. (2015). The collective Black and *Principles to Actions. Journal of Urban Mathematics Education, 8*(1), 17–23.

National Council of Teachers of Mathematics. (2014). *Principles to actions.* Reston, VA: Author.

National Council of Supervisors of Mathematics, & TODOS: Mathematics for ALL. (2016). *Mathematics education through the lens of social justice: Acknowledgement, actions, and accountability.* Retrieved from http://www.mathedleadership.org/resources/position.html

National Research Council. (2001). *Adding it up: Helping children learn mathematics.* Washington, DC: National Academy Press.

Oslund, J. (2012). Teaching mathematics for equity and social justice in a mathematics methods course for future elementary teachers. In L. Jacobsen, J. Mistele, & B. Sriraman (Eds.), *Mathematics teacher education in the public interest* (pp. 213–230). Charlotte, NC: Information Age.

Sanchez, W., Kastberg, S., Tyminski, A., & Lischka, A. (2015). *Scholarly inquiry and practices (SIP) conference for mathematics education methods*. Atlanta, GA: National Science Foundation.

St. Pierre, E. A. (2002). "Science" rejects postmodernism. *Educational Researcher, 31*(8), 25–27.

Walshaw, M. (2004). Pre-service mathematics teaching in the context of schools: An exploration into the constitution of identity. *Journal of Mathematics Teacher Education, 7,* 63–86. doi:10.1023/B:JMTE.0000009972.30248.9c

SECTION VII

COMMENTARY

CHAPTER 22

A COMMENTARY WITH URGENCY

Looking Across Theoretical Perspectives to Put Relationship Building with Underserved Students at the Forefront of Our Work

Richard Kitchen
University of Wyoming

> *I learned from the assessment that she [a kindergarten student] can do a lot more than what we are learning in class.*
>
> —Naomi

> *This assessment helped me to realize that first graders can work with story problems and should be exposed to them early and often.*
>
> —Susanne

These quotes are from prospective teachers (PTs) enrolled in one of my elementary mathematics methods courses. The PTs had just completed an assessment with a student using problems from the cognitively guided instruction (CGI) framework (Carpenter, Fennema, Franke, Levi, & Empson,

Building Support for Scholarly Practices in Mathematics Methods, pages 343–358
Copyright © 2018 by Information Age Publishing
343

1999). As part of the assessment, the PTs meet with a student, preferably in an early elementary grade, and conduct a one-on-one interview with the student, asking the child to solve four to six CGI problems. Through this assignment, I want PTs to learn about children's thinking and the strategies they use to solve problems (e.g., direct modeling, counting strategies, how children use derived facts when computing). One of my primary goals for this assignment is that PTs begin to come to the realization, much like Naomi did, that students come to school with mathematical knowledge and that they are capable of doing mathematics.

Year after year, this assignment leads to "aha moments" for PTs, many of whom start the assignment believing that they need to be careful to only assign problem types such as Join Result Unknown to their students because this problem has already been "covered in class." One PT wrote, "In order for students to have a strong number sense, sometimes they just need to be able to work things out without being told how to." Realizations such as this one are not trivial and require that PTs interact regularly with students to begin to appreciate their students' mathematical thinking and to move away from deficit perspectives that they may hold about what students are capable of accomplishing in mathematics. PTs' responses to the CGI assignment continually remind me of how mathematics is different from other subjects. Many PTs have struggled in their mathematics classes, particularly prospective elementary teachers, developing the attitude that mathematics is difficult and only those with the math gene can be successful in it. I have learned the value of assignments such as this one that give PTs insights into what students can actually do in mathematics. One PT who interviewed a fourth grader who "struggles in math" learned, much to her surprise, that her student could solve a variety of problem types using concrete, direct modeling strategies. Another who interviewed a kindergartener found that her student was able to solve problems involving single-digit numbers. This was an unexpected result for the PT because this student had been previously assessed as "not knowing any numbers" because she did not respond when asked to count as high as she could.

The CGI assignment, grounded in the cognitive perspective, is intended to give PTs direct contact with a student to learn about that student's mathematical thinking and to learn about that student. PTs have told me that after completing an interview with one individual, they want to interview every student in their classroom to learn more about the students and their mathematical thinking. They have told me that the assignment not only gives them insights into a student's thinking but it also has given them a way to connect with a student in a fun and academically rich manner. PTs feel that after completing the assignment, they often have a new and positive relationship with the student they interviewed. They have a new appreciation for the student whom they interviewed. PTs' experiences in this activity speak to the thesis of this commentary, the value of PTs developing meaningful relationships with their students that are grounded in the intellectual work they do as teachers.

After reading the chapters contained in this volume, it is clear that we have much to learn from one another about what is important to consider as we develop our mathematics methods courses for PTs. As noted by the book editors in Chapter 1, there is great variation in how authors design and enact their courses because mathematics teacher educators (MTEs) have differing theoretical perspectives, labor in divergent contexts, and have heterogeneous course learning goals. The multiplicity of perspectives and activities elucidated in this book provides a rich array of ideas. These ideas can help us think through the challenges associated with designing mathematics methods courses to prepare PTs both to teach for understanding and to work to meet the complex needs of the diverse students they will be teaching (Kitchen, 2005).

A particular strength of the book is how the authors of the respective chapters frame their discussions within a well-articulated theoretical perspective; some do so through multiple perspectives. The Scholarly Inquiry and Practices (SIP) conference (Sanchez, Kastberg, Tyminski, & Lischka, 2015) drew from three theoretical perspectives that are primarily used by MTEs to guide the development of mathematics methods courses: the cognitive, situative, and sociopolitical perspectives. There is strong agreement among the book's authors about the value of drawing from all three perspectives in the design of methods courses. Even so, Smith, Taylor, and Shin found in their research study that "the most popular perspective identified by MTEs to design and implement their secondary mathematics methods courses was the situative perspective" (Chapter 20, p. 321). Perhaps a primary reason for the dominance of the situative perspective is that many MTEs have a great deal of background knowledge and experience with this frame. A cursory examination of the *Second Handbook of Research on Mathematics Teaching and Learning* (Lester, 2007) reveals how the situative as well as the cognitive perspectives are dominant in the mathematics education research community. There is no doubt at this point in history about the relevancy of both the cognitive and situative perspectives in mathematics education and the need to incorporate them into our methods courses. I believe, however, the time has come for sociopolitical frames to become mainstream in mathematics education. This is a necessity, given the historic inequities that have plagued mathematics classrooms.

In the next section, I provide a review of the research literature that shows how ethnic and racial minorities, the poor, and the disenfranchised (referred to as "underserved students" for the remainder of the chapter) have historically been denied access to a quality mathematics education. I then briefly argue for the need for the sociopolitical perspective to take

center stage in our work as MTEs. I continue by discussing and then providing examples of the value of PTs developing meaningful relationships
with their students specifically as a means to address historic inequities in
mathematics education. At first glance, relationship building appears to be
an example of an activity located in the sociopolitical viewpoint. However,
after reviewing the chapters contained in this volume, I now view it as an
activity that cuts across perspectives allowing greater opportunities for students to have equitable access to a challenging mathematics education.

HISTORIC INEQUITIES THAT PERSIST
IN MATHEMATICS EDUCATION

In mathematics education reform documents (National Council of Teachers
of Mathematics, 1989, 1991, 1995, 2000; National Governors Association, 2010;
National Science Foundation, 1996), much emphasis has been placed on students engaging in reasoning to solve problems to make sense of mathematics by
making connections, communicating, and representing ideas (Gamoran et al.,
2003; Hiebert & Carpenter, 1992). Unfortunately, students may not have opportunities to study mathematics with teachers who have the content and pedagogical expertise to support the development of their mathematical reasoning and
justification (e.g., Stylianou, Blanton, & Knuth, 2009). This is particularly the
case for underserved students (Kitchen, DePree, Celedón-Pattichis, & Brinkerhoff, 2007; Leonard, & Martin, 2013; Téllez, Moschkovich, & Civil, 2011). Moreover, the sort of instruction recommended in mathematics education reform
documents is generally not a priority at schools attended largely by underserved
students (Davis & Martin, 2008; Kitchen, 2003; Martin, 2013).

Race, ethnicity, English language proficiency, and socioeconomic status
all influence the access students have to a challenging education in mathematics (Diversity in Mathematics Education [DiME-, 2007; Gutiérrez, 2008;
Martin, 2013). Educators of underserved students who teach in low-track
classes tend to make the memorization of math facts, algorithms, vocabulary, and procedures the focal point of their instruction rather than teaching students through the use of complex, challenging problems (Davis &
Martin, 2008; Kitchen et al., 2007; Lattimore, 2005). Haberman (1991)
named this type of instruction the *pedagogy of poverty*. Essentially, in schools
that are largely populated by poor students of color, pedagogy is highly
authoritarian and directive, tasks are skills-based, expectations are low, and
students are expected to comply and passively follow instructions. Racism
and classism have become so accepted as part of the very fabric of U.S. society that it essentially has become normative to expect less of certain student
groups in school. Students themselves may not resist the pedagogy of poverty because they become accustomed to it and can "succeed" at it without

having to do much (Haberman, 1991). At this time of heightened awareness of the racial injustices being perpetrated on the streets, in prisons and
jailhouses, and in courtrooms throughout the nation, there should also be
outrage about the continuing crisis nationally to educate ethnic and racial
minority children and those who grow up in poverty.

A question that I believe the mathematics education community needs
to address at this time is: What ethical role do we have as MTEs to put an
end to the historic legacy of students of color and low-income students in
the U.S. being denied access to a quality education in mathematics? For
too long, public education has often failed underserved students in general
(Darling-Hammond, 1996; Massey, 2009; Milner, 2013) as well as in mathematics education (see, e.g., DiME, 2007; Flores, 2008; Jacobsen, Mistele, &
Sriraman, 2013; Kitchen et al., 2007; Leonard & Martin, 2013; Martin, 2000;
Secada, 1992; Téllez et al., 2011). Ladson-Billings (2006) has argued that
the U.S. owes an "education debt" to the African American community and
other communities that have historically been marginalized and oppressed.
This debt includes historic, economic, sociopolitical, and moral elements.

The time has come to pay this debt by doing whatever it takes to guaran-
tee that quality schools exist in every community in the U.S., not just in the
nation's most privileged communities. Now is the time for the mathematics
education community to step up and to deal explicitly and collectively with
the injustices taking place in schools. There needs to be urgency among
MTEs to do our part to eliminate the educational disparities that continue
in U.S. classrooms. We have an essential role to play in our work to prepare
PTs to teach in equitable and just ways (Rodriguez & Kitchen, 2005). As
MTEs, we have an ethical obligation to implement the ideas in this book
first and foremost for the good of students who have historically been denied access to a quality mathematics education in the U.S.

WHY THE SOCIOPOLITICAL PERSPECTIVE MUST BECOME
MAINSTREAM IN MATHEMATICS EDUCATION

The sociopolitical lens affords a concentrated examination of structural inequities and injustices in education in general, as well as specific classroom-
level arrangements that may limit students' opportunities to learn at high
levels (Kitchen, 2003). In Chapter 21, McCloskey, Lawler, and Chao note
a legitimate issue regarding the widespread implementation of the sociopolitical perspective by the mathematics education community: "A final,
and maybe most debilitating obstacle, is that many MTEs admittedly do
not know enough about sociopolitical issues to adequately address them
or make them a part of their class" (Chapter 21, p. 336). I agree but believe that it is simply not acceptable for MTEs to be either uninformed or

unwilling to address the historic legacies of racism, classism, sexism, and other "isms" in public education, and in mathematics education in particular. Preparing MTEs to address these historic legacies is analogous to the notion of preparing PTs to teach for understanding—similar to how PTs need to experience mathematics as learners in which mathematical sense making is paramount. MTEs need to have experiences, at a minimum as doctoral students, to become familiar with educational injustices as well as with privilege and power and how they operate in schooling (Taylor & Kitchen, 2008).

As part of our work in the mathematics education community, we can use the sociopolitical perspective to derive practical steps and strategies to address historic injustices in our field with PTs in our methods classes. I am not convinced that MTEs must have an advanced understanding of the sociopolitical perspective. It may simply be sufficient that MTEs understand the historic role that mathematics education has played as gatekeeper, allowing only a few to have success at the expense of many (D'Ambrosio, 1983). Addressing this historical fact requires a willingness on our collective part to examine how to prepare and support PTs to take equity seriously so that students have opportunities to genuinely engage in the study of mathematics. Of course, we also need to be concerned about some significant pitfalls, such as essentializing the behaviors of particular groups and not taking structural racism seriously in our work.

An important consideration for MTEs is that differences based on race, as well as class, religion, gender, and sexual orientation, most likely translate into a PT not loving every one of her students equally (Martin, 2013). Recognizing that many of our PTs may have deficit perspectives for particular students on the basis of these differences compels us in our work as MTEs to examine issues related to racism, sexism, classism, power, and privilege in our work with PTs. As challenging as it can be for students who come from the dominant culture, they need to grapple with the fact that their "Whiteness" has granted them countless privileges. "Whiteness is the act of characterizing what is natural, real, or customary around the white experience" (Leonardo, 2002, cited in Joseph, Haynes, & Cobb, 2016, p. 3). Unconsciously viewing themselves as normal translates for many PTs into viewing the culture of "the other" from a deficit perspective, or as less than normal. Because many PTs in my methods classes come from the dominant culture but will teach in classrooms primarily populated by students of color, it is vital that they interrogate their Whiteness and privilege to develop more inclusive and equitable perspectives for their students who may not look like or act like them.

WORKING TO DEVELOP MEANINGFUL RELATIONSHIPS
AMONG PROSPECTIVE TEACHERS AND THEIR STUDENTS

A potentially impactful activity that some book authors discuss using a socio-political lens involves MTEs working to help their PTs develop meaningful professional and personal relationships with their students. To help frame this discussion, I turn to a quote by Stephanie, who is the third author of Chapter 8 by Chao, Hale, and Cross:

> A lot of my PTs' frustration comes from being up in front of the class and trying to teach for the first time and not feeling prepared. There's something about jumping from classes into full-time teaching while you're student teaching that's just not working. We don't provide them space to sit down with a child one-on-one as the classroom teacher in order to understand their students as people first, to listen to what they're thinking about mathematics. (p. 121)

Stephanie's insight concerning PTs having the opportunity to get to know their students first as people seems so fundamental, and simple as well, yet it may not be primary in our work. For schools to be places where every child is valued and cared for, spaces where teachers are genuinely vested in their students' well-being and education, it makes sense for schools to be places where teachers get to know their students and make the development of meaningful relationships with their students a priority (Kitchen et al., 2007).

Kitchen and colleagues (2007) found in a study conducted at public secondary schools that serve underserved students and have distinguished themselves nationally in mathematics that (a) high academic expectations at a school need to be backed with sustained support for academic excellence, (b) challenging academic content and high-level instruction must be present, and (c) the development of meaningful relationships among teachers with students is valued. Academic excellence is also among the three defining features of Ladson-Billings's (2009) characterization of culturally relevant pedagogy or teaching; the other two are cultural relevancy and challenging the status quo. In culturally relevant classrooms (Ladson-Billings, 1995, 2009), teachers embrace the cultural assets that students bring to the classroom while working to use culturally appropriate and meaningful instructional approaches that engage and inspire learners. As culturally relevant teachers, working to validate students for who they are and engaging them in the study of challenging academic content entails considering and respecting the experiences and knowledge that they bring to the educational enterprise. Developing meaningful relationships with students grounded in the intellectual work they do is primarily how teachers come to respect students' experiences and the knowledge that they bring to school (Kitchen et al., 2007). Students also need to develop a positive and meaningful relationship with at least one caring adult as a means to grow emotionally, socially, and academically (Dubois & Silverthorn, 2005).

It is difficult to imagine a school known for academic excellence that does not prioritize the development of positive, meaningful, and impactful relationships (Kitchen et al., 2007). Certainly, the development of meaningful relationships with youth is highly valued in many cultures; Mexican culture is one such example (e.g., Suárez-Orozco, 1989; Valenzuela, 1999). Much significance is ascribed to the family in Mexican culture, as is respect for the inherent dignity of the individual as a means to educate youth to become *bien educada/o* (well-educated). To be *bien educada/o*, one knows "how to live in the world as a caring, responsible, well-mannered, and respectful human being" (Valenzuela, 1999, p. 23). Teachers who have developed positive and trusting relationships with their students are able to press students to work harder with little or no student resistance (Kitchen et al., 2007). As students develop strong and healthy relationships with their teachers, they are more willing to trust their teachers and are more willing to work harder for them while being less likely to resist them as the academic rigor of their classes increases (Kitchen et al., 2007). It should be noted that relationship building may not necessarily address a central principle of the sociopolitical perspective, that it should provoke PTs "to reflect on the social, cultural, and political dimensions of mathematics teaching and learning" (Harper, Herbel-Eisenmann & McCloskey, Chapter 7, p. 100). However, I do believe that relationship building is a powerful way to enhance equity and access for students, and specifically for underserved students.

For the remainder of this commentary, I will give examples of how the development of meaningful relationships among PTs and their students is an activity that I would like to think Harper, Herbel-Eisenmann, and McCloskey (Chapter 7) would characterize as both feasible (for MTEs) and accessible (for PTs). In the examples that follow, I position relationship building using the frame of culturally relevant pedagogy (Ladson-Billings, 1995, 2009) and the intellectual work that teachers do to argue that relationship building has a central role to play in activities orchestrated across theoretical perspectives for the benefit of students. In Chapter 4, Kazemi identifies relationship building in two of the five principles that guide her in the design of the mathematics methods courses that she teaches. Although Kazemi's work is grounded in the situative perspective, her chapter helps to illustrate how cultivating relationships is an essential element of our work and is valuable across theoretical perspectives. To be sure, understanding the value that relationship building has for promoting equity and access in education in general and in mathematics education in particular needs to be studied through multiple theoretical lenses. From my work in schools, though, I have found that there is wide agreement among practitioners for the need to build positive relationships with diverse youth who populate their schools. In what follows, I will attempt to highlight relationship building and address its potential for advancing equity and access for underserved students by piggybacking on a few example activities elucidated in the book. Implicit in these attempts is my belief that relationship building should be front and center in all aspects of our work as MTEs.

Before proceeding, I would like to apologize for not being able to give more consideration to the many ideas presented in this book. It is beyond the scope of this commentary to consider all of the exceptional ideas and activities discussed here. After reviewing all of the chapters in this book, I have been reevaluating much of what I do when I teach mathematics methods courses. For instance, across the book's chapters, I noticed that many of us are thinking about relationship building across the theoretical perspectives that guide our work. This inspired me to consider the vital role that relationship building can play as a powerful tool for supporting equity and access for underserved students in mathematics.

CONNECTING TO A FEW EXAMPLES IN THE BOOK

In this section, I shift my attention to consider the central role that relationship development can play to help promote equity in mathematics classrooms. To do this, I revisit a few sample activities elucidated in the book and how the advancement of meaningful relationships with and among students is already an integral part of these activities and certainly could be an explicit focus of them as well. Throughout, I will emphasize the opportunities that we should leverage with our PTs during these activities to develop meaningful relationships with underserved students, the students that many PTs are generally the least prepared to teach (Rodriguez & Kitchen, 2005).

Chapter 15, written by Virmani, Taylor, and Rumsey with colleagues, inspires an initial example. They describe why mathematics methods courses should be embedded in actual schools. They write, "Central to these experiences are the opportunities to develop meaningful relationships with students and practicing teachers and learn from and with young people in the classroom" (Chapter 15, p. 234). Working with students as part of their methods courses provides a way for PTs to intimately connect with students to begin to develop caring relationships with them (Noddings, 2005). These relationships can only help to inspire PTs to identify with children who may be very different from them. I personally have taught methods courses in schools and found that the experience of PTs getting to directly implement "minilessons" that we have been exploring in methods class to be quite helpful for them. Students' responses or lack of response to their minilessons affords immediate feedback to PTs, which is a much more authentic form of evaluation than any response I give them. I have also witnessed how these experiences humanize teaching for PTs because they start examining their instruction in terms of how students responded to their instruction. Students are no longer abstractions for PTs after they have interacted with them and received honest feedback from them; they are real people with real needs and ideas.

For my second example, I return to Chapter 8 written by Chao, Hale, and Behm Cross. In this chapter, the authors promote clinical interviews within the sociopolitical perspective as a means for PTs to learn about their students before facing them in an impersonal classroom setting. Clinical interviews also serve the vital role of introducing PTs to the diverse students with whom they will be working closely on a daily basis for much of an academic year. Theodore, the lead author of the chapter, writes,

> What I want our PTs to realize is that it's really hard to know what's inside a child's head until you develop a relationship with the child, until you learn about who they are outside of the classroom environment, until you learn about their interests, until you talk to them as a fellow human being, as a person. (Chapter 8, p. 128)

The clinical interview allows for intimate interactions with students that promote PTs learning about their students' lives, connecting with their students both professionally and personally, and hopefully even coming to care about students as individuals with needs and personal stories. The authors situate clinical interviews within the sociopolitical frame mainly because of the potential of such interviews to help PTs listen to their students and to provoke reflection among PTs about deficit perspectives they may hold of students. This example points to the possibility of PTs promoting equity and access for students as they begin to identify and empathize with them through emerging relationships.

I now turn to the use of lesson rehearsals in methods courses as described by Arbaugh, Adams, Teuscher, Van Zoest, and Wieman in Chapter 9. In their chapter, the authors demonstrate how lesson rehearsals grounded in the situative perspective on learning are easily adaptable for a variety of purposes. For instance, the authors use this activity to extend PTs' mathematical content knowledge, to prepare PTs to attend to a mathematical practice found in the Common Core State Standards, and to help PTs focus more on the fundamental mathematical concept embedded in a lesson. Arbaugh and colleagues explain that their primary purpose for designing rehearsals is for secondary PTs to learn how to enact inquiry-based instruction (e.g., using a launch, explore, and summarize lesson sequence). From a sociopolitical perspective, it seems quite apropos that an explicit focus of lesson rehearsals could be to try out instructional strategies with underserved students with the goal of engaging every student in learning at high levels. For example, drawing on Harper, Sanchez, and Herbel-Eisenmann's ideas presented in Chapter 17, a potential focal point of these rehearsals could be on the role of language to support the mathematical learning of students learning English as an additional language (EALs). Moreover, these rehearsals could take into consideration other aspects of instruction that specifically support EALs, and every student in general, such as the development of collaborative learning spaces in which students feel comfortable sharing their ideas and taking mathematical risks (Staples, 2007). In such classrooms, students are positioned as competent (Turner, Celedón-Pattichis, & Marshall, 2008), and their mathematical ideas are paramount and form the core for instruction as the teacher and class challenge one another, while building on one another's mathematical ideas. As Arbaugh and coauthors note, MTEs need to be intentional to prepare secondary PTs to have students work in collaborative ways to support their peers' learning.

A final example of an activity in the book that has great potential to help foster meaningful relationships among PTs and students comes from Chapter 14. In this chapter, Ward uses the sociopolitical perspective to engage PTs in a community exploration to learn about potential connections that they can make between home and school. The PTs that Ward focuses on are early childhood majors. This activity engages early childhood majors as researchers of the funds of knowledge (Moll & Ruiz, 2002) found in the communities of the students that they serve. This is a powerful activity that is designed to help PTs move beyond deficit perspectives that they may hold about particular communities, to learn about the assets found in these communities, and to begin to identify with their students and the assets that they bring to school as community members. Once PTs begin to personally identify with students, they begin to humanize them, moving PTs to want to invest their time and energy into helping their students learn. Using mathematical parlance, there is a one-to-one correspondence between identifying in personal ways with students and coming to believe that students can learn and want to learn. Ward's activity not only allows PTs to see the assets that their students bring to the classroom, it promotes the development of the sort of relationships that inspire teachers to give a little more of themselves to their students.

To summarize, in the sample activities just reviewed, I have attempted to draw attention to how prioritizing the development of meaningful professional and personal relationships among PTs and their students can lead to PTs beginning to identify with their students. Because relationship building can foster identification with and empathy for students, it has tremendous potential to advance equity and access in mathematics, particularly for underserved students. I have also highlighted how we should consider the needs of underserved students first and foremost in every methods class activity. Giving precedence to the needs of underserved students can contribute to interrupting the legacy of underserved students being denied access to a challenging education in mathematics in the U.S. (Secada, 1992). Finally, I have tried to demonstrate that relationship building is an example of an activity that is located within the situative and sociopolitical frames, yet its use cuts across all three perspectives. In my opinion, this points to the versatility of relationship building and its potential utility across perspectives.

FINAL REMARKS

In this commentary, I argued for the sociopolitical perspective going mainstream in the mathematics education community in general, and specifically in our work as MTEs. Some may contend that it is already an overwhelming task to prepare PTs to teach mathematics for understanding and is just too much to ask to also prepare PTs to identify with their students. I agree that teaching for understanding is a challenging endeavor, particularly with secondary PTs who have generally experienced mathematics in traditional lecture-based courses. Once again, though, I believe that at this point in history we have an ethical duty to address historical inequities in mathematics education at more than a surface level in our work with PTs. Frankly, we have much work to do in this area. Sociopolitical framings that seek to address inequities and injustices in mathematics classrooms are becoming well established but remain unknown by many and may even intimidate some. And although sociopolitical perspectives are broad and may be difficult to articulate, I argue that the most important contribution this perspective offers is its explicit focus on issues related to equity. Specifically, every student, not just the most privileged students, should have the opportunity to access challenging mathematical content.

We would be remiss in our work with PTs if we did not prepare them for the many very real challenges they may face when they begin to identify with, care for, and even advocate for their students. As Gutiérrez argues in Chapter 2, teaching is a political act and teachers are not being prepared to deal with unjust educational practices that are being imposed upon them. An example of one such practice in this era of accountability and teacher shaming in P–12 education (Ravitch, 2010) is how the work of teachers has been reduced to focusing on the instruction of low-level skills to prepare students for "the test" (see, e.g., Burch, 2009; Valenzuela, 2005). The excessive focus on preparing students for success on the test is contributing to the educational legacy of denying poor students and students of color access to a rigorous mathematics education (Kitchen, Anderson Ridder, & Bolz, 2016).

In addition to preparing teachers to have viable strategies to contend with unreasonable and potentially discriminatory educational policies and practices, I believe PTs also need our support after they develop meaningful relationships with students and then begin to advocate for them. For instance, several students in the teacher education program at my university took strong stances for underserved students who were regularly being treated unfairly during mathematics classes at the school where they were student teaching. In each situation, the PTs played an important role in exposing unjust educational practices that led to positive outcomes for the students involved. I have little doubt that the PTs' emerging relationships with their students and the corresponding positive identification with them inspired the PTs to take positive professional steps to the benefit of their students.

In our work to transform the mathematics education of students who have historically been denied access to a challenging mathematics education, advancing equitable instruction must be a primary objective in our methods courses. As I noted earlier, there needs to be much more urgency in our work to eliminate the educational disparities that continue in mathematics classrooms in the U.S. It is simply not acceptable that so many students, particularly underserved students, are experiencing the subject that we love in such negative and damaging ways. A potential starting point for our work with PTs is to listen to them and their concerns and to understand their fear of not being "successful" as they initiate their teaching careers. I believe that a potentially impactful strategy to consider while actuating activities in our classes, aside from whether these activities are situated in the cognitive (e.g., the CGI assignment that I give PTs), situative, or sociopolitical perspective, is to explicitly ensure PTs have consistent opportunities to develop meaningful relationships with students. We need to do more than engage PTs in the study of mathematics in which we model inquiry-based instruction. We also need to help PTs see the potential that every student has to learn mathematics. One way to do this is to provide PTs opportunities to interact in meaningful ways with students. The goal is that through promoting understanding and empathy in PTs for the students they serve, PTs will work on behalf of every one of their students, not just their most privileged students.

To close, I would like to thank the book's editors, Signe Kastberg, Andrew Tyminski, Alyson Lischka, and Wendy Sanchez for inviting me to write this commentary for a book that will certainly be a very important and valuable contribution to the mathematics education community. I learned a great deal from reviewing the book's chapters and, as previously mentioned, am inspired to implement many of the activities elucidated in this book in future methods courses. This book has already had a large positive effect on how I look at my work as a MTE and will undoubtedly inspire many others as well as we continue to share and learn from one another about how to prepare PTs to teach mathematics in ways that promote equity and access in diverse classroom settings.

REFERENCES

Burch, P. (2009). *Hidden markets: The new education privatization.* New York, NY: Routledge.

Carpenter, T., Fennema, E., Franke, M. L., Levi, L., & Empson, S. B. (1999). *Children's mathematics: Cognitively guided instruction.* Portsmouth, NH: Heinemann.

D'Ambrosio, U. (1983). Successes and failures of mathematics curricula in the past two decades: A developing society viewpoint in a holistic framework. *Proceedings of the Fourth International Congress of Mathematics Education* (pp. 362–364). Boston, MA: Birkhaüser.

Darling-Hammond, L. (1996). The right to learn and the advancement of teaching: Research, policy, and practice for democratic education. *Educational Researcher, 25*(6), 5–17.

Davis, J., & Martin, D. B. (2008). Racism, assessment, and instructional practices: Implications for mathematics teachers of African American students. *Journal of Urban Mathematics Education, 1*(1), 10–34.

Diversity in Mathematics Education (DiME) Center for Learning and Teaching. (2007). Culture, race, power and mathematics education. In F. K. Lester (Ed.), *Second handbook of research on mathematics teaching and learning* (pp. 405–433). Charlotte, NC: Information Age.

Dubois, D. L., & Silverthorn, N. (2005). Natural mentoring relationships and adolescent health: Evidence from a national study. *American Journal of Public Health, 95*(3), 518–524.

Flores, A. (2008). The opportunity gap. In R. S. Kitchen & E. Silver (Eds.), *Promoting high participation and success in mathematics by Hispanic students: Examining opportunities and probing promising practices* [A Research Monograph of TODOS: Mathematics for All], *1,* (pp. 1-18). Washington, DC: National Education Association.

Gamoran, A., Anderson, C. W., Quiroz, P. A., Secada, W. G., Williams, T., & Ashman, S. (2003). *Transforming teaching in math and science.* New York, NY: Teachers College Press.

Gutiérrez, R. (2008). A "gap gazing" fetish in mathematics education? Problematizing research on the achievement gap. *Journal for Research in Mathematics Education, 39,* 357–364.

Haberman, M. (1991). Pedagogy of poverty versus good teaching. *Phi Delta Kappan, 73,* 290–294.

Hiebert, J., & Carpenter, T. P. (1992). Learning and teaching with understanding. In D. Grouws (Ed.), *The handbook of research on mathematics teaching and learning* (pp. 65–97). New York, NY: Macmillan.

Jacobsen, L. J., Mistele, J., & Sriraman, B. (Eds.). (2013). *Mathematics teacher education in the public interest: Equity and social justice.* Charlotte, NC: Information Age.

Joseph, N. M., Haynes, C. M., & Cobb, F. (Eds.). (2016). *Interrogating Whiteness and relinquishing power: White faculty's commitment to racial consciousness in STEM classrooms.* New York, NY: Peter Lang.

Kitchen, R. S. (2003). Getting real about mathematics education reform in high poverty communities. *For the Learning of Mathematics, 23*(3), 16–22.

Kitchen, R. S. (2005). Making equity and multiculturalism explicit to transform the culture of mathematics education. In A. J. Rodriguez & R. S. Kitchen (Eds.), *Preparing mathematics and science teachers for diverse classrooms: Promising strategies for transformative pedagogy* (pp. 33–60). Mahwah, NJ: Erlbaum.

Kitchen, R., Anderson Ridder, S., & Bolz, J. (2016). The legacy continues: "The test" and denying access to a challenging mathematics education for historically marginalized students. *Journal of Mathematics Education at Teachers College, 7*(1), 17–26. Retrieved from http://www.tc.edu/jmetc

Kitchen, R. S., DePree, J., Celedón-Pattichis, S., & Brinkerhoff, J. (2007). *Mathematics education at highly effective schools that serve the poor: Strategies for change.* Mahwah, NJ: Erlbaum.

Ladson-Billings, G. (1995). Toward a theory of culturally relevant pedagogy. *American Education Research Journal, 32*(3), 465–491.

Ladson-Billings, G. (2006). From the achievement gap to the education debt: Understanding achievement in U.S. Schools. *Educational Researcher, 35*(7), 3–12.

Ladson-Billings, G. (2009). *The dreamkeepers: Successful teachers of African-American children* (2nd ed.). San Francisco, CA: Jossey-Bass.

Lattimore, R. (2005). African American students' perceptions of their preparation for a high-stakes mathematics test. *The Negro Educational Review, 56*(2 & 3), 135–146.

Leonard, J., & Martin, D. B. (Eds.). (2013). *Beyond the numbers and toward new discourse: The brilliance of black children in mathematics.* Charlotte, NC: Information Age.

Leonardo, Z. (2002). The souls of White folk: Critical pedagogy, Whiteness studies, and globalization discourse. *Race, Ethnicity & Education, 5*(1), 29–50.

Lester, F. K. (Ed.). (2007). *Second handbook of research on mathematics teaching and learning.* Charlotte, NC: Information Age.

Martin, D. (2000). *Mathematics success and failure among African-American youth: The roles of sociohistorical context, community forces, school influence, and individual agency.* Mahwah, NJ: Erlbaum.

Martin, D. B. (2013). Race, racial projects, and mathematics education. *Journal for Research in Mathematics Education, 44,* 316–333.

Massey, D. S. (2009). The age of extremes: Concentrated affluence and poverty in the twenty-first century. In H. P. Hynes & R. Lopez (Eds.), *Urban health: Readings in the social, built, and physical environments of U.S. cities* (pp. 5–36). Sudbury, MA: Jones and Bartlett.

Milner, H. R. (2013). Analyzing poverty, learning, and teaching through a critical race theory lens. *Review of Research in Education, 37*(1), 1–53. doi:10.3102/0091732X12459720

Moll, L. C., & Ruiz, R. (2002). The schooling of Latino children. In M. M. Suárez-Orozco & M. M. Páez, (Eds.), *Latinos: Remaking America* (pp. 362–374). Berkley, CA: University of California Press.

National Council of Teachers of Mathematics. (1989). *Curriculum and evaluation standards for school mathematics.* Reston, VA: Author.

National Council of Teachers of Mathematics (NCTM). (1991). *Professional standards for teaching mathematics.* Reston, VA: Author.

National Council of Teachers of Mathematics. (1995). *Assessment standards for teaching mathematics.* Reston, VA: Author.

National Council of Teachers of Mathematics. (2000). *Principles and standards for school mathematics.* Reston, VA: Author.

National Governors Association Center for Best Practices, Council of Chief State School Officers. (2010). *Common core state standards for mathematics.* Washington, DC: Author.

National Science Foundation (NSF). (1996). *Indicators of science and mathematics education 1995.* Arlington, VA: Author.

Noddings, N. (2005). *The challenge to care in schools* (2nd ed.). New York, NY: Teachers College Press.

Ravitch, D. (2010). *The death and life of the great American school system: How testing and choice are undermining education.* Philadelphia, PA: Basic Books.

Rodriguez, A. J., & Kitchen, R. S. (Eds.). (2005). *Preparing mathematics and science teachers for diverse classrooms: Promising strategies for transformative pedagogy.* Mahwah, NJ: Erlbaum.

Sanchez, W., Kastberg, S., Tyminski, A., & Lischka, A. (2015). *Scholarly inquiry and practices (SIP) conference for mathematics education methods.* Atlanta, GA: National Science Foundation.

Secada, W. G. (1992). Race, ethnicity, social class, language, and achievement in mathematics. In D. A. Grouws (Ed.), *Handbook of research on mathematics teaching and learning* (pp. 623–660). New York, NY: Macmillan.

Staples, M. (2007). Supporting whole-class collaborative inquiry in a secondary mathematics classroom. *Cognition and Instruction, 25*(2), 161–217.

Stylianou, D., Blanton, M., & Knuth, E. (Eds.). (2009). *Teaching and learning proof across the grades: A K–16 perspective.* Mahwah, NJ: Taylor & Francis.

Suárez-Orozco, M. M. (1989). Psychosocial aspects of achievement motivation among recent Hispanic immigrants. In H. T. Trueba, G. Spindler, & L. Spindler (Eds.), *What do anthropologists say about dropouts?* (pp. 99–116). New York, NY: Falmer.

Taylor, E., & Kitchen, R. S. (2008). Doctoral programs in mathematics education: Diversity and equity. In R. Reys & J. Dossey (Eds.), *U.S. doctorates in mathematics education: Developing stewards of the discipline* (pp. 111–116). Washington, DC: Conference Board of the Mathematical Sciences.

Téllez, K., Moschkovich, J., & Civil, M. (Eds.). (2011). *Latinos and mathematics education: Research on learning and teaching in classrooms and communities.* Charlotte, NC: Information Age.

Turner, E., Celedón-Pattichis, S., & Marshall, M. (2008). Cultural and linguistic resources to promote problem solving and mathematical discourse among Hispanic kindergarten students. In R. S. Kitchen & E. Silver (Eds.), *Promoting high participation and success in mathematics by Hispanic students: Examining opportunities and probing promising practices* [A Research Monograph of TODOS: Mathematics for All], *1* (pp. 19–40). Washington, DC: National Education Association.

Valenzuela, A. (1999). *Subtractive schooling: U.S.–Mexican youth and the politics of caring.* Albany: State University of New York Press.

Valenzuela, A. (Ed.). (2005). *Leaving children behind.* Albany: State University of New York Press.

ABOUT THE EDITORS

Signe E. Kastberg is a professor of mathematics education in the College of Education at Purdue University. She engages in the scholarship of teaching and learning with a focus on the development of relational practice. Her central goal is to gain insight into relational practice by exploring and illustrating listening, trust, care, and empathy, and the ways in which these sustain and motivate student–teacher collaborations. She teaches undergraduate prospective teachers and graduate students with a focus on supporting evolving understandings of the beauty and complexity of learning to teach mathematics learners.

Andrew M. Tyminski is an associate professor in the College of Education and the department of mathematical sciences at Clemson University. His research interests are focused on elementary prospective teacher learning and the design and impact of mathematics methods courses. He studies the design, implementation, and learning outcomes of mathematics content and methods course activities. He teaches content and methods courses for prospective elementary and middle grades mathematics teachers at the undergraduate and graduate level.

Alyson E. Lischka is an assistant professor of mathematics education in the department of mathematical sciences at Middle Tennessee State University. Her research focuses on the implementation of ambitious teaching practices among prospective and practicing teachers, specifically explorations of effective feedback practices in the teaching of mathematics and mathematics methods courses. She teaches undergraduate courses in content

Building Support for Scholarly Practices in Mathematics Methods, pages 359–360
Copyright © 2018 by Information Age Publishing

and methods for prospective secondary teachers along with courses in the mathematics and science education doctoral program.

Wendy B. Sanchez is a professor of mathematics education in the department of secondary and middle grades education at Kennesaw State University. Her research focuses on the content, nature, and effectiveness of mathematics methods courses for prospective teachers. She teaches undergraduate and graduate mathematics content and pedagogy courses for prospective and practicing mathematics teachers. She also facilitates professional learning experiences for practicing teachers.

Printed in the United States
By Bookmasters